Comparative Physiology:
Primitive Mammals

Papers from the Fourth International Conference on Comparative Physiology held at Crans-sur-Sierre, Switzerland, 4–9 June 1978.

The conference was sponsored by the Interunion Commission on Comparative Physiology representing the International Unions of Biological Sciences, Physiological Sciences, and Pure and Applied Biophysics. The conference was made possible through the generous support of the Fonds National Suisse de la Recherche Scientifique; the University of Messina; the Swiss Bank Corporation; Lirca Laboratory; the Community of Chermignon; Winterthur Assurance, Crans; and Crans Golf Club.

Comparative Physiology: Primitive Mammals

EDITED BY

KNUT SCHMIDT-NIELSEN
Duke University

LIANA BOLIS
University of Messina

C. RICHARD TAYLOR
Harvard University

CO-EDITED BY

P. J. BENTLEY
Mount Sinai School of Medicine

C. E. STEVENS
Cornell University

CAMBRIDGE UNIVERSITY PRESS

Cambridge
London   New York   New Rochelle
Melbourne   Sydney

Published by the Press Syndicate of the University of Cambridge
The Pitt Building, Trumpington Street, Cambridge CB2 1RP
32 East 57th Street, New York, NY 10022, USA
296 Beaconsfield Parade, Middle Park, Melbourne 3206, Australia

Printed in the United States of America
Typeset by Huron Valley Graphics, Ann Arbor, Michigan
Printed and bound by The Book Press, Brattleboro, Vermont

Library of Congress Cataloging in Publication Data
Main entry under title:
Comparative physiology, primitive mammals.
1. Mammals – Physiology   2. Monotremata –
Physiology.   3. Marsupialia – Physiology.   4. In-
sectivora – Physiology.   5. Physiology, Comparative.
I. Schmidt-Nielsen, Knut, 1915–   II. Bolis,
Liana.   III. Taylor, Charles Richard, 1939–
IV. Title: Primitive mammals.
QL739.2.C65   599'.01   79-9245
ISBN 0 521 22847 6

CONTENTS

Contents

Contents

CONTRIBUTORS

R. McN. Alexander: Department of Pure and Applied Zoology, University of Leeds, Leeds LS2 9JT, England

R. B. Armstrong: Departments of Health Science and Biology, Boston University, Boston, Massachusetts, U.S.A.

R. V. Baudinette: School of Biological Sciences, Flinders University of South Australia, Bedford Park, South Australia 5042, Australia

P. J. Bentley: Departments of Pharmacology and Ophthalmology, Mount Sinai School of Medicine of The City University of New York, New York 10029, U.S.A.

J. R. Blair-West: Howard Florey Institute of Experimental Physiology and Medicine, University of Melbourne, Parkville, Victoria 3052, Australia

J. Bligh: Institute of Arctic Biology, University of Alaska, Fairbanks, Alaska 99701, U.S.A.

P. H. Burri: Department of Anatomy, University of Berne, Bühlstrasse 26, CH-3000 Berne 9, Switzerland

G. A. Cavagna: Istituto di Fisiologia Umana dell' Università di Milano, 20133 Milan, Italy

G. Citterio: Istituto di Fisiologia Umana dell' Università di Milano, 20133 Milan, Italy

H. Claassen: Department of Anatomy, University of Berne, Bühlstrasse 26, CH-3000 Berne 9, Switzerland

E. T. Clemens: Department of Veterinary Physiology, University of Nairobi, Nairobi, Kenya, East Africa

A. W. Crompton: Museum of Comparative Zoology, Harvard University, Cambridge, Massachusetts 02138, U.S.A.

T. J. Dawson: School of Zoology, University of New South Wales, Kensington, New South Wales 2033, Australia

D. W. Dellow: Department of Biochemistry and Nutrition, University of New England, Armidale, New South Wales 2351, Australia

J. F. Eisenberg: National Zoological Park, Smithsonian Institution, Washington, D.C. 20008, U.S.A.

S. M. Evans: Department of Zoology, Australian National University, Canberra, Australia

P. Gehr: Department of Anatomy, University of Berne, Bühlstrasse 26, CH-3000 Berne 9, Switzerland

A. Gibson: Howard Florey Institute of Experimental Physiology and Medicine, University of Melbourne, Parkville, Victoria 3052, Australia

T. R. Grant: School of Zoology, University of New South Wales, Kensington, New South Wales 2033, Australia

C. L. Guard: Department of Physiology, Biochemistry, and Pharmacology, New York State College of Veterinary Medicine, Cornell University, Ithaca, New York 14853, U.S.A.

C. J. F. Harrop (deceased): School of Zoology, University of New South Wales, Kensington, New South Wales 2033, Australia

N. C. Heglund: Concord Field Station, Museum of Comparative Zoology, Harvard University, Cambridge, Massachusetts 02138, U.S.A.

H. C. Heller: Department of Biological Sciences, Stanford University, Stanford, California 94305, U.S.A.

A. J. Hulbert: Department of Biology, University of Wollongong, Wollongong, New South Wales 2500, Australia

I. D. Hume: Department of Biochemistry and Nutrition, University of New England, Armidale, New South Wales 2351, Australia

P. Jacini: Istituto di Fisiologia Umana dell' Università di Milano, 20133 Milan, Italy

S. L. Lindstedt: Department of Physiology, College of Medicine, University of Arizona, Tucson, Arizona 85724, U.S.A.

I. R. McDonald: Department of Physiology, Monash University, Clayton, Victoria 3168, Australia

O. T. Oftedal: National Zoological Park, Smithsonian Institution, Washington, D.C. 20008, U.S.A.

P. Poczopko: Polish Academy of Sciences, Institute of Animal Physiology and Nutrition, 05-110 Jablonna near Warsaw, Poland

M. B. Renfree: School of Environmental and Life Sciences, Murdoch University, Western Australia 6153, Australia

S. Sehovic: Department of Anatomy, University of Berne, Bühlstrasse 26, CH-3000 Berne 9, Switzerland

C. Sernia: Department of Physiology, Monash University, Clayton, Victoria 3168, Australia

A. Shkolnik: George T. Wise Center for Life Sciences, Zoology Department, Tel-Aviv University, Tel-Aviv, Israel

C. E. Stevens: Department of Physiology, Biochemistry, and Pharmacology, New York State College of Veterinary Medicine, Cornell University, Ithaca, New York 14853, U.S.A.

C. R. Taylor: Museum of Comparative Zoology, Harvard University, Cambridge, Massachusetts 02138, U.S.A.

C. H. Tyndale-Biscoe: Division of Wildlife Research, CSIRO, Canberra, A.C.T. 2602, Australia

P. Vogel: Institut de Zoologie et de l'Ecologie Animale, Université de Lausanne, Place du Tunnel 19, CH-1005 Lausanne, Switzerland

E. R. Weibel: Department of Anatomy, University of Berne, Bühlstrasse 26, CH-3000 Berne 9, Switzerland

M. Weiss: Department of Physiology, Monash University, Clayton, Victoria 3168, Australia

EDITORIAL PREFACE

The papers contained in this volume were presented at the Fourth International Conference on Comparative Physiology, held at Crans-sur-Sierre in June 1978 and organized through the untiring efforts of Liana Bolis. It brought together scientists from various fields, who, from the perspective of comparative physiology, discussed the widely scattered information on primitive mammals.

What is meant by "primitive" mammals? Most of us think of marsupials and monotremes and perhaps an assortment of insectivores and prosimian primates that have retained certain morphological traits possessed by the early mammals evolving from reptilian ancestors. The first chapter in this book helps reconstruct a picture of what the earliest mammals were like, and subsequent chapters bring together our knowledge of the physiology of many living primitive mammals.

In ordinary language the word "primitive" often implies simple and perhaps even inadequate or inferior. How wrong this is in the context of mammalian physiology! The implication that primitive should mean simple certainly cannot stand close scrutiny. For example, marsupials have modes of reproduction that in many ways are more sophisticated than those of eutherian placentals, and, although marsupials tend to have a somewhat lower body temperature than eutherians, they use the same mechanisms for temperature regulation and use them just as effectively.

To avoid the unwarranted implication of inferiority or superiority, it is better to refer to "conservative" as opposed to more evolved or "derived" traits. The fact that an animal displays certain conservative traits in no way guarantees that in other respects it is not highly advanced. More important, conservative traits do not imply inferiority or inadequacy. On the contrary, the long persistence of conservative traits in animal evolution assures us that they are quite adequate for survival, and their persistence may in fact indicate some superiority of these traits.

This is the first book that attempts to give a common perspective to a wide variety of functional information pertaining to these problems. The result appears to be that the mere possession of some conservative morphological traits has little bearing on how complex or advanced the animal's functional characteristics are.

Knut Schmidt-Nielsen

1

Biology of the earliest mammals

A. W. CROMPTON

In a volume on "primitive mammals," it is appropriate to review our knowledge of the earliest known mammals and to speculate on some aspects of their biology. This is necessary because it is difficult to distinguish clearly between advanced and conservative features in those living mammals often designated as "primitive" – the monotremes, didelphoid and dasyurid marsupials, edentates, and, among the insectivores, the tenrecs, solenodons, and hedgehogs.

As a result of a series of remarkable discoveries, starting with Walter Kühne's finds in England shortly before the outbreak of World War II (see Parrington, 1971 for review), a fairly good record of the earliest mammals dating from the late Triassic, about 180 million years ago, is now available. Intensive searching and improved collecting techniques have provided a reasonable documentation of the first 100 million years of mammalian evolution. Recent discoveries in South America, in addition to the classic finds in southern and eastern Africa, have helped to document the transition from advanced mammal-like reptiles to the earliest mammals. It is against this background that I wish to review some aspects of the biology of the earliest mammals.

If we view the radiation of advanced mammal-like reptiles and early mammals (Figure 1), we find a clear separation between the mammalian and mammal-like reptile radiations. The most mammalian of the mammal-like reptiles are the cynodonts. It is from a single group of cynodont, or at the most from a group of closely related cynodonts, that mammals appear to have arisen (Crompton and Jenkins, in press). Representatives of the transitional forms have not as yet been discovered, but fossil forms slightly older and younger than this hypothetical group are now fairly well known. One way to determine the features that probably characterized these transitional stages is to compare the earliest mammals with the most advanced mammal-like reptiles. Clearly, many of the features possessed by mammals evolved earlier

1

within the mammal-like reptiles before this point was reached, such as secondary bony palate and double occipital condyle. However, new innovations were introduced at this stage and clearly separated mammals from the mammal-like reptiles and opened up a new adaptive zone. The earliest mammals were represented by two groups (Figure 1): one that appears to be related to therian mammals (marsupials and placentals and several extinct groups), the Kuehneotheriidae, and another that appears to be related to nontherian

Figure 1. Phylogenetic relationships of the principal groups of Mesozoic mammals and their suggested relationships to the most advanced group of mammal-like reptiles, the cynodonts. There is a clear separation between cynodonts and mammals, and it is suggested that this reflects the acquisition of numerous new features which are basic to the adaptive radiation of mammals at the transition between reptiles and mammals.

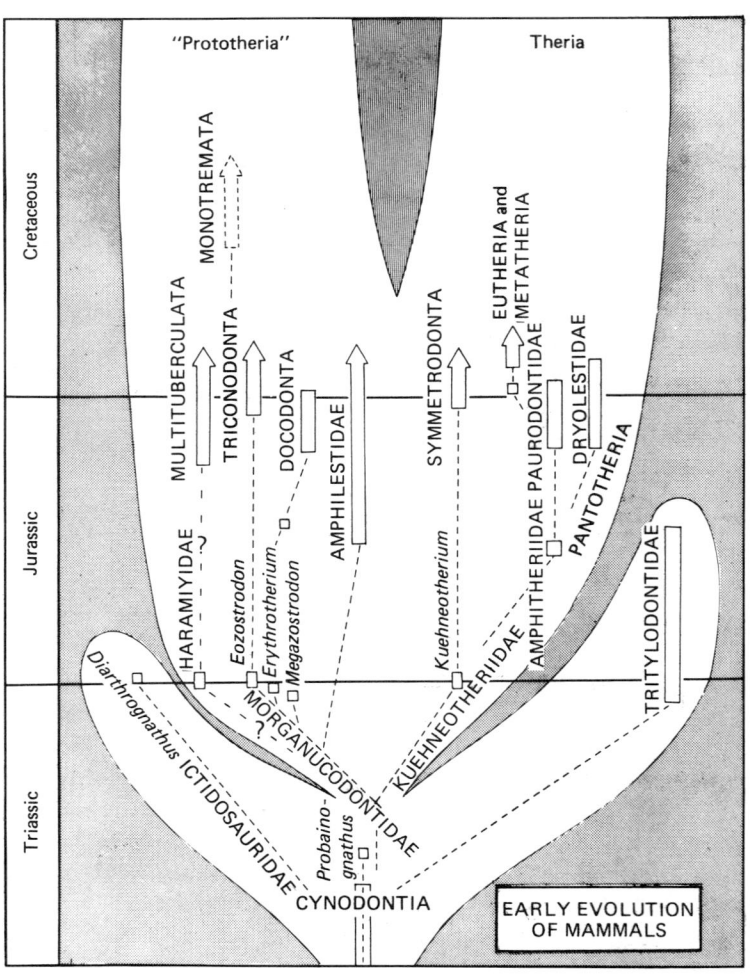

mammals (monotremes and several extinct groups), the Morganucodontidae. We know the Morganucodontidae much better than the Kuehneotheriidae.

The morganucodontids were rather small, about 20 to 30g in weight. Adaptions of the postcranial skeleton seems to suggest that they were adapted for foraging in a terrestrial and/or arboreal environment or on the interface between these two environments (Jenkins and Parrington, 1976). The first stages of the division of the vertebral column (Figure 2) into thoracic and lumbar regions with a transitional or anticlinal vertebra between these two divisions is present in these forms. This indicates that dorsoventral flexure of the vertebral column, a feature typical of mammals, was already present in these early forms. This feature is not present in the known mammal-like reptiles, in which lateral undulations of the vertebral column took place during locomotion (Jenkins, 1971). Jenkins (1974) has shown that dorsoventral flexure is a useful adaption, not only for cursorial locomotion but also for locomotion in small arboreal animals such as tree shrews. Structural modifications of the foot, which permitted the hallux to secure an active grip by opposing the remaining digits, were present in these early mammals. This feature is found in many arboreal mammals, such as the tree shrew and didelphoid marsupials. Jenkins and Parrington (1976) have listed numerous other features of the skeletons of early mammals which clearly separate them from the advanced mammal-like reptiles.

The dentition (Figure 3A, B) consists of relatively small incisors (I) and canines (C), but relatively large premolars (PM) and molars (M). This is in contrast to those mammal-like reptiles that appear to be closely related to mammals such as *Probainognathus* (Crompton, 1972a) where the canines

Figure 2. Reconstruction of the skeleton of a late Triassic morganucodontid, one of the earliest known mammals. (After Jenkins and Parrington, 1976.)

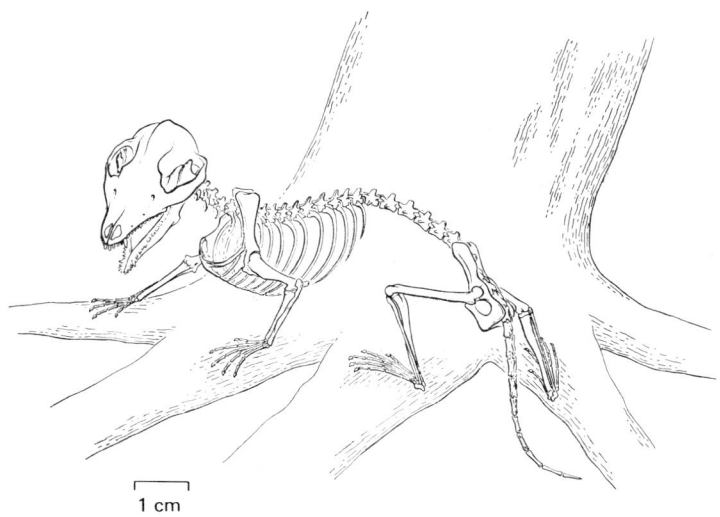

1 cm

tend to be large and the teeth behind relatively much smaller (Figure 4A). The early mammal postcanine teeth and especially the molars possess matching shearing surfaces (Figure 3C, D), and during occlusion only one side of the jaw was actively used an any one time (Crompton, 1974). This pattern of jaw movement required a high degree of musculature control so that the shearing surfaces could be brought precisely together when major forces were generated between the teeth. This represents a clear departure from the condition in the reptiles thought to be ancestral to or closely related to the first mammals. These forms did not occlude their teeth; the teeth bypassed one another so that a vertical space separated matching upper and lower teeth when the jaws are closed (Figure 4B); tearing rather than shearing characterized mastication (Crompton, 1972b).* Precise dental occlusion in the early mammals permitted a more effective breakdown of food in the mouth. The structure of the tall shearing molars and premolars, the potential for large gape, together with the small size of the animals suggests that they fed principally on arthropods and other invertebrates, but could also have taken young reptiles, amphibians, and birds. This was also true of the other group of early mammals, the Kuehneotheriidae.

In the mammal-like reptiles ancestral to mammals new teeth erupt between older ones (Figure 5C, D), and tooth replacement seems to have continued throughout life (Osborn and Crompton, 1973). In early mammals the molars occlude with two teeth in the matching upper or lower jaw. Consequently, alternate tooth replacement would disrupt occlusal relationships. For this reason, in all mammals, including the earliest, the molars tend to be added sequentially so that precise occlusal relationships will not be disturbed (Figure 5A, B).

In the earliest mammals and in the majority of living mammals, the postcanine row is divided into premolars and molars (Figure 3). The premolars replace the milk molars, and the molars are added sequentially, from front to back, and not replaced (Figure 5B). This limited replacement and division of the postcanine row into premolars and molars indicates a growth pattern different from that of the mammal-like reptiles. Living mammals suckle when they are young and do not require teeth during early growth stages for the processing of food. Consequently, a substantial part of total skull and body growth can take place before the complete eruption of the first set of deciduous teeth (Ewer, 1963; Hopson, 1973; Pound, 1977). A single replacement of the milk teeth and the addition of three or four molar teeth is all that is required to accommodate for the remaining growth after weaning. This is in contrast to reptiles or the mammal-like reptiles where hatchlings require a set of functional teeth to process food by mastication as soon as they leave the egg. Because teeth crowns, once formed, cannot enlarge,

* Precise dental occlusion was present in some specialized mammal-like reptiles that were not related to later mammals. However, the jaw movements and occlusion in these forms were fundamentally different from those of the early mammals.

Figure 3. Dentition of a late Triassic morganucodontid, *Eozostrodon*. (A) Snout and jaw in external view. (B) Lower jaw in medial view. (C) Matching shearing surfaces on the internal surface of the uppers and external surface of the lowers. (D) Matching shearing surfaces on the anterior surface of upper and a posterior surface of the lower molars, illustrating how the lowers move in a dorsomedial direction across the uppers during occlusion to shear the foods.

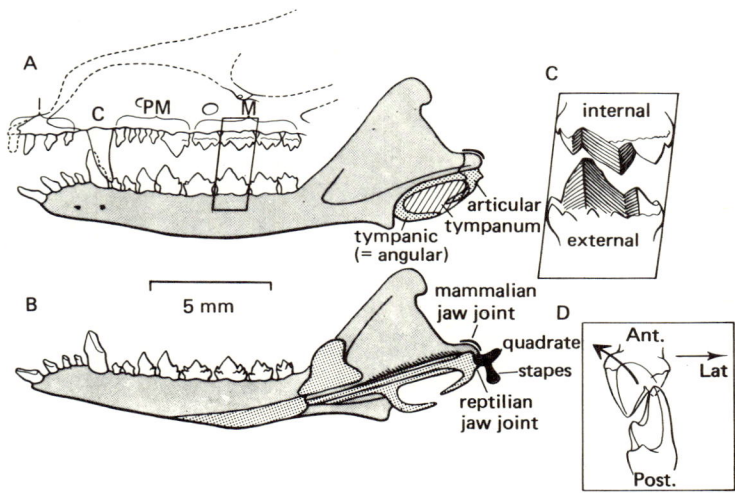

Figure 4. (A) Skull of *Probainognathus*, an advanced mammal-like reptile. (B) Posterior view of opposing upper and lower postcanine teeth to illustrate how, during the closing of the jaw, a space separates the teeth. Therefore, the teeth are capable of tearing, rather than shearing food.

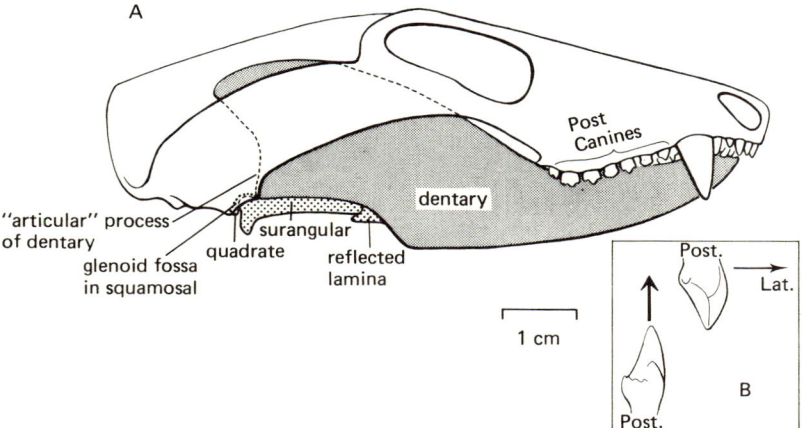

numerous replacements are necessary to accommodate the extensive growth range from hatchlings to adult. In mammal-like reptiles there may be at least six generations of teeth at any one tooth position (Osborn, 1973; Osborn and Crompton, 1973). The eruption and replacement pattern of the earliest mammals, therefore, suggests that maternal care and lactation were features of these forms but not of their immediate ancestors. A lack of replacement of the molars in the earliest and living mammals is accompanied by a clear termination of growth in the adult form. This is in contrast to their reptilian ancestors where a gradual, rather than an abrupt, termination of growth appears to have taken place.

The other group of early mammals, the Kuehneotheriidae, were slightly smaller than the Morganucodontidae. They appear to have had a mammalian tooth replacement pattern, indicating that lactation and maternal care characterized their early development. The Morganucodontidae appear to have been ancestral to several extinct orders as well as to the monotremes (Figure 1). It is, therefore, possible that reproduction in this group of early mammals was similiar to that of living monotremes. Here the hatchling, or neonatus, is poorly developed, and the nutrition for growth and development is derived principally from milk. Prior to birth, nutrition for the limited amount of growth that takes place is derived either from yolk or from

Figure 5. Tooth succession in an advanced mammal-like reptile (C,D) and an early mammal (A,B). In a mammal-like reptile, the postcanine tooth row is not divided into molars and premolars. Replacement continues throughout life, with new teeth erupting alternately between older teeth. The diagram (D) shows the successive teeth which are added at each tooth position during growth. In mammals, the tooth row is divided into premolars and molars. The premolars replace milk molars, and the molars are added sequentially and not replaced.

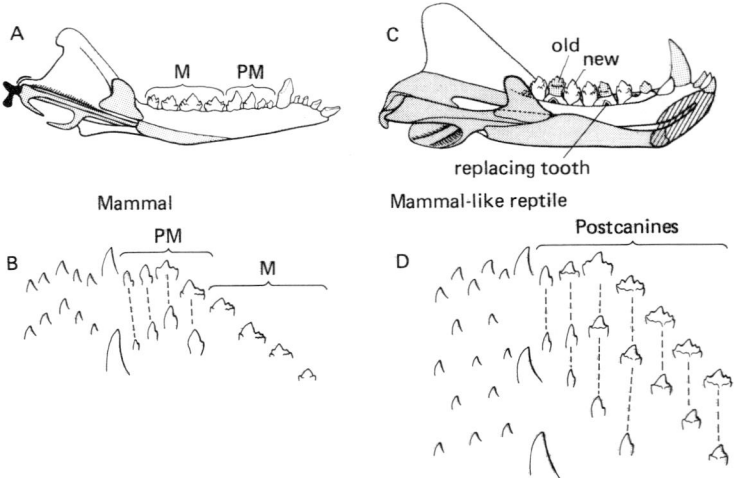

uterine secretions absorbed by the yolk sac. In marsupials, the young, at birth, are at about the same stage of development as the monotreme hatchling, and here also nutrition, for most of the growth and development, is supported by milk. It is relatively easy to derive the marsupial reproductive pattern from that of monotremes. Both the marsupials and the eutherians arose from a common stock which can be traced back to the other group of early mammals – the Kuehneotheriidae. This suggests that the birth, or hatching, of immature young and the heavy reliance on milk for growth and development probably also characterized this group of early mammals. Eutherian mammals increased the amount of nutrition derived from the uterine wall, and accordingly decreased the amount of growth dependent on nutrition derived from the milk.

A clearly defined feature that separates mammal-like reptiles from the earliest mammals is the jaw articulation (Crompton and Parker, 1978). In mammals an articulation exists between the dentary bone and the temporal bone, or the squamosal (Figure 6C); in mammal-like reptiles, it lies between the quadrate and articular bones (Figure 6A). Once the mammalian jaw articulation was established in early mammals (Figure 6B), the bones, which previously formed the reptilian jaw joint, separated from the lower jaw and

Figure 6. Jaw articulation of an advanced mammal-like reptile (A) an early mammal (B), and a living mammal, the opossum (C). In reptiles, the jaw articulation lies between the articular bone of the lower jaw and the quadrate attached to the skull. In early mammals a new articulation is established between the dentary bone and the temporal bone and lies alongside the old reptilian jaw joint between the quadrate and the articular. In modern mammals the old reptilian jaw joint bones have lost their contact with the lower jaw and are isolated in the middle ear, and they are involved in conducting vibrations from the middle ear to the inner ear (D). The malleus is the same as the articular, and the incus is the same as the quadrate.

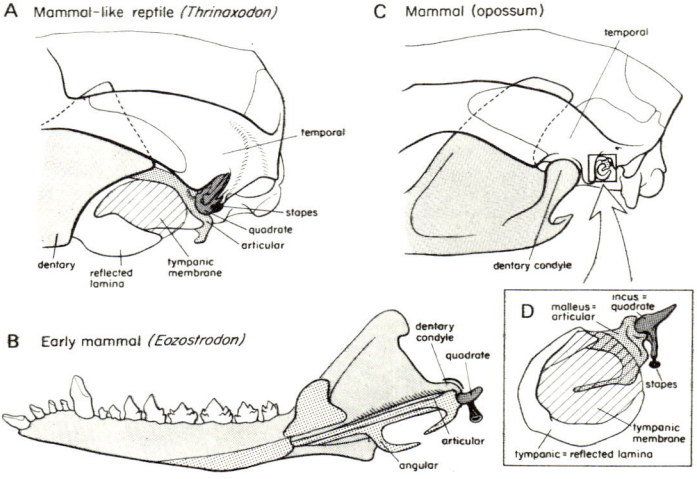

were isolated in the middle ear and became specialized for the transmission of vibrations from the tympanic membrane to the inner ear (Figure 6D). It has recently been suggested that the freeing of the reptilian jaw joint bones from a suspensory function improved their ability to conduct sound (Allin, 1975; Crompton and Parker, 1978). This view seems to be substantiated by changes in the cochlea housing. This is a feature that clearly separates the most advanced mammal-like reptiles from the earliest mammals. In the mammal-like reptiles, the cochlea housing is relatively small and is not visible on the ventral surface of the skull, in contrast to the condition in the earliest mammals where the cochlea housing is relatively very large and forms a dominant feature of the ventral surface of the skull. The improvement of the efficiency of the middle ear appears to have been correlated with a refinement and enlargement of the cochlea. This organ is designed to distinguish between different sound frequencies transmitted by the middle ear, and its enlargement presumably reflects an improvement in auditory acuity.

We are not yet in the position to determine the exact brain size of the earliest mammals, but the braincase volume of these forms relative to body size appears to have been three to four times larger than that of cynodont mammal-like reptiles such as *Probainognathus*, which were closely related to the first mammals. This enlargement of the brain was in part due to the enormous increase in cranial space anterior to the pituitary fossa as a result of the vertical suppression of the interorbital septum. Jerison (1973) has suggested that this rapid increase may be correlated with the ability of the earliest mammals to process more sensory information than their ancestors, especially from the nasal capsule and inner ear. We can observe a refinement of the structure of the ear, but a similar morphological refinement of the nasal capsule is difficult to document. This will require the discovery of ossified ethmoturbinals in early mammals. Jerison (1973) has argued that the increase in brain size and improvement of the auditory and olfactory senses were correlated with the invasion of a nocturnal environment by the earliest mammals. This is not a new idea, and the structure of the earliest mammals is not inconsistent with this view.

The fauna found contemporaneously with the earliest mammals tends to support the view that the earliest mammals were nocturnal (Crompton and Jenkins, 1978). The dominant terrestrial forms found together with the earliest mammals are a wide variety of saurischian dinosaurs (herbivores and carnivores) and ornithischian dinosaurs (herbivores). Even the smallest of these is several orders of magnitude larger than the early mammals (Santa Luca, et al., 1976). Another well-represented reptile group is the thecodonts, which were principally large to medium-sized carnivores (this is the group that gave rise to the first crocodiles some time in the middle to late Triassic times).

Some specialized survivors of the earlier vast radiation of mammal-like reptiles are found together with the earliest mammals. They were either insectivores cum carnivores or small herbivores, and were small to medium sized, although considerably larger than the Triassic mammals. Robinson (1967a, 1967b, 1973) has shown that the latter half of the Triassic was characterized by an adaptive radiation of lizards, some of them smaller than the earliest mammals, some about the same size, and many quite a bit larger. As a whole, they were small, lightly built, agile forms well suited for moving over uneven terrain, and some were also adapted for gliding flight. Also present were a series of small rhynchocephalians (related to the modern Tuatara of New Zealand). Many of these small reptiles, especially the lizards, were like the earliest mammals and probably insectivorous. Therefore, in late Triassic time, both lizards and mammals seem to have produced a significant radiation of insectivorous forms. If the improvement of auditory and olfactory senses and the enlargement of the brain in the early mammals were associated with a predominately nocturnal way of life, it is possible that the insectivore niche may have been partially divided between mammals and reptiles on a nocturnal–diurnal basis. The well-established fauna of large, medium, and small and presumably diurnal reptilian carnivores and herbivores living alongside the mammals appear to have prevented mammals from evolving into this niche. This situation was to continue for the next hundred million years, and the successful adaptive radiation of mammals during this time appears to have taken place in a nocturnal niche.

The most common of the early mammals (late Triassic), namely the Morganucodontidae, are found in deposits in southern Africa, China, and Europe. Morganucodontids from these countries, despite their extensive dispersal, are almost identical in structure and level of organization. This suggests a rapid dispersal of mammals shortly after their origin. This rapid dispersal and lack of diversity would be expected if the characteristics associated with the earliest mammals had enabled them to exploit a vacant or partially vacant niche for nocturnal insectivores. Diversity within this niche in different countries and regions would be expected in later times, and it is adequately documented by the fossil record.

One encounters a similar phenomenon with the late Triassic dinosaurs, the prosauropod dinosaurs. At the same time that the mammals were expanding in a nocturnal niche, these dinosaurs appear to have invaded and dominated the niche for terrestrial herbivores. Whether they are found in North America, South America, Europe, Africa, or China, these forms are almost identical in structure. The rapid dispersal and lack of an adaptive radiation, in both dinosaurs and mammals, may have been aided by the close connections that existed between the continents at that time and not only by the rapid dispersal in a new adaptive zone.

We (Crompton et al., 1978; see also Taylor, this volume) have argued that the successful exploitation of a nocturnal niche requires the ability to maintain a constant body temperature. Diurnal habitats present a wide mosaic of microenvironments suitable for the behavioral control of temperature (i.e., direct solar radiation, shaded areas, burrows). By selecting one of these micro-environments or by moving from one to another, a diurnal poikilothermic reptile can control its temperature within narrow limits. This mosaic does not exist at night, and if nocturnal temperatures fall significantly below a preferred temperature, poikilothermic animals are not able to forage effectively. We have suggested that this could be achieved under certain conditions without a significant increase in the basal metabolic rate. We have argued that the temperature regulation of early mammals was probably similar to modern Tenrecinae and Erinaceinae. These forms have a lower basal metabolic rate than "typical" mammals of similar size. Our hypothesis is that the earliest mammals achieved a constant body temperature by (1) restricting their periods of activity to twilight or night hours to avoid high external heat loads; (2) opting for a low body temperature (probably between 25 and 30 °C); (3) adding insulation (fur or superficial fat); and (4) retaining a basal metabolic rate similar to that of a "typical" reptile of the same size and at the same temperature. The main innovation of the earliest mammals was presumably their ability to increase metabolic rate, probably by shivering when ambient temperatures dropped below body temperature, and to control the loss or retention of heat by adding fur and fat and by improved peripheral vascular control.

The Triassic mammals can safely be referred to as primitive mammals. Unfortunately, there are no living survivors of these early mammals, and it is inadvisable to refer to some of the living mammals as "primitive." We know nothing about the phylogenetic history of monotremes other than that they appear to be related to a group of Triassic mammals different from the therian mammals. Therefore, at least 180 million years of independent evolution separates them from all of the living therian mammals. Both the marsupials and eutherians appear to have arisen from a common stock (Crompton and Kielan-Jaworowska, 1978) sometime in the early to middle Cretaceous, therefore, living marsupials and eutherians have had at least 70 to 80 million years of independent evolution. Little is known about edentate evolution other than a phylogenetic history dating back some 60 million years to the early Tertiary. Dilambdodont insectivores, such as the hedgehogs, arose before the end of the Cretaceous, and although it is probable that the zalambdodont insectivores, such as the Tenrecidae and Solenodontidae, arose from dilambdodont insectivores (Mills, 1966; Butler, 1972), their origin is unknown and they may be survivors of an early radiation of eutherian mammals (Eisenberg, this volume). Consequently all the so-called primitive mammals have had a long and independent history with more than adequate time to develop numerous specialized features. Some of these living "primitive" mammals retained several primitive or conservative features that characterize the earliest mammals. But without exception, they are found alongside highly specialized features in living forms. The most we can hope to do is to recognize conservative and advanced features; but when we move from the skeleton to the soft anatomy or to physiology, any decision will be highly speculative.

References

Allin, E. F. (1975). Evolution of the mammalian middle ear. *J. Morphol. 147*:403–38.

Butler, P. M. (1972). The problem of insectivore classification. In *Studies in Vertebrate Evolution*, eds. K. A. Joysey and T. Kemp, pp. 253–65. Edinburgh: Oliver and Boyd.

Crompton, A.W. (1972a). The evolution of the jaw articulation of cynodonts. In *Studies in Vertebrate Evolution*, eds. K. A. Joysey and T. Kemp, pp. 231–54. Edinburgh: Oliver and Boyd.

Crompton, A. W. (1972b). Postcanine occlusion in cynodonts and the origin of the tritylodontids. *Bull. Br. Mus. (Nat. Hist.) Geol. 21.2*:29–71.

Crompton, A. W. (1974). The dentition and relationships of southern African Triassic mammals *Erythrotherium parringtoni* and *Megazostrodon rudnerae*. *Bull. Br. Mus. (Nat. Hist.) Geol. 24.7*:397–437.

Crompton, A. W., and Jenkins, F. A., Jr. (1978). African Mesozoic mammals. In *African Fossil Mammals*, eds. H. B. S. Cooke and V. J. Maglio, pp. 46–55. Cambridge, Mass.: Harvard University Press.

Crompton, A. W., and Jenkins, F. A., Jr. (In press.) Orgin of mammals. In *Mesozoic Mammals*, eds. J. Lilligraven, W. Clemens and Z. Kielan-Jaworowska. Berkeley: University of California Press.

Crompton, A. W., and Kielan-Jaworowska, Z. (1978). Molar structure and occlusion in Cretaceous therian mammals. In *Studies on the Development, Structure and Function of Teeth,* eds. P. M. Butler and K. A. Joysey,

Crompton, A. W., and Parker, P. (1978). Evolution of the mammalian masticatory apparatus. *Am. Sci. 66*:192–201.

Crompton, A. W., Taylor, C. R., and Jagger, J. A. (1978). Evolution of homeothermy in mammals. *Nature 272*:333–6. pp. 249–87. New York: Academic Press.

Ewer, R. F. (1963). Reptilian tooth replacement. *News Bull. Zool. Soc. South Afr. 4* (2): 4–9.

Hopson, J. A. (1973). Endothermy, small size and the origin of mammalian reproduction. *Am. Nat. 107*:446–51.

Jenkins, F. A., Jr. (1971). The postcranial skeleton of African cynodonts. *Bull. Peabody Mus. Nat. Hist. 36*:1–216.

Jenkins, F. A., Jr. (1974). Tree shrew locomotion and the origin of primate arborealism. In: *Primate locomotion*, ed. F. A. Jenkins, Jr., pp. 85–115, Chap. 3. New York: Academic Press.

Jenkins, F. A., Jr. and Parrington, F. R. (1976). The postcranial skeletons of the Triassic mammals *Eozostrodon, Megazostrodon* and *Erythrotherium. Phil. Trans. R. Soc. London Ser. B. 273*:387–431.

Jerison, H. J. (1973). *Evolution of Brain and Intelligence.* New York: Academic Press.

Mills., J. R. E. (1966). The functional occlusion of the teeth of Insectivora. *J. Linn. Soc. Zool. 47*:1–24.

Osborn, J. W. (1973). The evolution of dentitions. *Am. Sci. 61*:548–59.

Osborn, J. W., and Crompton, A. W. (1973). The evolution of mammalian from reptilian dentitions. *Mus. Comp. Zool. Breviora 399*:1–18.

Parrington, F. R. (1971). On the Upper Triassic mammals. *Phil. Trans. R. Soc. London Ser. B 261*:231–72.

Pound, C. M. (1977). The significance of lactation in the evolution of mammals. *Evolution 31*:177–99.

Robinson, P. L. (1967a). Triassic vertebrates from lowland and upland. *Sci. Cult. 33*:169–73.

Robinson, P. L. (1967b). The evolution of the Lacaetilia. In *Problèmes actuels de Palèontologie (Evolution des vertébrés). 163*:395–407. Paris: Editions du Centre National de la Recherche Scientifique.

Robinson, P. L. (1973). A problematic reptile from the British Upper Triassic. *J. Geol. Soc. 129*:457–79.

Santa Luca, A. P., Crompton, A. W., and Charig, A. J. (1976). A complete skeleton of the Late Triassic ornithischian *Heterodontosaurus tucki. Nature 264*:324–8.

2

Biological strategies of living conservative mammals

JOHN F. EISENBERG

Toward a definition of a primitive mammal

There exists some difficulty among biologists concerning the definition of the word "primitive." For my purposes, I consider primitive to be synonymous with the possession of a set or sets of conservative (or plesiomorph) characters. Such conservative characters include not only aspects of morphology but also conservative physiological and behavioral traits. It is axiomatic then that the possession of such a conservative set of phenotypic characters implies that the animals very probably are occupying habitats and exploiting them in a manner similar to those mammals that existed at the end of the Cretaceous.

All living mammals also possess numerous derived (or apomorphic) characters and living mammals are in fact "mosaics" of characters. Crompton (this volume) has outlined the morphological features of the first mammals. We may conclude that they were small (30 to 50 g), had a long tail, were insectivorous, relied on olfaction and audition, secreted milk as nourishment for their young, and were adapted for climbing (scansorial). We may deduce that they were nocturnal, constructed nests, possessed long vibrissae, and were capable of endothermic temperature regulation during their nocturnal forays.

Table 1 compares the genera I will discuss with respect to mode of reproduction, occupancy of a conservative niche, and certain key morphological features. Although it is possible to discern forms that possess many conservative characters and thereby most nearly resemble in morphology and activity an early mammal (e.g., *Microgale dobsoni*), one may equally select for study a living mammal that possesses many derived characters, if it carries with it the conservative set of characters that is of interest. For example, *Tachyglossus* has many specialized or derived features, but it has a conservative reproductive tract and is oviparous.

Table 1. *Some conservative mammals compared*

Taxon	Ovi-parous	Ovovivi-parous (no trophoblast)	Viviparous (trophoblast)	Small size (30–150g)	Long tail	Coracoid bone	Epipubic bones	Moderate eye size	Cloaca	Internal testes	Occupies a conservative niche
Monotremata											
Tachyglossus	+	−	−	−	−	+	+	−	+	+	−
Ornithorhynchus	+	−	−	−	+	+	+	−	+	+	−
Metatheria											
Didelphis	−	+	−	−	+	−	+	+	−	−	+
Marmosa	−	+	−	+	+	−	+	+	−	−	+
Antechinus	−	+	−	+	+	−	+	+	−	−	+
Eutheria:											
Edentata											
Choloepus	−	−	+	−	−	−	−	+	+	+	−
Dasypus	−	−	+	−	+	−	−	+	−	±	±
Tenrecimorpha											
Hemicentetes	−	−	+	+	−	−	−	−	+	+	±
Tenrec	−	−	+	−	−	−	−	−	+	+	−
Microgale	−	−	+	+	+	−	−	+	+	±	+
Insectivora:											
Erinaceidae											
Echinosorex	−	−	+	−	+	−	−	+	−	±	+
Erinaceus	−	−	+	−	−	−	−	+	−	±	+
Solenodontidae											
Solenodon	−	−	+	−	+	−	−	−	−	±	+

Key to symbols: + = character present; − = character absent; ± = character partially expressed but modified.

The reproductive trichotomy

The three major stems of the class Mammalia still living – the Prototheria, Metatheria, and Eutheria – were already separated during the late Cretaceous. The living order and the two supercohorts representing this three-way division include the Monotremata, Marsupialia, and Eutheria, respectively (McKenna, 1975). Although many morphological differences discriminate these three major divisions from one another, the main feature separating these groups concerns the morphology and function of the reproductive tracts (Griffiths, 1968).

In spite of the possession by living forms of many derived characters, the Montremata still reproduce by means of laying a cleidoic egg, the Marsupialia show a simple placentation with no true trophoblast, and the Eutherians have evolved the trophoblast and thereby have a more efficient transfer of nutrients and wastes between the fetal and the maternal circulation. As a consequence the Eutheria place more emphasis on intrauterine nutrition of the fetus, and the monotremes and marsupials spend the major portion of their energy during the rearing cycle in lactation (Eisenberg, 1975a).

The living monotremes, although showing extremely conservative reproductive trends, by and large have accreted numerous derived characters in the course of their evolution and may be considered extremely specialized with respect to their mode of trophic exploitation and their life history. Both living forms produce small litters, are long-lived, and are examples of extreme K-selection,* thus indicating adaptation to their niches which presupposes extreme resource stability from the standpoint of their life history strategy (Burrell, 1927; Griffiths, 1968, 1978).

The Metatheria

The methatherians show a bewildering adaptive radiation which took place on two continental land masses, South America and Australia. The origins of the stem form and the history of this peculiar zoogeographical distribution remain to some extent obscure (Kirsch, 1977). Suffice it to say that two families deserve consideration under my definition of "primitive": the Neotropical family Didelphidae and the Australian family Dasyuridae.

If we look at the living Didelphidae, we see a family that shows an enduring trend toward arboreality. In comparison with other conservative mammals, the eye is quite large. They range in size from the small *Marmosa parvida,* which weighs less than 15 g as an adult, to *Didelphis virginiana* which may exceed 1.5 kg in weight when full grown. The possession of a prehensile tail dominates in most species, although several genera have lost this trait. Almost all species are quite strictly nocturnal. A number of niches have been invaded, ranging from aquatic through terrestrial to completely arbo-

* See Glossary at end of this chapter for definition of words marked with an asterisk.

real (see Hunsaker and Shupe, 1977). Much work remains to be done concerning thermoregulation, but it appears from the efforts of McNab (1978) that the basal metabolic rate averages higher in the Didelphidae than the average rates found for the family Dasyuridae (MacMillen and Nelson, 1969; Dawson and Hulbert, 1970). Some didelphids have relatively high core temperatures during their active phase. *D. virginiana* averages around 35 °C (Hunsaker, 1977). The genus *Marmosa* shows great variability. It is

Figure 1. Rectal or cloacal body temperatures during moderate activity for some morphologically conservative mammals. Arrows indicate ranges: black circles, the mean. D.v. = *Didelphis virginiana;* M.r. = *Marmosa robinsoni;* S.p. = *Solenodon paradoxus;* M.t. = *Microgale talazaci;* M.d. = *M. dobsoni;* H.n. = *Hemicentetes nigriceps;* S.s. = *Setifer setosus.* (Data for *Didelphis* and *Marmosa* from Hunsaker, 1977; data on insectivores from Eisenberg and Gould, 1966, 1970.)

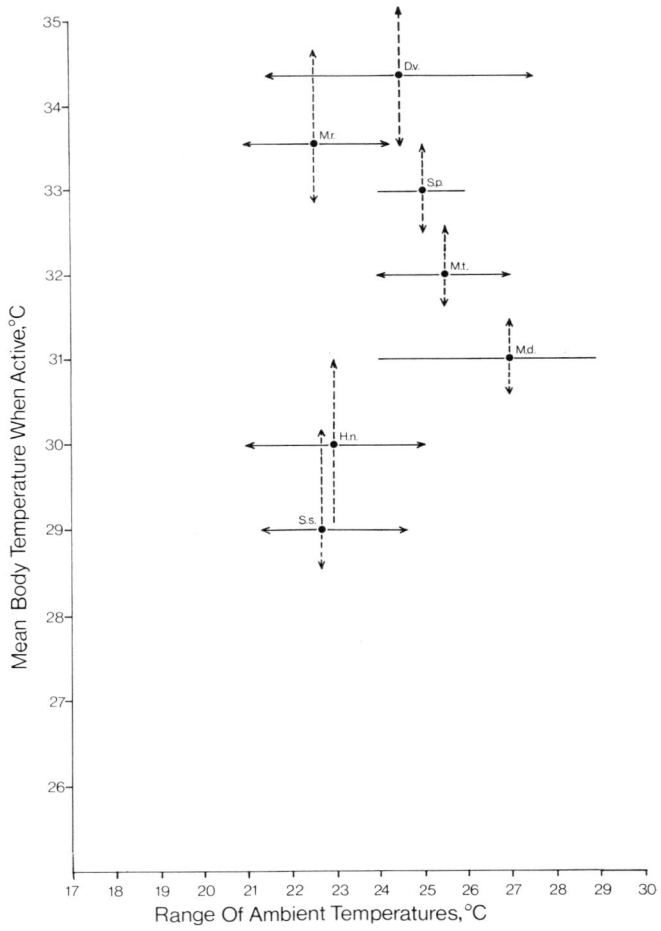

strongly suspected that *some* species of *Marmosa* become semitorpid during part of the year (Morrison and McNab, 1962); other species appear to show a pronounced 24-h rhythm in their thermoregulatory patterns, but maintain core temperatures in the range of 33 to 35 °C when active (Hunsaker, 1977).

It is wise at this point to distinguish between thermoregulatory patterns that show seasonal variation and those that show variation during a 24-h cycle and, furthermore, to distinguish groups of organisms that when active show lability in core temperatures from those that show constancy. Finally, we should examine those cases where distinct differences exist between taxa with respect to the core temperature when active. The point had been made by Crompton et al. (1978) that didelphids show a higher oxygen consumption when exercising than is the case with *Tenrec ecaudatus* and *Erinaceus europaeus*. This presupposes that the optimum temperature of enzymatic activity is lower in the latter two species when compared with that of *Didelphis*. One may further conclude that the didelphids operate at a higher core temperature when active than do the hedgehogs and the common tenrec. I will return to this point later but, suffice it to say that it would appear that the core temperature of at least *D. virginiana* and *M. robinsoni* when active are at 33 °C or above (see also Figure 1).

The didelphids, like all marsupials, are characterized by giving birth to extremely altricial* young. For those species that have been studied, the gestation period is less than or equal to 14 days. Extrauterine maturation is accomplished while the young is attached to a teat in a teat area or pouch. In common with the Dasyuridae in Australia, the Didelphidae show variation in the form of pouch development from virtually no pouch in *Marmosa* to a very well-developed pouch in *Didelphis*.

Feeding habits in the Didelphidae range from those of crustacivores, such as *Chironectes*, through insectivore-frugivores, such as *Marmosa*, to omnivores, such as *Didelphis*. Prey capture in the Didelphidae involves primarily a slow approach, rush, grasping the prey with the mouth and shaking it, or pinning the prey with the forepaw while administering a killing bite. Large prey are generally bitten and tossed to be returned to while in a crippled state. The bite and shake process are repeated (see Roberts et al., 1967; Eisenberg and Leyhausen, 1972).

Social structure in the Didelphidae appears to be simplified. Females generally show fidelity to a home range during the rearing cycle. Adult male home ranges are much larger and overlap those of several females. Most of the Didelphidae are geared for reproduction during a single season. Litters tend to be large, and there is a tendency toward semelparity.* Of course, there are exceptions, and *Caluromys* appears to be the most exceptional among the New World genera (see Eisenberg, 1975a, in press; Collins, 1973).

The family Dasyuridae shows many similarities to the Didelphidae but some exceptions. The Dasyuridae have evolved several genera that have departed from the conservative pattern and have adapted for life in extremely arid regions. These species, such as *Dasyuroides byrnei* and *Dasycercus cristicaudata*, have also departed from semelparity* and, for their size, may be rather long lived and have small litters. These two species may reproduce successively over many seasons. On the other hand, the genus *Antechinus* has a number of species that are rather conservative, are adapted to moist mesophilic or tropical forests, and show the same tendency toward semelparity* as has been exemplified in the genera *Marmosa* and *Didelphis* (Eisenberg, in press; Lee et al., 1977).

In feeding habits, the Dasyuridae range from carnivores through omnivores to insectivores; no member of the Dasyuridae has evolved into an aquatic form. Social structure within the Dasyuridae shows a wide range of variation, as might be anticipated from a consideration of the variety of trophic and demographic strategies. Some form of pair bonding may be shown within the genus *Dasyurus* (Settle, personal communication). On the other hand, reproductive patterns within the genus *Antechinus* appear to parallel those of most of the family Didelphidae, namely, parental care falling entirely to the female and a strong tendency toward semelparity,* especially in the male sex class (Lee et al., 1977).

In summary, a great deal of variation is shown within the families Didelphidae and Dasyuridae with respect to feeding patterns, and degrees of arboreality. Nevertheless, there is an enduring trend toward nocturnality, a simplified social structure, and a tendency to be either insectivores or small carnivores. In conformity with the small size of most of the species, there is a tendency toward almost a near-semelparous* mode of reproduction. Thermoregulation in the Dasyuridae appears to be a highly developed form of endothermy but with a slightly lower core temperature and resting metabolic rate than is the case with many of the genera in the Didelphidae (MacMillen and Nelson, 1969; Dawson and Hulbert, 1970).

The Edentata (Cingulata and Pilosa)

Turning to the Eutherians, four taxa show a rather conservative body plan. These include the Edentata, the Tenrecomorpha, the Erinaceomorpha, and the Soricomorpha. I will treat them in turn.

The Edentata are one of the most conservative groups of living eutherians, although, as is the case with the Monotremata, the majority of the living genera show many derived characters. There are three living families: the Bradypodidae, Dasypodidae, and Myrmecophagidae. Only the Dasypodidae show partial fidelity to a stem body plan. The Bradypodidae and Myrmecophagidae are extremely specialized. As was found with the Monotremata, the Myrmecophagidae and Bradypodidae have been strongly influenced by K-selection* (Eisenberg, 1975a). Litter size is usually one, and potential longevity may exceed 25 years. The sloths are very specialized for feeding on foliaceous vegetation, and the anteaters are specialized for feeding on the social insects of the order Isoptera and Hymenoptera. The armadillos or family Dasypodidae show a wide range of specializations from forms that are near-obligate myrmecophages (*Cabassous* and *Priodontes*) to generalized omnivores, such as *Chaetophractus* and *Euphractus*. The entire order is confined to the New World.

With such a diverse order, one could expect great variations in activity patterns. Suffice it to say, however, that most members of the order are crepuscular or nocturnal. *Bradypus*, the three-toed sloth, is an exception, being primarily diurnal in its activity patterns (Sunquist and Montgomery, 1973). Thermoregulation is characterized by a pronounced 24-h variation in deep body temperature and a tendency in the genus *Dasypus* to show seasonal semitorpor (Taber, 1945). Core temperatures may show an average lower value for the Bradypodidae than would be anticipated for a mammal of their size. In fact, the low metabolic rate of the sloths has been a subject of enduring interest and study by physiologists (see Goffart, 1971, for a review). On the other hand, there are many edentates that at maximum activity show rather high body temperatures; thus generalizations for the whole order would be extremely improper.

Reproductive trends indicate a long history of adaptation to their specific niches, and many of them are very strongly K-selected* forms. The genus *Dasypus* is an exception, having recently undergone r-selection* as it has adapted to numerous omnivore niches over its wide range from Argentina to the state of Oklahoma in North America (Eisenberg, 1975a). Complex cohesive social organizations beyond the mother–young unit are generally not shown by any member of this order.

The Tenrecomorpha

The tenrecoid insectivores (family Tenrecidae *sensu strictu*) present an extraordinary adaptive radiation on the island of Madagascar (Eisenberg and Gould, 1970; Figure 2). It is clear that they have an African origin, but on the island of Madagascar in the near absence of competitors, this stem group of insectivores was able to radiate in a manner unparalleled by insectivores anywhere else. They range in size from *Microgale cowani*, comparable to the holarctic shrews of the genus *Sorex*, to *Tenrec ecaudatus*, which may exceed 1.5 kg in weight. There are two subfamilies: the Tenrecinae and the Oryzorictinae. The latter are not spinescent, have a long tail, and in general more nearly resemble a "primitive" mammal. The Tenrecinae are larger and have shown an enduring tendency to develop spinescent coats as antipredator behavior. The least spinescent, as an adult, is *Tenrec ecaudatus*, which, by virtue of its large size, can offer effective antipredator behavior through biting and counterattack.

Most of the tenrecoid insectivores are nocturnal or crepuscular. There are several examples among the genus *Microgale* that suggest cryptic activity during daylight hours. The genus *Hemicentetes* is unique among the Tenrecinae in that one species, *Hemicentetes semispinosus,* is active at midday (Eisenberg and Gould, 1970). Thermoregulatory behavior shows wide variation within the family. Although all species show 24-h variation in their core temperature, the genus *Microgale* tends to show the least. Many species of the family Tenrecidae show an annual torpor, but some species of *Microgale* show no torpor and maintain a high core temperature. Whereas many species of the subfamily Tenrecinae tend to show maximum activity at rather low core temperatures, the Oryzorictinae appear to have a higher core temperature when active (see Figures 1 and 3).

Figure 2. Four insectivores. (A) *Solenodon paradoxus.* (B) *Tenrec ecaudatus.* (C) *Microgale talazaei.* (D) *Hemicentetes semispinosus.*

Feeding habits vary from omnivory in *Tenrec ecaudatus* to insectivory in most species of the genus *Microgale* to crustacivore or piscivore adaptations in the genus *Limnogale*. *Hemicentetes* is exceptional in its specialization for feeding on earthworms (Eisenberg and Gould, 1970).

The social structure shows as much variation as one would anticipate from such a variety of feeding niches as have been filled by the Tenrecidae. Even the most complex social structures are, however, derivatives of a basic family structure, and the most complex is shown within the genus *Hemicentetes*, where multigenerational groups of females may occupy the same communal burrow system. In the genus *Hemicentetes* the evolution of a stridulating organ or group of quills that produces ultrasonic sounds in the middle of

the back is rather unique within the Mammalia. The organ serves to coordinate the movements of mother and young when they form a cohesive foraging unit during the period approximately 7 days following the emergence of the young from their burrow (Eisenberg and Gould, 1970).

Reproductive trends within the tenrecomorphs also show considerable variation. Some species have been strongly influenced by r-selection* and have approximated a semelparous* mode of reproduction analogous to that shown for the Didelphidae and the edentate genus *Dasypus*. On the other hand, some species, such as *Microgale talazaci*, have shown an extreme tendency toward iteroparity* and, for their size class, are exceedingly long-lived small mammals. Figure 4 exemplifies the trends in comparison with selected Marsupialia (see also Eisenberg 1975b).

Figure 3. Some body temperatures expressed as degrees above the ambient for a series of "insectivores" during various phases of diel and annual activity patterns. Abbreviations as in Figure 1 with the additions of: H.s. = *Hemicentetes semispinosus;* T.e. = *Tenrec ecaudatus;* E.t. = *Echinops telfairi.* (Data from Eisenberg and Gould, 1966, 1970.)

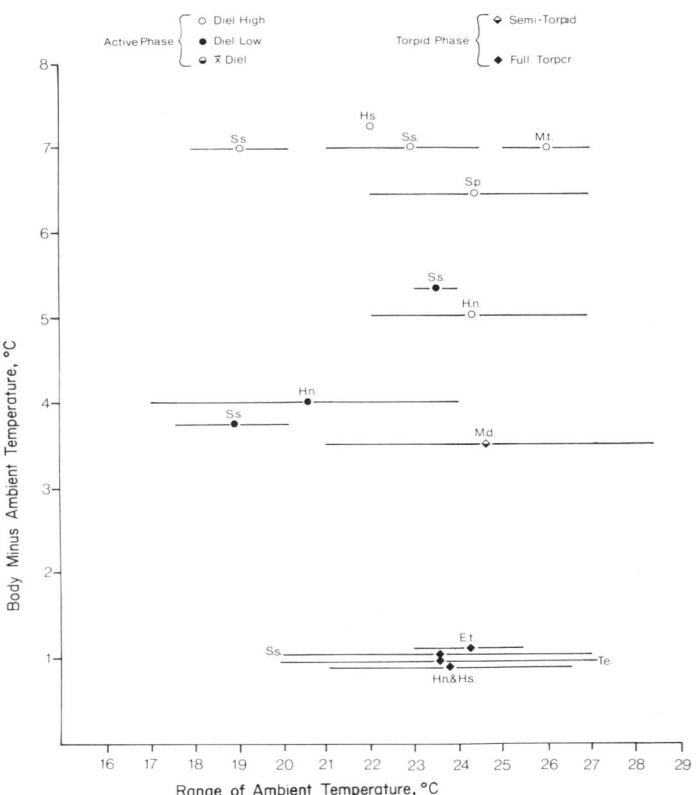

The Echinosoricinae

The Erinaceomorpha include the hedgehogs of the family Erinaceidae. There are two subfamilies: the Erinaceinae and the Echinosoricinae. *Erinaceus europaeus* is one of the best-studied insectivorous mammals, and I will not attempt to review its life history here except to say that it has a highly specialized antipredator defense mechanism convergent with that of the hedgehog tenrecs of the genera *Setifer* and *Echinops*. Of greater interest is the more conservative genus *Echinosorex* of the second subfamily. *Echinosorex*

Figure 4. The relation between maximum age at last reproduction (abcissa) and mean litter size (ordinate) for some marsupials, "insectivores," and rodents. Note the gradation from semelparity to iteroparity. Dark bands separate three groups of contrasting reproductive strategies. Those at the top show an extreme semelparous strategy, those in the middle an intermediate strategy, and those at the bottom an iteroparous reproductive strategy. M.r. = *Marmosa robinsoni;* A.s. = *Antechinus stuarti;* tenrecs (black dots) as in Figure 3; L.c. = *Lagurus sp.;* R.m. = *Reithrodontomys megalotis;* N.l. = *Neotoma lepida;* P.c. = *Peromyscus californicus;* S.a. = *Sorex araneus.*

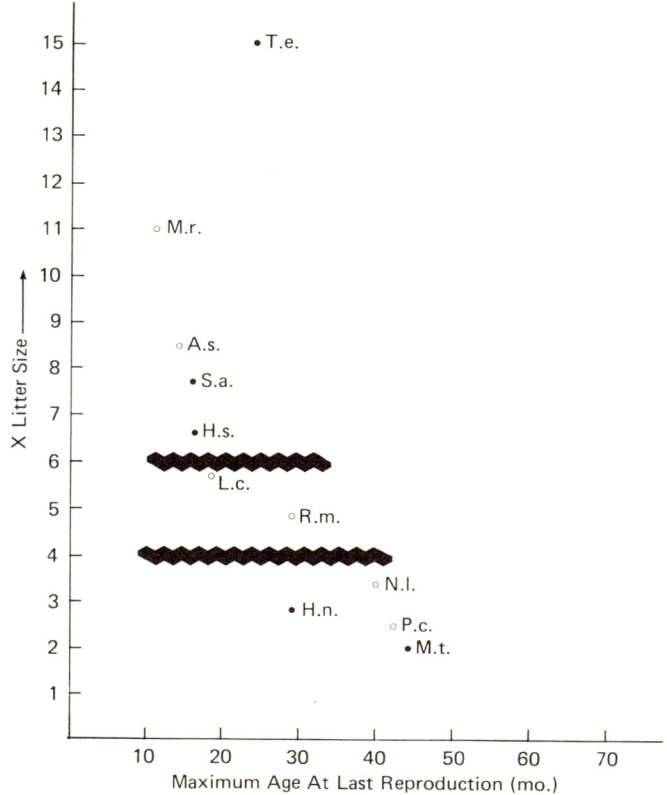

gymnura, which I and my colleague Dr. Edwin Gould have studied (Gould, 1979), is confined to Malaya, Sumatra, and Borneo. It is approximately the same size as *Didelphis virginiana* and shows a respectable core temperature when active. Living in the environment that it does, it does not exhibit any extreme variations in the annual cycle of its activity. Its reproductive pattern suggests extreme K-selection* because the modal number of young is one. In feeding habits, it is somewhat specialized for feeding along the shores of streams, having special prey-capture movements suggestive of capturing fish and crustaceans (Eisenberg and Leyhausen, 1972). In its social structure, it is in conformity with other conservative members discussed previously, showing no enduring social unit beyond the mother–young. In its interaction patterns and form of aggressive behavior, it is strongly reminiscent of the genus *Didelphis* (Gould, 1978).

The Solenodontidae

Concerning the Soricomorpha, I will confine my remarks to a single family, the Solenodontidae, and defer remarks on the Soricidae, which are discussed elsewhere in this volume.

The genus *Solenodon* is presently confined to the Greater Antilles. There are two species: *Solenodon paradoxus* on Hispaniola and *Solenodon cubanus* found on Cuba (see Eisenberg and Gould, 1966). The animals are slightly smaller than *Didelphis*, weighing slightly under 1 kg when adults. They are unique in the possession of a mobile proboscis which has a cartilaginous ball-and-socket joint inserting at the tip of the nasals. The claws on the forepaws of *Solenodon* are highly suggestive of a burrowing habitus. The eye is minute in *Solenodon* and is the smallest eye in proportion to body size of any of the genera heretofore described. The gait and the structure of tail and limbs are strongly suggestive of a sustained terrestrial habitus (which likewise is true for the Erinaceidae and most members of the Tenrecidae.) In activity *Solenodon* is more or less nocturnal. Thermoregulation shows some diel variation, but the core temperature when active is near 33 °C (see figure 1.) Given the habitat they occupy, there is very little indication of an annual cycle in activity, nor is there any indication that profound torpor occurs. A pronounced molt occurs, however, in August (Eisenberg and Gould, 1966).

In terms of reproduction, the modal litter size is one, although they may have two young for a maximum. This size is strongly indicative of K-selection.* Prior to the introduction of the mongoose on Hispaniola and Cuba, the only natural predators of *Solenodon* would have been raptorial birds and boas. In feeding habits, it would appear that *Solenodon* is specialized for probing in leaf litter and soft soil for arthropods and other invertebrates; however, experiments in captivity suggest that they are adept at killing ground-nesting birds (and especially young birds) as well as lizards (Eisenberg and Leyhausen, 1972).

The social structure of *Solenodon*, as can be deduced from captive studies and wild collections, suggests the basic unit of a mother–family. Communal use of burrow systems by several generations is strongly implied. Tolerance in captivity among members of the same family is high. A male may be tolerated with a female, but apparently the female retires to a separate denning area at the time of partus.

Discussion

Primitive mammals or mammals retaining conservative characters have remained conservative in form very probably because they continue to occupy an ecological niche that is in itself "conservative." On the other hand, the major reproductive modes are part of "phylogenetic inertia" (Wilson, 1975). As indicated earlier, the monotremes retain primitive characters in the form of their anatomy and retain a conservative mode of reproduction. On the other hand, they have acquired a great many derived characters and, in terms of their reproductive strategy, indicate extreme specialization.

The reproductive system of the Marsupialia involves a condition of ovoviviparity (Luckett, 1975). A shell membrane is retained throughout most of the intrauterine phase, no trophoblast develops, and exchange of nutrients between the uterine wall and the developing embryo occurs for the most part through the yolk sac. Gestation is extremely short, rarely exceeding the inter-estrous interval. At birth the young is extremely undeveloped and immediately attaches itself to a teat which may or may not be present in a pouch. An extended period of development occurs outside of the mother's body, and the young animal is transported for a considerable time prior to a nest phase. The nest phase was probably primitive in the Marsupialia and has been avoided by the cursorial Macropodidae (see also Tyndale-Biscoe, 1973).

The eutherians specialization their reproductive system by the evolution of an embryonic trophoblast which permitted retention of the fetus for a longer period of time within the uterus and increased efficiency of gaseous and nutrient exchange between the developing embryo and the mother. Primitively the young eutherian was born in a much more advanced state than the marsupial but was probably not precocial at the time of its birth. Precociality* is probably an adaptation and thus a specialization away from the stem form. By the same token, giving birth to very large litters of extremely altricial* young may be considered an alternative specialization from the intermediate primitive condition. A nest phase for the young was probably obligatory in the original eutherian rearing cycle.

Most of our living conservative mammals from the Metatheria and Eutheria continue to dwell in the tropics or subtropics. There are exceptions, with the opossum, *Didelphis virginiana,* adapting to the Temperate Zone, and several species from the family Dasyuridae adapting to extreme arid conditions in Australia or the Temperate Zone of Tasmania. One species from the order Edentata, *Dasypus novemcinctus*, has invaded Temperate Zone conditions in the north, whereas the genus *Chaetophractus* has shown a similar adaptive trend in the Southern Hemisphere. The hedgehog, *Erinaceus*, has of course adapted to a wide range of temperate and arid conditions. Given these exceptions, however, the least specialized conservative mammals still tend to occupy tropical habitats and are, for the most part, nocturnal.

When they are active, the mean body temperatures of didelphid marsupials and the Solenodontidae appears to be high (>32 °C). Some genera of the Tenrecidae, such as *Microgale*, also have a high body temperature when active. However, many of the Edentata and the tenrecine tenrecomorphs show lower body temperatures (26 to 29 °C) when active than would be predicted from a study of less conservative eutherians. Lower body temperatures imply lower basal metabolic rates (BMR). Most eutherian mammals show a decreasing metabolic rate as body size increases. The regression of BMR against body weight yields an intercept value higher than a comparable regression performed for marsupials (Dawson and Hulbert, 1970). Some conservative eutherians also show low BMR values for their size class, but there is considerable scatter. Deviations from taxonomic averages may represent specific adaptations and may not indicate conservative tendencies in endothermy (for a review see Poczopko, 1971).

The reproductive patterns of morphologically conservative mammals range from extreme K-selection* to extreme r-selection.* In and of themselves, these attributes are not indicative of a conservative character but rather are the end points of selection tuning reproductive rates to the optimum. In feeding habits all of the morphologically conservative mammals tend as a basic mode of exploitation to be small carnivores or insectivores.

The extent to which learning plays a role in the ontogeny of the young may be reflected in part by the relative brain size of the species in question. Many morphologically conservative mammals show rather low encephalization quotients (Jerison, 1973; Eisenberg, 1975a); however, some have surprisingly large brains in relation to their body size. Inevitably, the larger-brained forms tend to be those that have undergone moderate K-selection, are rather long lived, and apparently exploit a feeding niche that is to some extent unpredictable. Relatively large brain size and complex social organizations form one alternative subset of an outcome of extreme K-selection (Eisenberg, in press).

The behavioral repertoires of the conservative eutherians and marsupials can be conveniently compared (Eisenberg and Golani, 1977). In both the Marsupialia and the eutherians, the unspecialized forepaw was utilized in grooming the fur of the face and the vibrissae. In the primitive condition, the animal sits upright and wipes with the wrist while grasping at the muzzle and vibrissae on the down stroke. Primitively this face-washing movement probably also served to spread saliva and glandular secretions associated with the major sensory hairs on the anterior and lateral parts of the animal's body; thus, it is also a self-marking movement (Eisenberg and Kleiman, 1972).

In the Marsupialia, feces were typically not buried, probably because the primitive marsupial line was already specialized for arboreality and lived in tree cavities. On the other hand, the conservative eutherian generally buries its feces, which is probably indicative of living in burrows and represents an adaptation to preserve nest sanitation. In both the Marsupialia and the Eutheria, urine may be deposited in droplets in the course of movements and thus has a secondary marking function. Locus-specific urination either from a branch or in a special chamber of the burrow is probably a primitive trait. Marking with glandular secretions from the perineal region or from the cloaca or anus is probably a conservative trait in both the Marsupialia and the eutherians.

Prey capture in the Marsupialia and the eutherians was primitively accomplished by seizing small prey with the mouth or by pinning with the forepaws while biting. Although salivary glands specialized for producing venom are to be found in some species of the insectivore family Soricidae and in the Solenodontidae, this may be a specialized character rather than conservative. No poison bite has been detected in the Marsupialia.

Feeding on small prey involves bracing with the forepaws and side-to-side or up-and-down movements of the head thus dismembering larger prey. The forepaws are used to pin the prey. Smaller prey may be chewed directly (Eisenberg and Leyhausen, 1972). Holding foodstuffs in the forepaws while feeding is a conservative trait in the Marsupialia, but the conservative Eutheria typically do not hold food in the forepaws or sit upright while feeding. Caching food is a specialization evolved many times within the Mammalia, but it is not considered to be a conservative trait. Caching is absent in the Didelphidae, Tenrecidae, and Erinaceidae.

Burrow construction is probably a conservative trait within the Eutheria but may not be so in the Marsupialia. Nest construction from leaves or grass is present in both taxa. In both lines, transport of nesting materials in the mouth is a conservative trait. The transport of nesting material in the tail as well as in the mouth appears to be a conservative behavioral character in both the Monotremata and the Marsupialia. Transport of nesting material using the tail is unknown in the eutherians.

Envoi

There are several misconceptions concerning the behavior patterns of morphologically conservative mammals, and I would like to correct them for the record. It has been stated that "primitive" mammals when compared with "advanced" mammals have a simplified behavioral repertoire, especially with respect to their social behavior. I think this generalization is somewhat premature because a behavioral inventory or ethogram for primitive mammals is often rather incomplete, the reason being that the great variety of chemical signals employed in coordinating their behavior patterns is generally not counted in as part of their repertoire. Furthermore, the vocalizations employed in intraspecific interactions are rarely enumerated completely for morphologically conservative mammals because very often they are high frequency sounds above the range of human perception (Poduschka, 1977). Often the behavioral repertoires of interacting small mammals are not as finely subdivided as in larger mammals by a human observer because temporal resolution is very difficult and perhaps can only be accomplished with the aid of high speed cinematography (Gould, 1969).

It has often been said that "primitive" mammals have a limit with respect to the social complexity that they exhibit. This misconception may occur because the durations of complex social interactions, for example in the mother–young unit, are often brief but not necessarily less complex (Eisenberg and Gould, 1970). Most morphologically conservative mammals still alive today are of rather small size with a concomitant shortened life expectancy and an early sexual maturation. Because sexual reproduction occurs without the formation of sterile castes, there is no selective premium for a long-term, interdependent social organization such as one finds among the higher social insects (Wilson, 1975). Thus a mechanistic social system based on neuter castes and permissive of only small amounts of learning, such as exemplified by the societies of the Hymenoptera and Isoptera, is a closed pathway in the evolution of social structures not only in small mammals but in all of the Mammalia. There is instead a premium for selection favoring preprogramming of certain basic pathways in the nervous system of those small mammals that exhibit a tendency toward semelparous* reproduction regardless of whether they are conservative or advanced. As a result, the relative brain size is small, and the need to learn complex patterns is reduced. Nevertheless, complex interdependent social structures could have evolved in organisms with relatively small brains only if selection also favored the formation of sterile nonreproducing castes. Because this has not happened in the sexually reproducing Mammalia, the social complexity that one finds in the social insects is absent, but this absence is not in and of itself a correlate of the small brain size.

It appears that the behavioral differences between morphologically conservative mammals having a low encephalization quotient and those mammals exhibiting more derived characters including high encephalization quotients are differences of degree rather than attributable to any absolute differences. Larger mammals with larger encephalization quotients often exhibit more plasticity in their behavior and a greater capacity to program new information during their growth and development, differences that tend to set them off from small mammals with low encephalization quotients whether or not they exhibit a conservative morphology.

Glossary

The definitions given here are taken from Wilson, 1975.

altricial: pertaining to young animals that are helpless for a substantial period following birth; used especially with reference to birds. (Contrast with precocial.)

iteroparity: the production of offspring by an organism in successive groups. (Contrast with semelparity.)

K-selection: selection favoring superiority in stable, predictable environments in which rapid population growth is unimportant. (Contrast with r-selection.)

precocial: referring to young animals who are able to move about and forage at a very early age; especially in birds. (Contrast with altricial.)

r-selection: selection favoring rapid rates of population increase, especially prominent in species that specialize in colonizing short-lived environments or undergo large fluctuations in population size. (Contrast with K-selection.)

semelparity: the production of offspring by an organism in one group all at the same time. (Contrast with iteroparity.)

References

Burrell, H. (1927). *The Platypus.* Waterloo: Eagle Press.

Collins, L. (1973). *Monotremes and Marsupials.* Washington D.C.: Smithsonian Institution Press.

Crompton, A. W., Taylor, C. R. and Jagger, J. (1978). Evolution of homeothermy in mammals. *Nature 272:*333–6.

Dawson, T. J., and Hulbert, A. J. (1970). Standard metabolism, body temperature, and surface areas in Australian marsupials. *Am. J. Physiol. 218:*1233–8.

Eisenberg, J. F. (1975a). Phylogeny, behavior, and ecology in the Mammalia. In *Phylogeny of the Primates: An Interdisciplinary Approach,* eds. P. Luckett and F. Szalay, pp. 47–68. New York: Plenum Press.

Eisenberg, J. F. (1975b). Tenrecs and solenodons in captivity. *Int. Zoo Yearb. 15:*6–12. London: The Zoological Society of London.

Eisenberg, J. F. (In press.) *The Mammalian Radiations: Studies in Evolution and Adaptation.* Chicago: University of Chicago Press.

Eisenberg, J. F. and Golani, I. (1977). Communication in Metatheria. In *How Animals Communicate,* ed. T. A. Sebeok, pp. 575–99. Bloomington: University of Indiana Press.

Eisenberg, J. F. and Gould, E. (1966). The behavior of *Solenodon paradoxus* in captivity with comments on the behavior of other Insectivora. *Zoologica, (N.Y.) 51:*49–58.

Eisenberg, J. F., and Gould, E. (1970). The tenrecs: A study in mammalian behavior and evolution. *Smthson. Contrib. Zool. 27:*1–137.

Eisenberg, J. F., and Kleiman, D. G. (1972). Olfactory communication in mammals. *Annu. Rev. Ecol. Syst. 3:*1–32.

Eisenberg, J. F., and Leyhausen, P. (1972). The phylogenesis of predatory behavior in mammals. *Z. Tierpsychol. 30:*59–93.

Goffart, M. (1971). *Function and Form in the Sloth.* London: Pergamon Press. 225 pp.

Gould, E. (1969). Communication in three genera of shrews (Soricidae): *Suncus, Blarina,* and *Cryptotis. Commun. Behav. Biol., Part A 3:*11–31.

Gould, E. (1979). The behavior of the moonrat, *Echinosorex gymnurus* (Erinaceidae) and the pentail shrew, *Ptilocercus lowi* (Tupaiidae) with comments on the behavior of other Insectivora. *Z. Tierpsychol. 48:*1–27.

Griffiths, M. (1968). *Echidnas*. London: Pergamon Press, 282 pp.

Griffiths, M. (1978). *The Biology of Monotremes*. London: Academic Press.

Hunsaker, D. (1977). Ecology of New World marsupials. In *The Biology of Marsupials*, ed. D. Hunsaker, pp. 95–158. New York: Academic Press.

Hunsaker, D. and Shupe, D. (1977). The behavior of New World marsupials. In *The Biology of Marsupials*, ed. D. Hunsaker, pp. 279–348. New York: Academic Press.

Jerison, H. J. (1973). *Evolution of the Brain and Intelligence*. New York: Academic Press.

Kirsch, J. A. W. (1977). The classification of marsupials. In *The Biology of Marsupials*, ed. D. Hunsaker, pp. 1–50. New York: Academic Press.

Lee, A. K., Bradley, A. J., and Braithwaite, R. W. (1977). Corticosteroid levels and male mortality in *Antechinus stuartii*. In *The Biology of Marsupials*, eds. B. Stonehouse and D. Gilmore, pp. 209–20. Baltimore: University Park Press.

Luckett, P. (1975). The ontogeny of the fetal membranes and placenta. In *Phylogeny of the Primates: An Interdisciplinary Approach*, eds. W. P. Luckett and F. Szalay, pp. 157–82. New York: Plenum Press.

McKenna, M. C. (1975). Toward a phylogenetic classification of the Mammalia. In *Phylogeny of the Primates: A Multidisciplinary Approach*, eds. W. P. Luckett and F. Szalay, pp. 21–46. New York: Plenum Press.

MacMillen, R. E. and Nelson, J. (1969). Bioenergetics and body size in dasyurid marsupials. *Am. J. Physiol. 217:*1246–51.

McNab, B. K. (1978). The comparative energetics of Neotropical marsupials. *J. Comp. Physiol. 125:*115–28.

Morrison, P. R., and McNab, B. K. (1962). Daily torpor in a Brazilian murine opossum (*Marmosa*). *Comp. Biochem. Physiol. 6:*57–68.

Poczopko, P. (1971). Metabolic levels in adult homeotherms. *Acta Theriol. 16:*1–21.

Poduschka, W. (1977). Insectivore communication. In *How Animals Communicate*, ed. T. A. Sebeok, pp. 600–33. Bloomington: Indiana University Press.

Roberts, W. W., Steinberg, M. T., and Means, T. W. (1967). Hypothalamic mechanisms for sexual, aggressive and other motivational behaviors in the opossum, *Didelphis virginiana*. *J. Comp. Physiol. Psychol. 64:*1–15.

Sunquist, M., and Montgomery, G. G. (1973). Activity patterns and rates of movement of two-toed and three-toed sloths (*Choloepus hoffmanni* and *Bradypus infuscatus*). *J. Mammal. 54:*946–54.

Taber, F. W. (1945). Contribution on the life history and ecology of the nine-banded armadillo. *J. Mammal. 26:*211.

Tyndale-Biscoe, H. (1973). *The Life of Marsupials*. New York: Elsevier Press.

Wilson, E. O. (1975). *Sociobiology*. Cambridge: Harvard University Press.

3

Milk and mammalian evolution

OLAV T. OFTEDAL

Milk in the evolution of mammalian infancy

Milk secretion is a characteristic of mammalian reproduction. Once adopted in the course of mammalian evolution, it was universally retained. Some birds, including pigeons and doves, the greater flamingo, and the emperor penguin also produce nutritive fluids for the young, but their secretions are of crop or esophageal origin (Fisher, 1972). Thus mammary milk is not only universal to mammals, it is unique to them.

The origin and early evolution of milk production is obscure. Several authors, including Charles Darwin, have speculated on the incremental changes that could have led from a presumably dilute sweat-like fluid to the nutrient-rich secretion we know as milk (Long, 1972). The evolution of milk must have been tied to the transformation of well-developed and self-reliant reptilian hatchlings to the more altricial mammalian young. Living representatives of the "primitive" orders Monotremata, Marsupialia, and Insectivora have altricial neonates as a rule. The monotreme hatchling and the marsupial newborn are virtually at an embryonic stage of development at birth (Griffiths et al., 1969; Tyndale-Biscoe, 1973). These mammals are highly dependent on a long lactation period. If, in fact, the altricial condition is a primitive or conservative feature in mammals, so must be a reliance on milk.

Hopson (1973) argued that the energy costs of producing precocial young would be too much for endothermic small mammals of the size prevalent in the late Triassic. According to this view, altriciality and a dependence on maternal milk were essential components of early mammalian evolution. Patterns of skull growth and teeth eruption in early mammalian fossils support the notion of an extended period of maternal care with infant suckling (Crompton, this volume). Did milk secretion then originate among the reptilian predecessors of mammals just as among some groups of living birds? By the development of milk secretion, the mammals or their reptilian forebears were able to store energy, minerals, and other nutrients for mobilization as needed. In contrast to the predominant avian mode of parental food collection and feeding of the young, the milk feeders became less dependent on the immediate availability and quality of food sources.

The maternal preprocessing of food made acquisition of adult masticatory and digestive function unnecessary for neonates. The digestive processes for fat, protein, and carbohydrates in mammalian young are geared to maternal milk (Oftedal, 1975); even precocial young are generally unable to cope with adult diets. Lactivory or milk feeding, coupled with altered digestive function, has allowed postnatal transfer of passive immunity (immunoglobulins) from mother to young in many species (Brambell, 1970). Enzyme activities associated with intermediary metabolism in the young also reflect reliance on milk nutrients (Oftedal, 1976). The evolution of lactation has been a key factor in the evolution of mammalian infancy.

Milk must represent a compromise between the need to conserve the resources of the female and the need to maintain nutrient intakes of the young at a level that permits optimal growth. The actual compromise achieved will presumably reflect the overall reproductive strategy. In species geared to produce many young in a short span of time (r-selection) the energy and nutrient drain on the female may be particularly severe. Thus in the masked shrew *Sorex cinereus* the five to nine young in a litter are reared nearly to adult size in just 20 days, at which time they are weaned (Forsyth, 1976). These lactating females often do not survive into the next breeding season. On the other hand, bats of a similar size characteristically have only one or two young per litter, but live for many years. It would be interesting to compare the maternal investment in milk and the nutrient intakes of the young in such extremes.

Milk and infant nutrition

The rapid growth that characterizes the early postnatal period implies high requirements for protein, minerals, and other materials needed for tissue formation. As the sole food of sucklings, milk must contain all nutrients in adequate amounts for normal development. The true digestibility of milk components is high (Roy, 1970). In some species milk proteins and triglycerides may be absorbed intact from the gastrointestinal tract, at least early in lactation (Oftedal, 1975).

Interspecies differences in nutrient requirements can stem from differences in growth rates, relative amounts of tissue deposited during growth, and patterns of energy utilization in the maintenance of body temperature. Altricial young typically are born with very limited nutrient stores and achieve a rapid rate of growth at a time when they are at least partially ectothermic, relying on maternal heat to support body temperatures. The ratio of growth needs (protein, minerals, etc.) to maintenance needs (especially energy) will be high. Where growth includes substantial fat deposition, as in aquatic and some arctic mammals, the relative proportion of fat in milk is elevated. On the other hand, some species such as golden hamsters deposit very little lipid prior to weaning (Adolph and Heggeness, 1971). Golden hamster milk is reported to be rather low in fat (Jenness and Sloan, 1970).

Translation of estimated nutrient requirements to interspecies differences in milk composition is complicated by the need to consider volumes of milk ingested. Rapid growth with associated high nutrient needs may be achieved by an elevated milk consumption, a more concentrated milk, or a combination of both. Linzell (1972) has demonstrated that peak milk production (both in volume and in energy output) of diverse species is a function of metabolic body size. Relative to body weight, a small species produces more milk than a larger one. This milk may also be more concentrated. Blaxter (1961) pointed out that energy requirements of mammalian young are proportional to their metabolic body size (body weight $^{0.73}$) whereas milk intake capacities (gastrointestinal volumes) are proportional to body weight (wt). The dry matter content should then be proportional to energy requirement divided by milk volume: wt $^{0.73}$/wt $^{1.0}$ = wt$^{-0.27}$. Payne and Wheeler (1968) obtained an exponent of about -0.28 in regression of both milk energy and milk protein contents against birth weights of a number of species. On average, the smaller the species the richer the milk.

The substantial scatter of values about the Payne and Wheeler (1968) regression lines suggest the influence of other parameters. For example, Ben Shaul (1962) noted a rough correlation of dry matter content to nursing interval: Those species that nursed their young less frequently produced a more concentrated (high dry matter content) milk. Two good illustrations, the echidna and the tree shrews, will be discussed further on. In any case the early and often quoted attempts of Bunge (1902) and Abderhalden (1908) to relate milk protein content to growth rate (as assessed by time to double birth weight) were overly simplistic.

The proportion of milk energy provided by protein is a more meaningful index of resources available for infant growth. Powers (1933) reevaluated the data of Abderhalden and others as percentages provided by protein, fat, and carbohydrate and found that for most species protein supplied about 20% of the energy, irrespective of body size or growth rate. On the basis of tabulated data for 132 mammalian species (Jenness and Sloan, 1970), similar calculations reveal that nearly two-thirds of the species produce milk with 15 to 30% of the energy as protein. A comprehensive review of the relation of milk composition to growth is clearly needed.

In considering particular nutrients in milk, account must be taken of the capacity for prenatal storage with subsequent mobilization after birth. For example, iron may be deposited in the prenatal liver for use during the suckling period. The milk of many mammals is low in iron. The rabbit has substantial stores of iron at birth and does not normally accumulate much additional iron during suckling, as its milk is poor in iron, although not as poor as some (Tarvydas et al., 1968). A different situation exists in those species born in a very altricial state. A significant growth period precedes the physical development that allows access to nonmilk iron sources, and size considerations seem to rule out the possibility of adequate prenatal iron stores. Newly hatched monotremes and newborn marsupials must obtain iron in milk just as physiologically comparable placental young obtain iron in utero. In fact, the milks of two species analyzed, the echidna *Tachyglossus aculeatus* and the quokka *Setonix brachyurus*, are rich in iron at least in the early part of lactation (Griffiths et al., 1969; Loh and Kaldor, 1973).

Interpretation of milk composition

Several comparative studies based on literature values for milk composition were quoted above. The validity of any conclusions rests on the accuracy of the data. Unfortunately, analytical values have often been reported without such essential information as stage of lactation, prior separation of mother from young, use of sedation and/or oxytocin, and volume collected. A brief discussion of the influence of these factors on milk composition data must precede any review of the milks of "primitive mammals."

The milk of any species exhibits substantial variation in gross composition, both within and among individuals (Jenness, 1974). In general, the lipid or fat fraction is most variable and carbohydrate the least. Protein content may vary with stage of lactation. Among well-studied domestic species and the human, the earliest milk (colostrum) is high in protein, followed by a decline (transitional phase) to the more or less stable level of mature milk. Late milk often has elevated protein levels, especially during the process of mammary involution. Lactose levels are usually low in colostral milk, while fat content is variable. Samples obtained early or late in lactation may not be typical of mature milk. Stage of lactation should by reported whenever milk analyses are published.

The method of milk collection may influence composition, especially in regard to fat content. Studies in several species of eutherians have shown that the fat content of milk rises during the course of milking. The first milk drawn (foremilk) is lower in fat than the last milk (hindmilk or strippings). The degree of mammary emptying achieved during milking thus assumes importance, as does the degree to which the mammaries may have been emptied by suckling young prior to sampling. Milk collection shortly after suckling may result in abnormally high fat values; incomplete sampling may give abnormally low values.

An effort may be made to minimize sampling bias through a combination of separation, sedation, and oxytocin administration. A period of separation allows the mammary glands to fill, thereby reducing the influence of the most recent suckling. Excessively prolonged separation may interfere with normal lactation, however. Injection of oxytocin induces contraction of the mammary myoepithelial cells and facilitates more complete gland emptying.

Sedation is necesary for wild or excitable species that may become highly stressed when restrained. In the absence of sedation, circulating adrenalin can inhibit the action of oxytocin, blocking the milk ejection reflex (Denamur, 1965).

The milk of primitive mammals

We can now ask: Is any particular type of milk representative of the earliest mammals? It there a "primitive" milk? If the strategy of growth and development of early mammals were known, one could speculate on the milk composition that would be consistent with this developmental scheme. The small size of early Mesozoic mammals, for example, suggests milk of a rather high dry-matter content.

An alternative approach is to review data on milk composition of existing mammals commonly considered primitive. The flaw of this method is in assuming "primitive" mammals to adhere to an evolutionarily conservative pattern of milk production. Published information on the major components of monotreme, marsupial, and insectivore milks illustrates the diversity of results obtained and emphasizes the questionable validity of much of this material. For present purposes both the elephant shrews (Macroscelididae) and the tree shrews (Tupaiidae) are considered to belong to the Insectivora, although their true systematic affinities are unclear (Patterson, 1965; Campbell, 1974). It may also be logical to subdivide the Marsupialia into several orders, not all of which can be considered primitive (Tyndale-Biscoe, 1973).

Monotremata

The monotremes produce a milk high in dry matter: 20 to 41% in the platypus, *Ornithorhynchus anatinus*, and 26 to 53% in echidna, *Tachyglossus aculeatus* (Griffiths, 1968; Griffiths et al., 1973). As with so many features of these creatures, rich milk must be seen as a specialized rather than a primitive trait. Echidnas suckle their young at the infrequent rate of once every 1 to 2 days, at least when the young are rather developed (Griffiths, 1968). The single young may ingest 7 to 10% of body weight per suckling. Whereas suckling intervals in the platypus are unknown, its semiaquatic habit may require fat deposition in the young for insulative purposes prior to leaving the nest. The fraction of milk dry matter contributed by fat has not been determined.

The gross composition of echidna milk has been reported for but a few samples (Table 1). The early report of Marston (1926) refers to a sample taken two weeks after the pouch young had been removed and is probably abnormal. The rather late lactation samples (young 2 to 3 weeks out of pouch) of Griffiths (1965) and the samples of undisclosed origin of Jenness and Sloan (1970) differ markedly in fat and protein content. Whether this is due to sampling method, stage of lactation, or real variability is not clear.

Table 1. *Values of milk composition for some "primitive" mammals*

Species	No. of samples	Dry matter(%)	Fat (%)	Protein (%)	Carbo-hydrate(%)	Ash (%)	Data source
Echidna							
Tachyglossus aculeatus	1	36.7	19.6	11.3[a]	2.8	0.78	Marston (1926)[b]
	1–2	47.0	14.8	16.6	–	–	Griffiths (1965)[b]
	2	–	9.6	12.5[a]	0.9	–	Jenness and Sloan (1970)[c]
Opossum							
Didelphis virginiana	13–14	23.2	11.3	8.4	1.6[d]	1.7	Bergman and Housley (1968)[b]
	1	24.4	7.0	4.8[a]	4.1	–	Jenness and Sloan (1970)[c]
Brush-tailed possum							
Trichosurus vulpecula	46	24.5	6.1	9.2	3.2	1.6	Gross and Bollinger (1959)[b]
Quokka							
Setonix brachyurus	4	13.4	0.9	4.0[a]	3.4	0.9	Jenness and Sloan (1970)[c]
Red kangaroo							
Megaleia rufa	?	12	4.0	3.9	4.7	0.75	Ben Shaul (1962)[c]
	16	22.8	4.9	6.7[e]	2.0[d]	–	Lemon and Barker (1967)[f]
	7	15.2	5.7	6.2[e]	1.9[d]	–	Lemon and Barker (1967)[g]
	1	20.0	3.4	4.6[a]	6.7	1.4	Jenness and Sloan (1970)[c]
Wallaroo							
Macropus robustus	2	19.5	3.8	5.7[a]	6.2	1.3	Bollinger and Pascoe (1953)[b]
Red-necked wallaby							
Protemnodon rufogrisea	?	13	4.6	4.0	4.5	0.77	Ben Shaul(1962)[c]

Hedgehog *Erinaceus europaeus*	?	20.6	10.1	7.2	2.0	2.3	Ben Shaul (1962)[c]
Water shrew *Neomys fodiens*	?	35.0	20.0	10.0	0.1	0.8	Ben Shaul (1962)[c]
Short-tail shrew *Blarina brevicauda*	?	19.9	6.5	11.0	3.2	0.8	Ben Shaul (1962)[c]
Musk shrew *Suncus murinus*	pooled	37.5	17.5	10.7	0.8	1.9	Dryden and Anderson (1978)[b]
Tree shrew *Tupaia belangeri*	1	40.5	25.6	10.4	1.5	–	Martin (1968)[b]

[a]Estimated from protein nitrogen, not total nitrogen.
[b]Discussed in text.
[c]No data available on source or collection of sample(s).
[d]Estimated by reduction method without hydrolysis.
[e]Recalculation using factor of 6.38, not 6.25, for conversion of nitrogen to protein.
[f]Pouch young 122–232 days postpartum.
[g]Young out of pouch, 245–304 days postpartum.

Griffiths (1968) reported that the first milk drawn during the milking of an echidna was higher in dry matter (47%) than the last milk (26%) – a reverse of the usual pattern observed in eutherians. The low carbohydrate content of echidna milk reported by Jenness and Sloan (1970) was confirmed by Messer and Kerry (1973). They determined the principal carbohydrate in echidna milk to be fucosyllactose and in platypus milk to be difucosyllactose, with only small amounts of lactose.

Marsupialia

In the prolonged period of marsupial postnatal development, the efficiency with which milk solids can be converted into infant weight gain must decline as energy is increasingly needed to support endothermy. The relative proportions of fat and protein required from milk may therefore change during development. Griffiths et al. (1972) have demonstrated a rise in fat content during lactation in the red kangaroo, *Megaleia rufa*, but concurrent levels of protein were not determined. With the birth of a second young at about the time the first leaves the pouch, the bizarre situation arises in which the older of the two receives milk that is 3.6 to 4.1 percentage points higher in fat than that received by the neonate (Griffiths et al., 1972). The milks supplied to the two young also differ in fatty acid composition and in the composition of the whey proteins (Lemon and Bailey, 1966). No eutherian is known to produce two different types of milk simultaneously. As with the echidna, the absence of detailed data on the method of collection and stage of lactation of some of the samples listed in Table 1 makes evaluation difficult.

Reported mean values for milk of the American opossum (*Didelphis virginiana*) encompass the entire period of 3 to 110 days postpartum (Table 1). They show a wide range of values: dry matter 9.4 to 31.3%, fat 4.9 to 17.9%, protein 6.7 to 11.8%, and carbohydrate 0.73 to 3.15% (Bergman and Housley, 1968). Unfortunately, the authors do not present their data by age category; the actual significance of the mean values is therefore unclear. By assessing carbohydrate as reducing sugar, without hydrolysis, the values obtained may be low, as the oligosaccharides present in at least some marsupial milks (Jenness et al., 1964) are not fully accounted for. The levels of carbohydrate reported by Jenness and Sloan (1970) for opossum, quokka, and red kangaroo can be considered more representative.

The most reliable estimates of marsupial milk are those of the brush-tailed possum, *Trichosurus vulpecula* (Table 1). Forty-six mature milk samples of 3 to 14 ml each were obtained from five females using the methods of separation, sedation, and oxytocin administration (Gross and Bolliger, 1959). Analyses of a few pooled early samples (18 to 65 days postpartum) suggested that early milk is lower in dry matter, protein, and perhaps fat; the significance in terms of relative proportions of energy and protein is not clear.

Only two of the samples of wallaroo (*Macropus robustus*) milk analyzed by

Bolliger and Pascoe (1953) are included in Table 1. The remaining samples were all collected 2 days or more after removal of the young and are probably not representative.

Insectivora

Shrews are remarkable for their small size, rapid growth, and large weaning weight (Forsyth, 1976; Dryden, 1968). In view of the observed trends of increased milk dry matter content and increased relative mammary size with diminishing body size (Blaxter, 1961; Payne and Wheeler, 1968; Linzell, 1972), reliable information on composition and amount of milk among the smallest mammals would be invaluable. Given the minute size of some shrews, special care must be taken to ensure representative sampling and analytical accuracy. The values reported for two shrew species (Table 1) by Ben Shaul (1962) are questionable, as no information on sample size, stage of lactation, method of collection, or analytical procedures was provided. Dryden and Anderson (1978) excised the stomachs of recently suckled young of musk shrews (*Suncus murinus*) to obtain milk samples during mid-lactation (9 to 13 days postpartum). Their results (Table 1) for this mouse-sized crocidurine shrew are remarkably similar to the water shrew (*Neomys fodiens*) values of Ben Shaul (1962) even though the sampling method was different.

The milk composition values for a tree shrew (*Tupaia belanger*) published by Martin (1968) are also derived from stomach contents of suckling young, as are less complete data reported for two other tree shrew species, *Tupaia minor* and *Lyongale tana* (D'Souza and Martin, 1974). Unfortunately stomach contents may not be representative of milk as secreted by the dams. Upon ingestion, the fat and casein proteins form a clot in the stomach, whereas lactose and whey proteins in the liquid fraction pass rapidly into the small intestine. Milk removed from the stomach of suckling rats immediately after suckling was significantly higher in dry matter than milk collected from lactating females (Naismith et al., 1969). Fat content relative to protein content was exaggerated, and lactose content was reduced.

The unusual nursing pattern of tree shrews (Tupaiidae) has been demonstrated by Martin (1966, 1968) and D'Souza and Martin (1974). Lactating females return to the nest only once every 48 hours to suckle their young, at which time the young may consume milk equivalent to 40% of the body weight. Tree shrew milk may well be high in dry matter and fat as expected from the long nursing interval, but the figures reported by Martin (1968) and D'Souza and Martin (1974) are probably not representative. The elephant shrews (Macroscelididae) provide an intriguing contrast to many small mammals; they are born precocial and are weaned early (Rathbun, 1976).

From these observations, a picture emerges of a diversity in milk composition among "primitive" mammals. Bias in sampling or analytical procedure may explain some of the disparities, especially among samples collected from the same species, but some of the interspecies differences are undoubtedly real. Unfortunately, in many cases the sum of fat, protein, carbohydrate, and ash does not approach the reported dry matter content, leaving a large part of the dry matter unaccounted for. Until more reliable data are obtained, it is inadvisable to attempt detailed comparisons between species. Any guess as to what might constitute a primitive or evolutionarily conservative milk would be hazardous at best.

Conclusion

No trait evolves in isolation from other facets of the biology of an organism: Milk and mammalian infancy are interrelated phenomena that have co-evolved. The selective advantage implicit in minimizing the drain of lactation on the nursing female suggests that nutrient levels in milk will meet but not exceed the requirements for normal infant growth. These needs of the young may be influenced by many factors, including metabolic body size, growth rate, relative amounts of tissue deposited, and the degree of endothermy. Attempts to correlate these aspects of infancy with interspecies differences in milk composition have been only partly successful. Much of the unexplained variation may reflect error or bias introduced by sampling or analytical procedures. The need for comparative studies based on careful and systematic methods is obvious.

Until a better grasp is attained of the causal factors underlying interspecies differences in milk composition, any attempt to outline the evolution of milk remains speculative. The limited data on monotremes, marsupials, and insectivores reveal little that suggests primitiveness. In fact, lactation among the macropod marsupials is as complex as that of any mammal, in that two disparate milks may be produced simultaneously by adjacent mammary glands. The high dry matter content of echidna milk is another example of what we must consider a specialized rather than a primitive feature. Milk secretion per se may be a primitive or conservative mammalian trait, but this need not imply that the milk of any particular species can be designated as primitive.

References

Abderhalden, E. (1908). *Textbook of Physiological Chemistry*. New York: Wiley.

Adolph, E. F., and F. W. Heggeness (1971). Age changes in body water and fat in fetal and infant mammals. *Growth* 35:55–63.

Ben Shaul, D. M. (1962). The composition of the milk of wild animals. *Int. Zoo Yearb.* 4:333–42.

Bergman, H. C., and Housley, C. (1968). Chemical analyses of American opossum (*Didelphis virginiana*) milk. *Comp. Biochem. Physiol.* 25:213–18.

Blaxter, R. L. (1961). Lactation and growth of the young. In *Milk: The Mammary Gland and Its Secretion*, vol. 2, eds. S. K. Kon and A. T. Cowie. New York: Academic Press.

Bolliger, A., and Pascoe, J. V. (1953). Composition of kangaroo milk (Wallaroo, *Macropus robustus*). *Aust. J. Sci. 15*:215–17.

Brambell, F. W. (1970). *The Transmission of Passive Immunity from Mother to Young.* New York: American Elsevier.

Bunge, G. V. (1902). *Text-Book of Physiological and Pathological Chemistry.* Philadelphia: P. Blakiston's Son.

Campbell, C. B. G. (1974). On the phyletic relationships of the tree shrews. *Mammal Rev. 4*:125–43.

Denamur, R. (1965). The hypothalamo-neuro hypophyseal system and milk-ejection reflex. *Dairy Sci. Abstr. 27*:193–224, 263–80.

Dryden, G. L. (1968). Growth and development of *Suncus murinus* in captivity on Guam. *J. Mammal. 49*:51–62.

Dryden, G. L., and Anderson, R. R. (1978). Milk composition and its relation to growth rate in the musk shrew, *Suncus murinus. Comp. Biochem. Physiol. 60A*:213–16.

D'Souza, F., and Martin, R. D. (1974). Maternal behavior and the effects of stress in tree shrews. *Nature 251*:309–11.

Fisher, Hans (1972). The nutrition of birds. In *Avian Biology,* vol. 2, eds. D. S. Farner, J. R. King, and K. C. Parkes. New York: Academic Press.

Forsyth, D. J. (1976). A field study of growth and development of nestling masked shrews (*Sorex cinereus*). *J. Mammal. 57*:708–21.

Griffiths, M. (1965). Rate of growth and intake of milk in a suckling echidna. *Comp. Biochem. Physiol. 16*:383–92.

Griffiths, M. (1968). *Echidnas.* Oxford: Pergamon Press.

Griffiths, M., McIntosh, D. L., and Coles, R. E. A. (1969). The mammary gland of the echidna, *Tachyglossus aculeatus,* with observations on the incubation of the egg and on the newly-hatched young. *J. Zool. 158*:371–86.

Griffiths, M., McIntosh, D. L., and Leckie, R. M. C. (1972). The mammary glands of the red kangaroo with observations on the fatty acid components of the milk triglycerides. *J. Zool. 166*:265–75.

Griffiths, M., Elliott, M. A., Leckie, R. M. C., and Schoefl, G. I. (1973). Observations of the comparative anatomy and ultrastructure of mammary glands and on the fatty acids of the triglycerides in platypus and echidna milk fats. *J. Zool. 169*:255–79.

Gross, R., and Bolliger, A. (1959). Composition of milk of the marsupial *Trichosurus vulpecula. Am. J. Dis. Child. 98*:102–9.

Hopson, J. A. (1973). Endothermy, small size and the origin of mammalian reproduction. *Am. Nat. 107*:446–52.

Jenness, R. (1974). The composition of milk. In *Lactation: A Comprehensive Treatise,* vol. 3, eds. B. L. Larson and V. R. Smith. New York: Academic Press.

Jenness, R., and Sloan, R. E. (1970). The compositions of milks of various species. *Dairy Sci. Abstr. 32*:599–612.

Jenness, R, Regehr, E. A., and Sloan, R. (1964). Comparative biochemical studies of milks – II. Dialyzable carbohydrates. *Comp. Biochem. Physiol. 13*:339–52.

Lemon, M., and Bailey, L. F. (1966). A specific protein difference in the milk from two mammary glands of a red kangaroo. *Austr. J. Exp. Biol. Med. Sci. 44*:705–8.

Lemon, M., and Barker, S. (1967). Changes in milk composition of the red kangaroo, Megaleia rufa (Desmarest), during lactation. *Austr. J. Exp. Biol. Med. Sci. 45*:213–19.

Linzell, J. L. (1972). Milk yield, energy loss in milk, and mammary gland weight in different species. *Dairy Sci. Abstr. 34*:351–60.

Loh, T. T., and Kaldor, I. (1973). Iron in milk fractions of lactating rats, rabbits and quokkas. *Comp. Biochem. Physiol. 44B*:337–46.

Long, C. A. (1972). Two hypotheses of the origin of lactation. *Am. Naturalist 106*: 141–4.

Marston, H. R. (1926). The milk of the montreme – *Echidna aculeata multi-aculeata*. *Austr. J. Exp. Biol. Med. Sci. 3*:217–20.

Martin, R. D. (1966). Tree shrews: Unique reproductive mechanisms of systematic importance. *Science 152*:1402–4.

Martin, R. D. (1968). Reproduction and ontogeny in tree shrews (*Tupaia belangeri*), with reference to their general behavior and taxonomic relationships. *Z. Tierpsychol. 25*:409–95, 505–32.

Messer, M., and Kerry, K. R. (1973). Milk carbohydrates of the echidna and the platypus. *Science 180*:201–3.

Naismith, D. J., Mittwoch, A., and Platt, B. S. (1969). Changes in composition of rat's milk in the stomach of the suckling. *Br. J. Nutr. 23*:683–93.

Oftedal, O. T. (1975). Gastrointestinal development in mammalian lactivorous young: A comparative approach. In *Literature Reviews of Selected Topics in Comparative Gastroenterology*, vol. 4. Ithaca, New York: New York State Veterinary College.

Patterson, B. (1965). The fossil elephant shrews (family Macroscelididae). *Bull. Mus. Comp. Zool. 133*:295–335.

Payne, P. R., and Wheeler, E. F. (1968). Comparative nutrition in pregnancy and lactation. *Proc. Nutr. Soc. 27*:129–38.

Powers, G. F. (1933). The alleged correlation between the rate of growth of the suckling and the composition of the milk of the species. *J. Pediatr. 3*:201–16.

Rathbun, G. B. (1976). The ecology and social structure of the elelphant shrews *Rhynchocyon chrysopygus* Gunther *Elephantulus refescens* Peters. Ph.D. thesis, University of Nairobi.

Roy, J. H. B. (1970). *The Calf. Nutrition and Health*. University Park, Pennsylvania: Pennsylvania State University Press.

Tarvydas, H., Fordon, S. M., and Morgan, E. H. (1968). Iron metabolism during lactation in the rabbit. *Br. J. Nutr. 22*:565–73.

Tyndale-Biscoe, H. (1973). *Life of Marsupials*. New York: American Elsevier.

4

The reptilian digestive system: general characteristics

CHARLES L. GUARD

The basic features common to the digestive system of modern reptiles may reasonably be considered to represent those of the earliest mammals. Although certain anatomical and physiological traits differ between reptiles and mammals, the requirements imposed on the digestive tract by a given diet and habitat are similar for both groups. If the earliest mammals, as Dr. Crompton has suggested, fed primarily on insects, then they shared this diet with many extant reptiles.

The gut of *Anolis carolinensis*, a small insectivorous lizard, resembles a simple tube. The stomach appears to be little more than an outpouching of the foregut. The intestine is clearly divided into a small and large bowel by a sphincter. The former is not greatly coiled; the later is short, straight, and larger in diameter. Other contemporary reptiles have digestive systems adapted to carnivorous, omnivorous, and herbivorous diets. Some of these structural adaptations and their functional correlates will be explored in this paper. The properties of the reptilian digestive system, from that of the structurally simple in carnivores to that of the more complex, derived state in hervibores, form a baseline to which comparative studies on mammals can be referred.

The oral region demonstrates a more diverse adaptive morphology in living reptiles than the gut itself. The dentition of all snakes and most lizards serves only to capture or prehend food items. In some mollusc-eating lizards several pairs of teeth in the caudal part of the mouth are modified for crushing (Edmund, 1969). Chelonians lack teeth and use their horny beaks for grasping and cropping. The salivary glands of reptiles secrete only mucus.

The digestive tracts of four species are shown in Figures 1 and 2. The caiman is a carnivore, the turtle is an omnivore, and the tortoise and iguana are herbivores. The tuatara, most snakes, the crocodilians, and some lizards are carnivores. Most turtles and lizards are omnivorous. About 40 of the 2500 species of lizards, the tortoises, and some marine turtles are truly herbivorous. Lizards that are herbivorous as adults are usually omnivorous until they reach body weights of 50 to 300 g (Pough, 1973).

The stomach of most reptiles is fusiform in shape and lined with only two types of glandular epithelium. The fundic glands, which secrete HCl and pepsinogen from a single cell type, line the rostral portion of the stomach. These glands may be homologous to the proper gastric glands of mammals which secrete HCl and pepsinogen from separate cells. Mucus-secreting pyloric glands line the aboral portion (Gabe and Saint Girons, 1972). The gastric mucosa of two species of turtles and five species of insectivorous lizards have been shown to secrete the enzyme chitinase (Jeuniaux, 1963). This enzyme can hydrolyze the arthropod cuticle to N-acetyl-*d*-glucosamine components, which are subject to further digestion by other enzymes. It was present in the gastric mucosa and pancreas of all fish and amphibians studied by Jeuniaux but absent in the herbivorous tortoises. The stomach of the Crocodilia is very muscular and greatly curved with the pylorus near the cardia. Stones have often been observed in the stomach in captured specimens. Many functions, including trituration, have been suggested for the stones, but none proved (Dandrifosse, 1974).

The small intestine is shorter in relation to body length and less coiled in reptiles than in mammals, but otherwise quite similar in form. Among reptiles there is a general relationship between small intestinal length and diet:

Figure 1. Gastrointestinal tracts of a spectacled caiman and a western painted turtle. Note the greatly curved stomach of the caiman with a small pyloric compartment. The ureters and urinary bladder are shown joining the cloaca. In this and the following figure, body length refers to distance from mouth to anus in the intact animal. (Drawings by Erica Melack.)

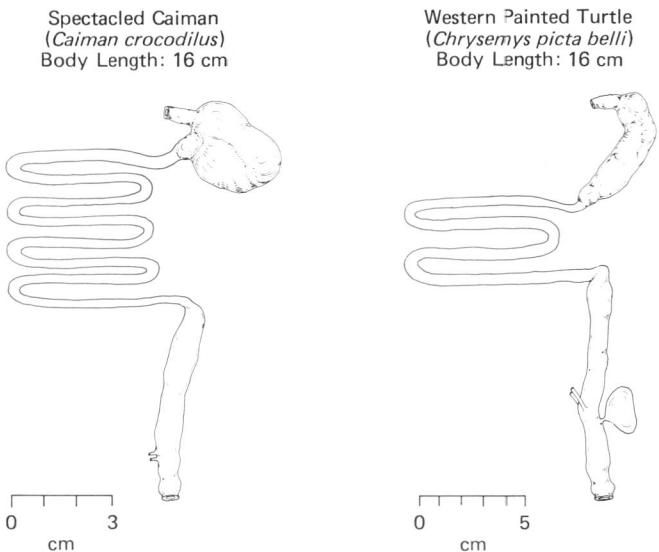

Spectacled Caiman
(*Caiman crocodilus*)
Body Length: 16 cm

Western Painted Turtle
(*Chrysemys picta belli*)
Body Length: 16 cm

0   3
cm

0   5
cm

The small intestine is longest in carnivores, intermediate in omnivores, and shortest in herbivores. This relationship is reversed for the relative length and, more important, the volume of the large intestine. Within the Iguanidae the large intestine varies from about 15 to 50% of the total intestinal length. It is shortest in the carnivorous gecko and longest in the herbivorous green iguana (Guibé, 1970). An ileocolonic valve always separates the small and large bowel. A cecum is variably developed. As indicated in Figure 2, it is merely an eccentric dilatation of the proximal colon in the tortoise. However, in the green iguana and other herbivorous lizards of the families Agamidae and Iguanidae it is compartmentalized by mucosal folds along with the remainder of the proximal colon (El-Toubi and Bishai, 1959). The mucosal folds that project into the colonic lumen of the iguana produce a compartmentalization similar to that provided by infoldings of the gut wall associated with bands of longitudinal muscle in the colon of many species of mammals.

The cloaca is continuous with the large intestine. The most proximal compartment is the coprodeum where urine is stored. The middle compartment is the urodeum where the ureters, reproductive tract, and bladder (if present) join the gut. The hindmost region or proctodeum is not really a

Figure 2. Gastrointestinal tracts of a red-footed tortoise and a green iguana. Note the short small intestine and voluminous large intestine in these herbivores. A longitudinal section of the cecum and proximal colon of the iguana is shown in the inset. The urogenital tract of this male specimen is also included. (Drawings by Erica Melack.)

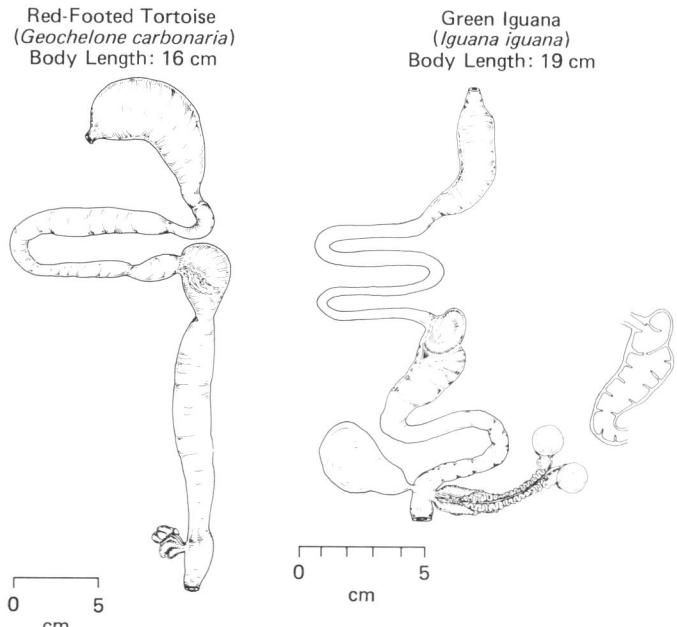

Red-Footed Tortoise
(*Geochelone carbonaria*)
Body Length: 16 cm

Green Iguana
(*Iguana iguana*)
Body Length: 19 cm

separate compartment, but it is lined with a cutaneous type of epithelium. The epithelial lining of the coprodeum and urodeum is identical to that of the colon.

Regulation of digesta movement within the gut is a complex physiological process. The episodic nature of feeding, digestion, absorption, and defecation requires efficient temporal coordination. The system may be thought of as a mechanical pump for digesta consisting of the gross structure of the gut with regulation effected by gastrointestinal muscle. The latter in turn is controlled by the basic electrical rhythm of the muscular tunic cells, which is modulated by extrinsic and intrinsic innervation. Digesta retention in a given segment of the gut allows for (1) storage, (2) enzymatic digestion, (3) absorption of nutrients, and water, and (4) microbial digestion. However, digesta transit time has been studied in relatively few species and seldom in a manner that facilitates interspecies comparisons. The rate of digesta transit in a given animal is dependent on diet, feeding schedule, body temperature, and particle size. Transit through the entire digestive tract is the simplest parameter to measure and can be defined as the time elapsed between oral or intragastric administration of a digesta marker and its appearance in the feces. Perhaps the best single measure is mean transit time. Transit through a given segment of the tract can be measured by reentry fistulae, radiographic techniques, or collection of gut contents following sacrifice at various times after feeding. The perfect digesta markers, both fluid and particulate, should be physically identical to normal digesta but chemically inert to digestive processes and neither secreted nor absorbed by the gastrointestinal tract. The perfect markers have not been found. However, digesta transit has been studied in a broad range of species under comparable conditions utilizing similar techniques (Argenzio et al., 1974a; Clemens et al. 1975a, b). Animals were acclimated to a fixed diet and regular feeding schedule. Fluid markers (polyethylene glycol, ^{51}Cr-EDTA, or $BaSO_4$) and particulate markers (lead-impregnated polyethylene tubing, 2-mm diameter and cut in various lengths) were administered via stomach tube at the time of a meal.

The mean transit times of fluid and particulate markers in the four reptiles illustrated in Figures 1 and 2 are shown in Table 1 for comparison with data on mammals to be discussed in the following chapters. *Caiman crocodilus* and *Chrysemys picta belli* were kept in an ambient temperature of 30 °C. *Geochelone carbonaria* and *Iguana iguana* were maintained in a thermal gradient with extremes of 20 and 55 °C. Liquid markers passed more quickly than particles in all species. Smaller particles also tended to pass more rapidly than larger ones. Radiological studies of the caiman showed that the particles remained in the stomach until about 24 h prior to their appearance as a bolus in the feces. Some of the 10-mm particles were regurgitated at approximately the same time as the remainder were moved into the intestine. In the turtle the stomach was also the main site of retention, though for a shorter time than in the caiman. The cecum and proximal colon were the major sites of the markedly prolonged particle retention observed in the herbivorous tortoise and lizard. This was demonstrated by both radiological examination and direct observation following euthanasia.

The mechanisms available for retention of digesta include stasis, sphincters or valves, compartmentalization, and antiperistalsis. In all four species but particularly in the caiman the pyloric region of the stomach was capable of retaining particles while passing liquids into the small intestine. The major difference observed between the mean transit time for particles in the caiman and turtle on one hand and in the tortoise and iguana on the other was related to the relative volume and compartmentalization of the large intestine. The slow passage of digesta through the hindgut of the iguana could be explained by both its volume and mucosal septa. However, retention of digesta in the mammalian large intestine is also accomplished by retrograde propulsion (antiperistaltic waves of contraction). This mechanism may be important for digesta retention in the four reptilian species included in this report. The large intestine of the tortoise with no anatomical compartmentalization may greatly depend on antiperistalsis for prolonged particle retention. Among lower vertebrates the phenomenon of large intestinal antiperistalsis has been described in the tortoise (Hukuhara, et. al., 1975) and turkey (Dzuik, 1971). In these two species the predominant patterns of motor activity of the large intestine were repetitive waves of contraction that began at the cloaca and moved orally. Aborally directed peristaltic waves were infrequent and always resulted in defecation.

The retention of digesta in the proximal large intestine by antiperistalsis is common to all species of mammals that have been studied in this regard. Elliott and Barclay-Smith (1904) described the motor events in the colon of a wide variety of mammals. They concluded that antiperistalsis in the proximal segment of the colon served to fill the cecum. The retention of digesta

Table 1. *Mean transit time (hours) for digesta markers through the entire gastrointestinal tract.*

Species	Liquid marker	Particulate markers		
		2 mm long	5 mm long	10 mm long
Caiman crocodilus	41	162	162	162
Chrysemys picta belli	35	56	57	60
Geochelone carbonaria	<48	270	285	363
Iguana iguana	<48	207	221	386

Note: Values represent two trials on four specimens of each species except *C. crocodilus* (three specimens). Liquid marker was polyethylene glycol or $BaSO_4$. Particulate markers were cut from 2.2-mm-diameter polyethylene tubing.

under anaerobic conditions and a near-neutral pH in the cecum and/or proximal colon is accompanied by a dense growth of bacteria to equal 25 to 50% of the wet weight of the digesta (McBee, 1970). These indigenous, symbiotic microorganisms are found in the hindgut of all terrestrial vertebrates that have been examined. They have been shown to serve a protective function in preventing overgrowth of pathogenic organisms and are capable of synthesizing protein, vitamins, and other metabolic cofactors. They also are capable of digesting carbohydrates, including those not susceptible to host-elaborated enzymes, into volatile fatty acids (VFA). These VFA are the major anions of the colonic content and have been shown to be readily absorbed by the large intestine of the pig (Argenzio and Southworth, 1975) and pony (Argenzio et al., 1974b). VFA produced in the hindgut have been shown to provide an important source of energy in mammalian herbivores. The coupled absorption of VFA and Na appears to be the primary determinant of water absorption in the hindgut of the mammals in which this has been studied (Argenzio et al., 1977; Argenzio et al., 1975).

Therefore, VFA levels were measured as an index of microbial digestive activity in the feces and segmental samples of gut contents of the four reptilian species reported in this study. Significant amounts of VFA were present in the large intestine of all four species (Table 2). In the turtle, the high level of VFA observed in the small intestine may have been derived from the intestinal contents of the earthworms in the diet, which were found to contain about 5 mmol liter^{-1}. VFA concentrations were highest in the ceca of the two herbivores. Lower levels found in aboral segments of the large intestine may have been due to VFA absorption. The possible role of VFA in the stimulation of sodium and water reabsorption by the reptilian hindgut has not been reported in the literature but is currently under investigation.

The ureteral urine that enters the cloaca may be markedly affected by the antiperistaltic movements of the hindgut. Radioopaque solutions adminis-

Table 2. *Concentrations of volatile fatty acids (VFA) in the gastrointestinal contents and feces.*

Species	Stomach	Small intestine	Cecum	Large intestine	Feces
Caiman crocodilus	0	3.8	–	15.1	7.3
Chrysemys picta belli	3.8	34.6	–	24.2	5.6
Geochelone carbonaria	0.8	6.8	62.7	22.9	9.1
Iguana iguana	0.6	10.1	50.8	15.7	12.7

Note: Values for the gut contents are the means of duplicate samples from two specimens of each species. Values for the feces were determined on six or more samples collected immediately after passage. The first two species lack a cecum, as indicated by the dash. All values are in millimoles per liter.

tered intravenously to the roadrunner were filtered by the kidneys and observed to flow retrogradely through the colon to the tips of the ceca (Ohmart et al., 1970). Similar experiments conducted with the green iguana have demonstrated retrograde flow into the middle portion of the large intestine (personal observation). Additionally, $BaSO_4$ placed in the coprodeum of this lizard reached the proximal colon in about 1 h. Retrograde passage of ureteral urine in the hindgut may be another important function of antiperistalsis in animals with a cloaca. Skadhauge (1977) recently reviewed the evidence for the involvement of the coprodeum, colon, and cecum in osmotic and ionic regulation in lower vertebrates. The membrane transport capabilities for Na, Cl, and water reabsorption and K excretion are now well established. In addition there is a good possibility that the urinary N is recycled by the microbes of the large intestine. In mammals, urea that diffuses into the cecum and colon via the mucosa is recycled (Knutson et al., 1977), and in birds the cecal flora is capable of metabolizing uric acid (Barnes, 1972). In reptiles recycling of urinary N would be most advantageous to herbivores subsisting on diets very low in protein.

The literature on primitive mammals contains very little information on rates of digesta passage. Specific groups are discussed by other authors in this volume, i.e., monotremes and marsupials (Harrop and Hume) and insectivores and primates (Clemens). Among the remaining primitive mammalian groups the insectivorous bats may be considered comparable to some reptiles in diet and diurnal fluctuation in body temperature. These bats have a very simple digestive tract with no externally visible separation of the intestine into a small and large bowel. Minimum transit times for identifiable parts of marker insects were 28 and 35 min (*Nyctalus noctula*, Cranbrook, 1965; *Myotis lucifugus*, Buchler, 1975). Such very rapid transit through digestive tracts that lack a distinct large intestine suggests an absence of intestinal mechanisms for digesta retention. However, this probably represents a secondary loss of function of the hindgut rather than the preservation of a conservative trait.

Christensen (1971) and colleagues have studied the electromyographic basis of gastrointestinal motility in mammals. Their findings support the earlier observations of Elliott and Barclay-Smith. Christensen defines a pacemaker as a site of initiation of electrical activity leading to the propagation both orally and aborally of contractile waves. The midcolon of the cat was shown to satisfy his definition; in the rabbit a pacemaker appears to be located at the junction of the sacculated proximal colon and the smooth middle portion. Antiperistaltic waves originating here move ileal effluent into the cecum. No comparable electromyographic studies have been performed on reptiles. However, the observations of antiperistaltic activity in the tortoise originating at the cloaca suggest this as a site worthy of further study.

Information on the digestive physiology of reptiles is based on relatively few studies on a handful of species. Yet, the gap between our understanding of this system in reptiles and mammals does not seem great. Current knowledge suggests that the requirements placed upon the digestive system of reptiles and mammals are similar. Furthermore, it seems likely that most of these requirements were met by the ancestors of both the current reptiles and the earliest mammals. However, contemporary representatives of the earliest mammals are composites of conservative and derived characters. As demonstrated in the following chapters, many of the mammals that are considered primitive because of some other prominent conservative character have structurally derived digestive tracts. For example, the secretion of chitinase by the gastric mucosa appears to be a conservative trait. It has been demonstrated in insectivorous reptiles, a hedgehog, a mole, a bat, a prosimian, and the domestic pig (Jeuniaux, 1963; Dandrifosse, 1975). The echidna *Tachyglossus aculeatus* feeds entirely on insects, yet it has a stomach lined largely or entirely with stratified epithelium. It therefore seems unlikely that the stomach of the echidna is capable of secreting chitinase.

Some members of the following mammalian orders have a true cloaca or a chamber common to the alimentary and urogenital systems: Monotremata, Marsupialia, Insectivora, Primates, Lagomorpha, and Rodentia. This list may be incomplete, but the inference is clear. The cloaca has not totally disappeared from contemporary mammals. The evolutionary history of the hindgut as an organ involved with osmotic and ionic regulation suggests that the mammals with functional cloacae be closely examined for the retention of this trait.

The information illustrated herein results from an investigation supported by Grant AM 09280 from the National Institutes of Health.

References

Argenzio, R. A., and Southworth, M. (1975). Sites of organic acid production and absorption in gastrointestinal tract of the pig. *Am. J. Physiol.* 228:454–60.

Argenzio, R. A., Lowe, J. E., Pickard, D. W., and Stevens, C. E. (1974a). Digesta passage and water exchange in the equine large intestine. *Am. J. Physiol.* 226:1035–42.

Argenzio, R. A., Southworth, M., and Stevens, C. E. (1974b). Sites of organic acid production and absorption in the equine gastrointestinal tract. *Am. J. Physiol,* 226:1043–50.

Argenzio, R. A., Miller, N., and von Engelhardt, W. (1975). Effect of volatile fatty acids on water and ion absorption from the goat colon. *Am. J. Physiol.* 229:997–1002.

Argenzio, R. A., Southworth, M., Lowe, J. E., and Stevens, C. E. (1977). Interrelationship of Na, HCO_3, and volatile fatty acid transport by equine large intestine. *Am. J. Physiol.* 233(6):E469–78.

Barnes, E. M. (1972). The avian intestinal flora with particular reference to the possible ecological significance of the cecal anaerobic bacteria. *Am. J. Clin. Nutr.* 25:1475–79.

Buchler, E. R. (1975). Food transit time in *Myotis lucifugus. J. Mammal.* 56:252–5.

Christensen, J. (1971). The controls of gastrointestinal movements: Some old and new views. *N. Engl. J. Med., 285:*85–98.

Clemens, E. T., Stevens, C. E., and Southworth, M. (1975a). Sites of organic acid production and pattern of digesta movement in the gastrointestinal tract of swine. *J. Nutr. 105:*759–68.

Clemens, E. T., Stevens, C. E., and Southworth, M. (1975b). Sites of organic acid production and pattern of digesta movement in the gastrointestinal tract of geese. *J. Nutr. 105:*1341–50.

Cranbrook, Earl of (1965). Grooming by vespertilonid bats. *Proc. Zool. Soc. London 145:*143–4.

Dandrifosse, G. (1974). Digestion in reptiles. In *Chemical Zoology. IX. Amphibians and Reptiles,* eds. M. Florkin and B. T. Scheer, pp. 249–75. New York: Academic Press.

Dandrifosse, G. (1975). Purification of chitinases contained in pancreas or gastric mucosa of frog. *Biochimie, Paris 57(6/7):*829–31.

Dzuik, H. E. (1971). Reverse flow of gastrointestinal contents in turkeys. *Fed. Proc. 30:*610, Abstr.

Edmund, A. G. (1969). Dentition, In *Biology of the Reptilia,* vol. 1, ed. C. Gans, pp. 117–200. New York: Academic Press.

Elliott, T. R., and Barclay-Smith, E. (1904). Antiperistalsis and other muscular activities of the colon. *J. Physiol., London 31:*272–304.

El-Toubi, M. R., and Bishai, H. M. (1959). On the anatomy and histology of the alimentary tract of the lizard *Uromastyx aegyptia* (Forskal). *Bull. Fac. Sci., Cairo Univ. 34:*13–50.

Gabe, M., and Saint Girons, H. (1972). Contribution á l 'histologie de l 'estomac des Lépidosauriens (Reptiles). *Zool. Jahrb. Anat. 89:*579–99.

Guibé, J. (1970). L'appereil digestif. In *Traité de Zoologie,* ed. P. P. Grassé, vol. 14, part 2, pp. 521–48. Paris: Masson.

Hukuhara, T., Naitoh, T., and Kameyama, H. (1975). Observations on the gastrointestinal movements of the tortoise (*Geoclemys reevsii*) by means of the abdominal window technique. *Jpn. J. Smooth Musc. Res. 11:*39–46.

Jeuniaux, C. (1963). *Chitine et Chitinolyse,* pp. 131–4. Paris: Masson.

Knutson, R. S., Francis, R. S., Hall, J. L., Moore, B. H., and Heisinger, J. F. (1977). Ammonia and urea distribution and urease activity in the gastrointestinal tract of rabbits (*Oryctolagus* and *Sylvilagus*). *Comp. Biochem. Physiol. 58A:*151–4.

McBee, R. H. (1970). Metabolic contributions of the cecal flora. *Am. J. Clin. Nutr. 23:*1514–18.

Ohmart, R. D., McFarland, L. S., and Morgan, J. P. (1970). Urographic evidence that urine enters the rectum and ceca of the roadrunner (*Geococcyx californianus*) Aves. *Comp. Biochem. Physiol. 35:*487–9.

Pough, F. H. (1973). Lizard energetics and diet. *Ecology 54:*837–44.

Skadhauge, E. (1977). Excretion in lower vertebrates: Function of gut, cloaca and bladder in modifying urine. *Fed. Proc., 36:*2487–92.

5

The gastrointestinal tract of mammals: major variations

C. E. STEVENS

Many of the basic characteristics of the mammalian digestive system are shared by lower vertebrates and some invertebrate species (Stevens, 1977). This is especially true of the motor, secretory, digestive, and neurohumoral control mechanisms associated with midgut functions. Some of the major variations among mammals, and between mammals and reptiles, are seen in the fore- and hindgut. Variations in dentition and jaw structure as they relate to the evolution of mammals are reviewed by Peyer (1968) and McKenna (1975). The esophagus varies in the presence or extent of smooth versus striated muscle and the nature of the gastroesophageal junction. The following discussion will be limited to variations in the stomach and large intestines as these may relate to taxonomical classification, diet or habitat.

The phylogenetic classification of mammals is largely based on fossil records of skeletal structure, including the teeth and jaws. Whereas there is general agreement on the classification of some orders (e.g., monotremes evolved very early and the ungulates appeared relatively late), there is disagreement over the sequential appearance and derivation of many other groups (Simpson, 1945; McKenna, 1975). The classification of Vaughn (1972) will be used as a basis for the present discussion, and the Monotremata, Marsupialia, Insectivora, Dermoptera, Chiroptera, Primates, and Edentata will be considered to represent orders that appeared early in the evolution of mammals. Therefore the discussion will begin with a brief description of gross structural characteristics of the gastrointestinal tract in these orders. This will be followed by a broader discussion of structural variations in mammals and their functional significance in an attempt to determine characteristics that may be conservative as opposed to derived.

There is relatively little species variation in the gastrointestinal tract within the orders Monotremata, Insectivora, and Dermoptera. Figure 1 illustrates the tracts of an echidna and a mole. The gut of the echidna typifies that of the monotremes, which will be discussed in much greater detail by Harrop and Hume (this volume). It has a simple (noncompartmentalized) stomach and a relatively simple large intestine, consisting of a cecum or appendix and a nonsacculated, relatively nonvoluminous colon. The cloaca at the terminus of the colon is not indicated in the drawing. The mole demonstrates the pattern of gastrointestinal structure seen in most Insectivora. It also has a simple stomach, but the intestine shows no externally visible evidence (sphincter, cecum, or change in diameter) of division into a small and large bowel.

There is considerable species variation in the gastrointestinal tract of bats, edentates, primates, and marsupials. The stomach of bats varies from the simple structure seen in insectivorous species to the more voluminous, partially compartmentalized stomach of some frugivores and nectivores and the extremely voluminous stomach of vampire species (Figure 2). There is usually no cecum or other distinct division of the intestine into small and large bowel. The stomach of edentates also varies from the simple structure seen in insectivores and armadillos to the complex stomach of sloths (Figure 3). The large intestine of edentates is relatively simple, although the rectum of the sloth is quite voluminous, serving to store feces during prolonged periods between defecations. Primates also show a relatively wide variation in gut pattern. The gastrointestinal tract of a prosimian and a monkey are illustrated by Clemens (this volume). The stomach of prosimians, apes, man, and most monkeys is simple in structure, but that of the colobus and langur monkey are voluminous and sacculated. A cecum is present in all primates.

Figure 1. Gastrointestinal tracts of an echidna and a mole. In this and following figures, body length refers to distance from mouth to anus in the intact animal. A photograph of the echidna specimen was supplied by Dr. I. D. Hume, University of New England, Armidale, Australia. (Drawings by Erica Melack.)

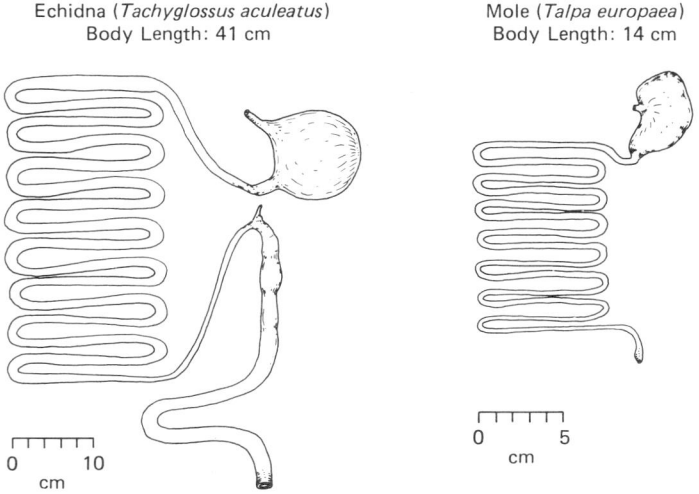

Echidna (*Tachyglossus aculeatus*)
Body Length: 41 cm

Mole (*Talpa europaea*)
Body Length: 14 cm

0   10
cm

0   5
cm

Figure 2. Gastrointestinal tracts of an insectivorous bat and a vampire bat.
Live specimens supplied by Dr. W. A. Wimsatt, Cornell University, Ithaca,
New York. (Drawings by Erica Melack.)

Insectivorous Bat (*Myotis lucifugus*)   Vampire Bat (*Desmodus rufus*)
     Body Length: 7 cm   Body Length: 7.5 cm

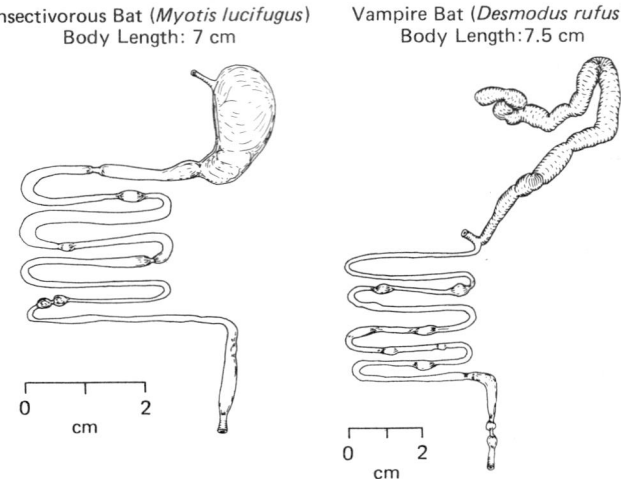

Figure 3. Gastrointestinal tracts of an armadillo and a sloth. Photographs of
specimens supplied by R. G. Parra, Instituto de Producción Animal, Facul-
tad de Agronomia, Maracay, Venezuela. (Drawings by Erica Melack.)

Armadillo (*Dasypus sevenicola*)   Sloth (*Bradypus infuscatus*)
     Body Length: 20 cm   Body Length: 55 cm

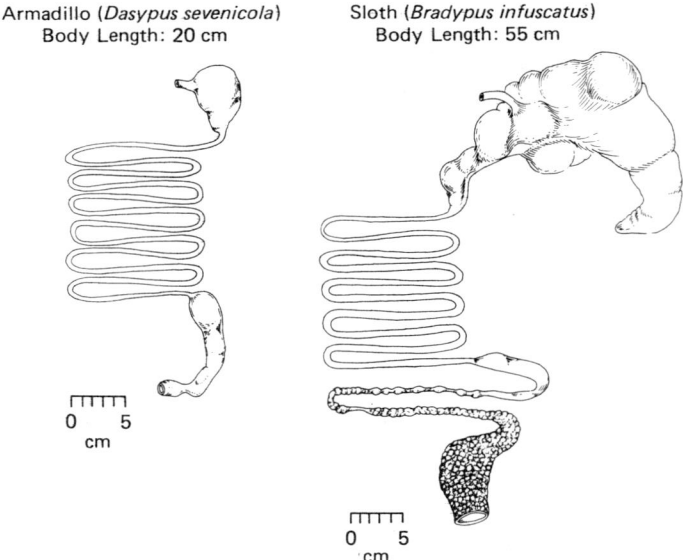

It is extremely small in man but relatively large in most species. The proximal colon of most primates other than prosimians is sacculated. In man, as well as some apes and monkeys, it is sacculated throughout much of its length. The marsupials show extremely wide species variation in gastrointestinal tract structure, from species with a simple stomach and no discernible division of the intestine into a small and large bowel to species with an extremely complex stomach and/or large intestine. These variations are well illustrated in the discussion of the marsupial digestive system by Hume and Dellow (this volume).

Figures 4 and 5 illustrate the gastrointestinal tract of species that are considered to have arrived relatively late in the evolution of mammals. They are included here to demonstrate some major variations in large intestinal structure and for later discussion of digesta transit in the dog, pig, and rabbit. The dog (Figure 4) demonstrates the simple stomach and relatively simple large intestine typical of most Carnivora. The mink demonstrates the extremely simple large intestine characteristic of the arctoid Carnivora. The large intestine of the latter lacks a cecum and anatomical ileocolonic sphincter, a characteristic similar to that of most Insectivora species. The domestic pig and the rabbit (Figure 5) each has a simple stomach and complex large intestine. However, the pig has a relatively nonvoluminous cecum and a colon that is voluminous and sacculated, in association with bands of longitudinal muscle throughout most of its length. Conversely, the cecum of the rabbit is extremely voluminous, whereas the colon is nonvoluminous and sacculated only in its proximal segment.

Table 1 gives Vaughn's classification of mammalian orders, categorized by the diet(s) of inclusive species. For the purpose of this discussion, the term "carnivore" applies to species that feed only on animals, and "herbivores" to species that can subsist largely on the fibrous portion of plants. The term "omnivore" will be used to indicate both those species that feed on animals and plants and those that feed on plant concentrates (fruit, seeds, nectar, etc.). Although this is not the accepted definition of the term omnivore, it provides a useful means of relating gastrointestinal tract structure to broad categories of diet.

The first group of orders includes only carnivorous species. Monotremes feed on insects (echidna) or other invertebrates (platypus). The Mysticeti (baleen whales) feed on invertebrates, and the Odontoceti (toothed whales) on invertebrates, fish, or aquatic mammals. Pholidota (scaly anteaters) and Tubulidentata (aardvarks) are both anteaters. The second group in Table 1 includes orders with both carnivorous and omnivorous species. As stated earlier, the Chiroptera include insectivorous, vampire, frugivorous, and nectivorous species. Although most Insectivora are carnivores, as their name would imply, hedgehogs are omnivorous. The Carnivora also contain omnivorous species such as the raccoons and most bears.

Figure 4. Gastrointestinal tracts of a mink and a dog. (Drawings by Erica Melack; from Stevens, 1977).

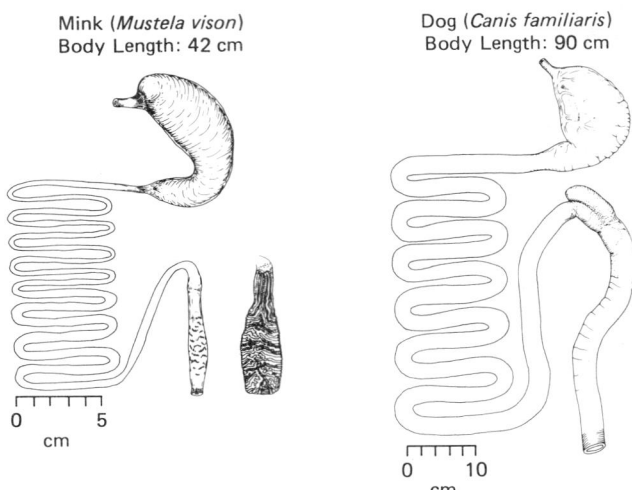

Mink (*Mustela vison*)
Body Length: 42 cm

Dog (*Canis familiaris*)
Body Length: 90 cm

Figure 5. Gastrointestinal tracts of a pig and a rabbit. (Drawings by Erica Melack; from Stevens, 1977.)

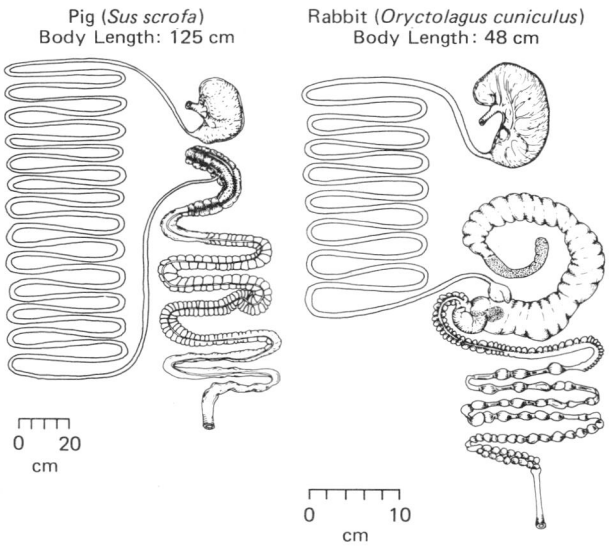

Pig (*Sus scrofa*)
Body Length: 125 cm

Rabbit (*Oryctolagus cuniculus*)
Body Length: 48 cm

Mammalian orders listed in the third group contain carnivorous, omnivorous, and herbivorous species. Edentates include anteaters, omnivorous armadillos, and the herbivorous sloths. The marsupials are represented by a range of species and diets almost as broad as that of all eutherian orders. The rodents, which account for approximately 40% of all mammalian species, also demonstrate a broad range of diets. The primates include insectivorous prosimians, a large number of omnivorous species, and strict herbivores such as the colobus and langur monkeys. The two orders Dermoptera and Hyracoidea represent a limited number of omnivorous species.

Haltenorth (1963) subdivided the Artiodactyla into Suina and Ruminantia. The former includes the pigs and hippopotamuses, and the latter the tylopods, camelids, and Pecora (chevrotains and advanced ruminants). The domestic pig is an omnivore, but most suiform species and all Ruminantia are herbivores.

The last group consists of orders that contain only herbivorous species:

Table 1. *Variations in diet and gastrointestinal structure in various mammalian orders*

Order	Diet Carnivorous	Omnivorous	Herbivorous	Structure of stomach[a] Complex	St. sq. ep.	Structure of large intestine Cecum	Sac. colon
Monotremata	+			−	+	+	−
Mysticeti	+			+	+	+	−
Odontoceti	+			+	+	−	−
Pholidota	+			−	+	−	−
Tubulidentata	+			−		+	+
Chiroptera	+	+		±	−	±	−
Insectivora	+	+		−	−	±	±
Carnivora	+	+		−	−	±	±
Marsupialia	+	+	+	±	±	±	±
Edentata	+	+	+	±	+	±	±
Rodentia	+	+	+	±	±	±	±
Primates	+	+	+	±	±	+	±
Dermoptera		+		−	−	+	+
Hyracoidea		+		−	+	+	−
Artiodactyla		+	+	±	+	±	±
Perissodactyla		+		−	+	+	+
Lagomorpha		+		−	−	+	+
Proboscidea		+		−	−	+	+
Sirenia		+		+	−	+	−

[a]St. sq. ep. = Stratified squamous epithelium.
Key to symbols: + = character present in all species; − = character absent in all species; ± = character present in some species, absent in others.

the Perissodactyla (equids, rhinos, and tapirs), Lagomorpha (rabbits, hares, and picas), Proboscidea (elephants), and Sirenia (manatees and dugongs).

The table shows that whereas the orders generally considered to contain the most primitive species can include omnivores and herbivores, all but the Dermoptera include carnivores. Conversely, orders that contain mostly or only herbivorous species are those considered highly advanced. It also is interesting to note that in each order containing carnivorous species except Odontoceti and Carnivora, some or all species feed primarily on small invertebrates.

Table 1 also lists variations in the structural characteristics of the stomach and large intestine. The stomach of most mammals consists of a simple outpocketing of the digestive tract which serves as a site for storage, maceration, and the initial stages of digestion. However, the stomach of many species is more structurally complex and voluminous. This is true for all whales and sirenians, most artiodactyls, and a number of marsupials, bats, primates, edentates, and rodents. In many of these species the stomach is divided into a complex series of permanent compartments. In others (e.g., kangaroos and the colobus monkey) the longitudinal muscle forms axial bands and the stomach is sacculated in a manner similar to that seen in the large intestine of numerous species. The compartmentalization or sacculation aids in the retention of gastric contents. A voluminous, compartmentalized stomach allows for greater and more prolonged storage of digesta. This allows for intermittent feeding in carnivores, such as the vampire bat, and provides herbivores with a site for microbial digestion of plant fiber.

The type and distribution of gastric epithelium can also show wide species variation. The distribution of gastric mucosa in man and dog typifies that of many simple-stomached mammals. The cardiac mucosa consists of a narrow band that blends with the associated proper gastric mucosa. The terminal segment of stomach is lined with pyloric mucosa. However, the stomach of some species contains an expanded area of cardiac mucosa. An additional area of nonglandular stratified epithelium is found in a wide range of species with simple or complex stomachs (Slijper, 1946). The latter include a number of simple-stomached, carnivorous species (monotremes, whales, scaly anteaters, and edentates) that feed on invertebrates, and most herbivores that have a complex stomach. Stratified squamous epithelium has long been assumed to be nonpermeable and serve as a physical barrier against damage from ingested material. However, in ruminants it has been shown to be very permeable to the volatile fatty acid (VFA) end products of carbohydrate digestion by microorganisms and capable of actively absorbing Na and Cl (Stevens, 1973). Little is known about its function in other species.

Although the intestine of some mammals is not distinctly divided into a small and large bowel — that is, there is no change in diameter or evidence of either an ileocolonic valve or cecum — a small intestine and a large intestine are evident in most species. Gross structural characteristics of the mammalian small intestine show little species variation other than differences in relative length, diameter, types of villi, degree of mucosal folding, and thickness of muscle layers. The first four characteristics help determine mucosal surface area, and all of the above can affect transit time; but passage of digesta through the small intestine is relatively rapid (a few hours) in comparison to either gastric emptying or large-intestinal transit time in species that have been studied.

The large intestine shows wide species variation in its relative length, diameter, volume, and complexity. It terminates in a cloaca in the monotremes and at least some species belonging to other orders, but the digestive and urinary tracts exit separately in most mammals. A cecum of varying dimensions may be present. The cecum and portions of the colon may be sacculated in association with longitudinal bands of muscle (teniae), and in the Perissodactyla and Proboscidea the colon is both sacculated and divided into distinct, permanent compartments. A complex and voluminous large intestine also allows for a more prolonged storage of digesta, providing the major site for microbial digestion in many herbivores.

Table 1 shows that there is a tendency for simplicity of the stomach and large intestine in carnivorous orders and complexity of the stomach and/or large intestine in herbivorous orders. As previously noted, this relationship between diet and complexity of the stomach does not hold for all species. The same is true for the large intestine, which must serve the additional function of absorbing inorganic electrolytes and water of both dietary and secretory origin. Soergel and Hofmann (1972) estimated that the daily secretions of the digestive system in a man, previously starved for 24 hours, amounted to about 40% of the extracellular fluid volume (ECFV) and contained approximately 40 to 45% of the Na, Cl, and HCO_3 present in the extracellular fluid space. Ninety-eight percent of the fluid and ions secreted by the gut was resorbed prior to excretion.

Herbivores secrete larger volumes of salivary, pancreatic, and biliary fluids, and the pony large intestine alone has been shown to provide additional secretions equivalent to approximately 40% of the ECFV (Argenzio et al. 1974). Therefore, the absorptive load of the large intestine is extremely large in the pony, and probably grossly underestimated in other species because of lack of data on large-intestinal secretion. The relationship between large-intestinal complexity and the need for water conservation is supported by the simplicity or absence of a large intestine in whales. Also, among the Artiodactyla, the hippopotamus has the simplest, least voluminous large intestine, whereas that of the camelids is one of the longest and most complex.

The transit of digesta through the large intestine can be delayed by retention in a cecal cul-de-sac. The comparative study by Elliott and Barclay-Smith (1904), mentioned in the previous chapter, provided early evidence that cecal filling was largely due to retrograde propulsion (antiperistalsis) by the proximal colon. More recent electromyographic studies (Christensen,

1971) indicate that the pacemaker for propulsive contractions of the colon is located some distance aboral to the junction of the small and large intestine. Review of a series of comparative studies of fluid and particulate marker passage through various segments of the gastrointestinal tract of the dog, pig, pony, and rabbit (Stevens, 1977) showed that the large intestine was the major site of digesta retention in each of these species. However, they also showed marked species variations in both the primary site and the total duration of digesta retention within the large intestine. In the rabbit, the cecum provided the major site of retention, but in the dog, pig, and pony the proximal colon was a more critical site. The rate of passage also was inversely related to the degree of colonic compartmentalization.

The capacity of the large intestine for retaining digesta has often been assumed to correlate with the relative volume of the cecum (e.g., simple-stomached herbivores are often referred to as "cecal-digesters"). The above studies suggest that in many species the retrograde propulsion and saccula-tion in the colon may be more critical determinants. Examination of the relationship between the weight of gastrointestinal contents and body weight in a wide range of herbivores (Parra, 1972) showed that the relative capacity of the digestive tract decreased with decreasing body weight. Because the rate of metabolism per unit mass tends to increase with a decrease in body weight, it was concluded that small animals require different strategies for compensation. This could account for the voluminous cecum and nonvolu-minous colon as well as the practice of coprophagy in lagomorphs and many rodents. Therefore, the gastrointestinal tract of the herbivorous species in these two orders, which constitute a large proportion of the laboratory ani-mals used in studies of the digestive system, may have derived a number of special characteristics that do not apply to most other mammals.

From the previous overview of variations in the gastrointestinal tract of mammals, it is apparent that they tend to demonstrate a closer correlation with diet and other factors, such as body weight and need for water conser-vation, than with taxonomical classification. The major variations involve the stomach and hindgut. Compartmentalization or sacculation of the stomach in some species appears to be one characteristic derived from the conserva-tive simple stomach of reptiles and early mammals. The presence of strati-fied epithelium in the stomach is another characteristic limited to mammals. Embryological studies of sheep (Bryden et al., 1972) indicate that the areas of stratified squamous epithelium were derived from gastric rather than esophageal tissue, at least in this specie. Oppel (1897) and Bensley (1902–3) proposed that the absence of stratified squamous epithelium in lower verte-brates and the decrease in glandular complexity from proper gastric to cardiac to nonglandular stratified tissue suggests a regressive evolutionary change. If this is true, the stomach of monotremes, which is lined largely with stratified squamous epithelium, is highly advanced. The same is true of edentates and herbivorous marsupials, in which it also is present.

The presence of a cloaca with a common exit for the excretion of urine and digesta is a conservative characteristic shared by reptiles and adult amphibians. Its disappearance in most mammals appears to correlate with an increase in the kidney's ability to conserve electrolytes and water (Stevens, 1977). Sacculation of the large intestine in association with bands of longitudinal muscle is seen only in mammals and primarily limited to advanced species. This appears to be a specialization that favors retention of digesta for conservation of electrolytes and water as well as microbial digestion in the larger terrestrial species. The lack of a distinct large intestine, evidenced by the absence of an intestinal sphincter or a cecum, in most Insectivora and Chiroptera might suggest that this is a conservative characteristic. However, its presence in reptiles, monotremes, and other primitive mammalian species would strongly indicate that this is a regressive derivation or specialization in Insectivora and Chiroptera as well as the arctoid carnivores and many whales.

Dr. Crompton (this volume) concluded from the available evidence that the earliest mammals were relatively small carnivores that fed on insects or other small invertebrates. The presence of species with this type of diet in most of the mammalian orders considered to contain primitive species and their absence from orders believed to have evolved later would support this hypothesis. Relatively little is known about the digestive system of presently living species that could be included in this category. Their study could provide valuable information on its evolution in mammals. However, most of the evidence relating to the dietary characteristics of early mammals is based on tooth and jaw structure. Although they give some important clues as to the diet, they are only an indication of how food was procured and initially triturated for the episodic, complex process of digestion. Furthermore, other strategies for maceration of food such as a gizzard or gizzard-like stomach have evolved in many invertebrate and vertebrate species. Therefore, a great deal of additional comparative information on the structure and function of the gastrointestinal tract is needed for reasonable speculation on its evolution.

References

Argenzio, R. A., Lowe, J. E., Pickard, D. W., and Stevens, C. E. (1974). Digesta passage and water exchange in the equine large intestine. *Am. J. Physiol. 226*:1035–42.

Bensley, R. R. (1902–3). The cardiac glands of mammals. *Am. J. Anat. 2*:105–56.

Bryden, M. M., Evans, H. E., and Binns, W. (1972). Embryology of the sheep. II. The alimentary tract and associated glands. *J. Morphol. 138*, 187–206.

Christensen, J. (1971). The controls of gastrointestinal movements: Some old and new views. *N. Engl. J. Med. 285*:85–98.

Elliott, T. R., and Barclay-Smith, E. (1904). Antiperistalsis and other muscular activities of the colon. *J. Physiol. 31*:272–304.

Haltenorth, Th. (1963). Klassifikation der Säugetiere: Artiodactyla. *Handb. Zool.* 8:1–167.

McKenna, M. C. (1975). Toward a phylogenetic classification of the mammalia. In *Phylogeny of the Primates*, eds. W. P. Luckett and F. S. Szalay, pp. 21–46. New York: Plenum Press.

Oppel, A. (1897). *Lehrbuch der vergleichenden mikroskopischen Anatomie der Wirbeltiere*; *Zweiter Teil, Schlund und Darm*. Jena: Gustav Fischer.

Parra, R. G. (1972). Comparative aspects of the digestive physiology of ruminants and non-ruminant herbivores. In *Literature Reviews of Selected Topics in Comparative Gastroenterology*, vol. 2. Ithaca, New York: New York State Veterinary College.

Peyer, B. (1968). *Comparative Edontology*, trans. and ed. R. Zangerl. Chicago: University of Chicago Press.

Simpson, G. G. (1945). The principles of classification and a classification of mammals. *Bull. Am. Mus. Nat. Hist.* 85:1–350.

Slijper, E. J. (1946). Die physiologische Anatomie der Verdauungsorgane bei den Vertebraten. In *Tabulae Biologicae*, eds. H. J. Vonk, J. J. Mansour-Bek, and O. E. J. Slijper, pars 1, vol. 21, pp. 1–81. Amsterdam: W. Junk.

Soergel, K. H., and Hofmann, A. F. (1972). Absorption. In *Pathophysiology: Altered Regulatory Mechanisms in Disease*, ed. E. D. Frohlich, pp. 423–53. Philadelphia: J. B. Lippincott.

Stevens, C. E. (1973). Transport across rumen epithelium. In *Transport Mechanisms in Epithelia*, eds. H. H. Ussing and N. A. Thorn, pp. 404–26. Copenhagen: Munksgaard.

Stevens, C. E. (1977). Comparative physiology of the digestive system. In *Dukes' Physiology of Domestic Animals*, ed. M. J. Swenson, 9th ed., pp. 216–32. Ithaca: Cornell University Press.

Vaughn, T. A. (1972). Classification of mammals. In *Mammalogy*, pp. 35–38. Philadelphia: W. B. Saunders.

6

Digestive tract and digestive function in monotremes and nonmacropod marsupials

C. J. F. HARROP* and I. D. HUME

Owen (1868) was one of the first to describe the general anatomy of the digestive tracts of the monotremes and of several nonmacropod marsupial species. Subsequently MacKenzie (1918) provided a more detailed description of the digestive tracts of both platypus (*Ornithorhynchus anatinus*) and echidna (*Tachyglossus aculeatus*), and of members of the marsupial families Dasyuridae (Tasmanian devil), Peramelidae (bandicoots), Vombatidae (wombat), Phascolarctidae (koala), and Phalangeridae (*Trichosurus* and *Pseudocheirus*). However, information on the digestive function and nutrition of the nonmacropod marsupials is only now being accumulated. The little we know about the nutrition of the monotremes is due almost entirely to the efforts of Griffiths (1968, 1978).

Monotremes
This review is restricted to a discussion of digestive physiology and nutrition. More complete accounts of the general biology of monotremes are found in the monographs *Echidnas* (Griffiths, 1968) and *The Biology of the Monotremes* (Griffiths, 1978).

Platypus (Ornithorhynchus anatinus)
The platypus is primarily carnivorous, and small shellfish, insects, and insect larvae constitute the major portion of its diet (Burrell, 1927). Food habit studies based on stomach content analysis are virtually useless in the platypus because of the small size of the stomach and the fine particle size of the digesta in the stomach. Most of the identifiable parts of insects and molluscs are ejected from the mouth; examination of the cheek pouches is thus perhaps the best sampling method for dietary analysis. The cheek pouches invariably contain mud and gravel ingested with the food, and this material

* Deceased.

no doubt helps in grinding of the food; it probably also supplies organic detritus, both animal and plant, to the platypus.

The digestive tract. The overall impression of the digestive tract of the platypus is one of simplicity (Figure 1), as is so characteristic of many eutherian carnivores. However, both the esophagus and the stomach of the monotremes are lined with stratified squamous epithelium; there are no proper gastric glands, and so no peptic digestion. The only glands found in the platypus stomach are Brunner's glands, which are confined to the submucosa of the distal stomach, and which secrete a neutral mucopolysaccharide. Brunner's glands are found in all mammalian species (Krause, 1972), and their secretions are thought to protect the proximal duodenal mucosa from the mechanical trauma of contents moving through the intestinal tract and from the ulcerating effects of acid and pepsin secreted by the stomach. This raises the question of the function of Brunner's glands in the platypus because no acid or pepsin is produced. Their location in the stomach of the platypus also differs from other mammals, in which they are confined to an area immediately distal to the pylorus, in the duodenum.

In the platypus there is no distinct pylorus, but there is an abrupt change in the mucosa from stratified squamous to glandular epithelium. The short small intestine is devoid of villi, but has numerous transverse folds (Krause, 1975). Crypts of Lieberkühn lying between these folds are lined with simple

Figure 1. Digestive tract of the platypus (*Ornithorhynchus anatinus*). Body length = 44 cm.

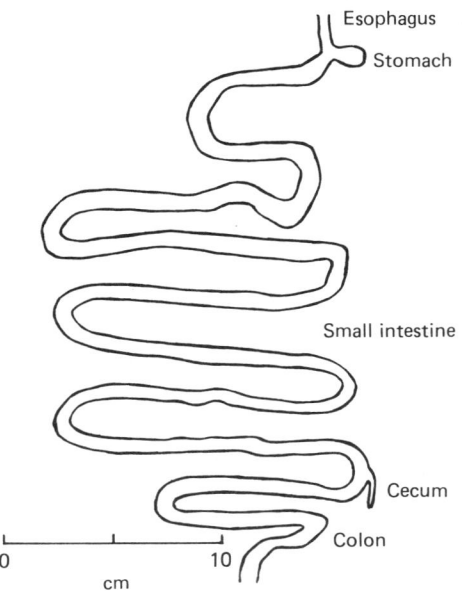

columnar epithelium. Nothing is known of the digestive enzymes of the intestinal glands or of the pancreas, though no doubt the proteolytic activity of both is high.

There is a tiny, nonfunctional cecum, and a short large intestine (Figure 1).

Digestive function. The short intestine suggests a fast rate of passage of food residues through the platypus gut and a readily digestible diet. This is an oversimplification, and neglects the important part played by the specialized mouth parts in the prehension, sorting, and comminution of food items. The role of the bill as an important peripheral sensory organ was proposed by early naturalists. The bill, particularly its upper border, is densely innervated, and remarkably large regions within the central nervous system are devoted to inputs from the bill (Bohringer, in press).

The cheek pouches apparently replace the stomach as a food storage area. When the pouches are full, the platypus rests on the surface of the water, and the contents of the pouches are transferred to the buccal cavity where they are comminuted by the grinding action of the horny pads on the maxillae and lower jaw. Exoskeletons of prey and other hard food items are ejected into the water through a series of horny serrations along the margin of the lower jaw. Food passing to the esophagous is finely ground, but almost nothing is known of the digestive processes that follow.

Metabolism. The standard metabolic rate of monotremes is low compared with eutherians and marsupials (Dawson and Hulbert, 1970). However, there is no information on daily food requirements and food intake, or on energy requirements of the wild platypus. At least toward the southern limit of its range, the platypus undergoes seasonal changes in body weight which have been shown to be due to changes in the amount of depot fat in the tail (Temple-Smith, 1973). Highest tail-fat indices are found in summer, and lowest values in spring, suggesting a decline in food supply in late winter relative to requirements (Temple-Smith, 1973). Grant and Carrick (1978) have calculated that energy requirements for maintenance of body temperature are increased by at least 18% at this time of the year.

Echidna (Tachyglossus aculeatus)
The food habits of the echidna have proved far easier to study than those of the platypus, for several reasons. The animal is terrestrial rather than semi-aquatic, food items are ingested whole without any sorting or ejection from the mouth, and comminution of ingesta is incomplete so that examination of the scats is a valid means of identifying prey species. Since the exoskeletons of the ants and termites pass through the digestive tract without being digested, the diet can be readily identified from the jaws of the chitinous exoskeletons of the prey.

From such examination Griffiths (1978) has found that in the cooler, wetter parts of the echidna's range ants are the major food item; this apparent preference for ants can be explained by low availability of termites. In the more arid parts of the echidna's range the preference is for termites even though ants are plentiful. There is no difference in the sodium content of termites and ants, but the water content of termites is higher (74% compared to 64%), which may explain the preference for termites (Griffiths, 1978).

The digestive tract. As in the platypus, the esophagus and the stomach of the echidna are nonglandular, the lining being stratified squamous epithelium. Unlike that of the platypus, however, the echidna's stomach is a large storage organ (Figure 2), capable of holding up to 200 g of termites. The stratified epithelium of the stomach continues into a relatively narrow tube, the "pseudoduodenum" (Griffiths, 1965), into which opens a prominent set of Brunner's glands. At the caudal end of the pseudoduodenum the stratified epithelium is replaced by villi and crypts of Lieberkühn lined with columnar epithelium; this is the cranial end of the small intestine. The distal portions of the villi are lined with simple columnar epithelium, the central

Figure 2. Digestive tract of the echidna (*Tachyglossus aculeatus*). Body length = 46 cm.

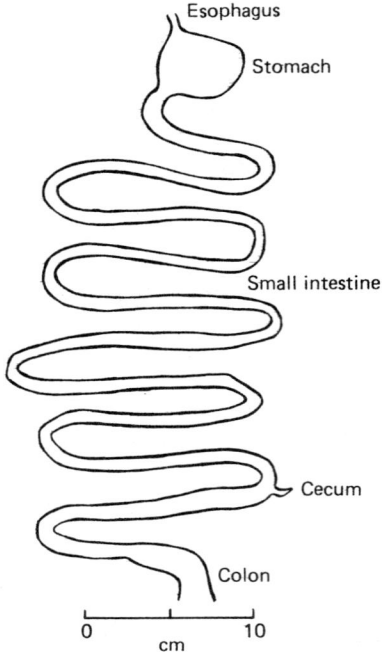

parts of the crypts with goblet cells producing mucin, and the deeper parts of the crypts with Paneth cells, which presumably produce trehalase (Krause, 1975) and other enzymes.

The echidna has a vestigial cecum, but both the small intestine and the large intestine are longer than in the platypus (Figure 2).

Digestive function. The principal feature of the echidna's mouth parts is the long vermiform tongue with which the animal both catches and masticates its food. The tongue is lubricated with a sticky secretion to which the insects are held while being drawn into the buccal cavity as the tongue is retracted. The insects are then ground between the dorsal surface of the base of the tongue, which has a set of keratinized spines, and sets of transverse spines on the palate. The lingual spines are lubricated by mucus secreted by glands in the dermis of the pad, and the musculature of the tongue is complex (Griffiths, 1978).

Although the echidna stomach secretes no digestive enzymes, there is evidence for autodigestion of the live insect prey within the stomach. Homogenates of termites exhibit strong amylase activity, and the pH in the stomach is normally close to the optimum for amylase. As yet there is no evidence that the stomach serves an absorptive role, but histologically and ultrastructurally the stratified epithelium lining the monotreme stomach closely resembles that of the bovine rumen, through which volatile fatty acids, water, and ions are absorbed; it is therefore possible that the monotreme stomach epithelium may have a similar absorptive function.

The small intestine has been shown to have maltase, isomaltase, and trehalase activity, but no sucrase activity. Trehalose is the principal sugar in insects, and upon digestion yields two molecules of glucose per molecule of trehalose. The glucose is absorbed and is no doubt the principal source of energy to the echidna. No lactase activity has been found in adult echidnas, but would probably be present in the young because echidna milk contains some lactose (Messer and Kerry, 1973). The combined bile–pancreatic secretion contains lipase, amylase, and proteinase activity, but ligation of the duct had little effect on experimental animals (Griffiths, 1968). This was explained by Griffiths (1968) on the basis of autodigestion of termites in the stomach, and the slow rate of passage of digesta through the relatively long intestine, at least 2 days being required for clearance of a meal of 100 to 200 g wet weight of termites.

Metabolism. Griffiths (1965) consistently found that the addition of glucose to a natural diet of termites improved growth and nitrogen retention in echidnas. From this it was concluded that the energy content of ants and termites may limit echidna growth in the wild. This may explain the high incidence of echidna attacks on mounds of the meat-ant *Iridomyrmes detectus* in spring, when virgin queen ants containing 45% fat are present; at this time echidnas emerge from hibernation (Griffiths and Simpson, 1966), and energy requirements are presumably high.

Long-beaked echidna (Zaglossus)

The food of *Zaglossus* is mainly, if not solely, earthworms and scarab beetle larvae. The digestive tract is similar to *Tachyglossus*, so presumably the rate of passage of digesta is also slow, allowing digestion to be carried out in the small intestine by enzymes of pancreatic and intestinal origin; there is no evidence for self-digestion of the earthworms in the stomach.

The principal difference between *Zaglossus* and *Tachyglossus* is in the mouth parts. The tongue of *Zaglossus* has spines housed in a deep groove in the distal one-third of the tongue. The spines point backward, and the tongue is extended only 2 cm from the end of the beak during feeding. The prey is hooked onto the exposed spines by a slight forward and upward movement of the beak, then jerked back into the mouth where it is ground as in *Tachyglossus*. The musculature of the tongue is again complex, but differs from *Tachyglossus* in that it lacks striated circular muscles. This has been explained by Griffiths (1978) in relation to the different feeding habits of the two genera.

Nonmacropod marsupials

Food habits within the nonmacropod marsupial group cover much the same range as in eutherians, from virtually strict carnivory (e.g., *Sarcophilus harrisii*, Tasmanian devil) through omnivory (Peramelidae, bandicoots) to strict herbivory (e.g., *Phascolarctos cinereus*). All the nonmacropod marsupial herbivores are hindgut fermenters.

Carnivores

The three families Dasyuridae, Notoryctidae, and Thylacinidae contain all the carnivores and insectivores among the marsupials. The dasyurids are all predators, the smaller members eating mainly invertebrates, the larger members in addition eating many vertebrates including mammals, birds, and reptiles. The Tasmanian devil also eats carrion (Guiler, 1970), and all probably eat some vegetable matter.

The tiger cat (*Dasyurus maculatus*) has a digestive tract very similar to that of most eutherian arctoid carnivores, with a simple stomach, a short but wide small intestine, and no cecum. Therefore there is no anatomical distinction between a small and a large intestine (Figure 3). Owen (1868) considered that the stomach of carnivorous marsupials was "relatively much more capacious" and "better adapted for the retention of food" than members of the eutherian order Carnivora, but detailed information is lacking. Rate of passage of digesta through the gut can be expected to be rapid because the stomach is often found empty with little digesta in the rest of the tract. Cowan et al. (1974) found that in the small carnivorous marsupial *Antechinus swainsonii* 90% of food residues were eliminated in 12 hours, and 99% in 18 hours.

Omnivores

Among the eutherians, the small intestine of omnivores tends to be longer than that of carnivores. The cecum and colon also tend to be longer and larger in diameter. The same principle holds in the marsupials, as seen in the family Peramelidae, the bilbies, and bandicoots (Figure 4). The diet of

Figure 3. Digestive tract of the tiger cat or tiger quoll (*Dasyurus maculatus*). Body length = 55 cm.

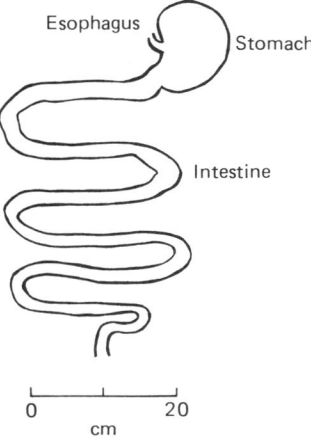

Figure 4. Digestive tract of the long-nosed bandicoot (*Perameles nasuta*). Body length = 36 cm.

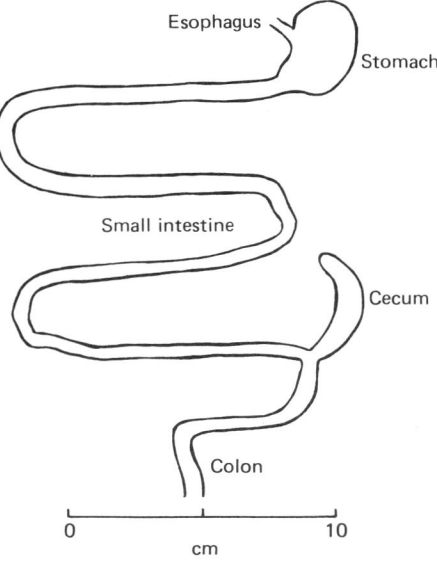

members of this family includes invertebrates such as insects and their larvae, spiders, earthworms, snails, occasional small vertebrates such as frogs
and lizards, roots, herbs, and fruit.

The rate of passage of digesta through the bandicoot gut would be expected to be slower than in the carnivores. Griffiths (see Waring et al., 1966)
used different species of termites as passage markers in *Isoodon macrourus*
(short-nosed bandicoot) by examining feces for undigested exoskeletons.
Marker termites first appeared in the feces 7 h after ingestion, reached a
peak in concentration at 9 h, and were completely eliminated in 29 h. This
indicated a longer passage time than in some eutherian carnivores such as
the mink (*Mustela vison*) (Sibbald et al., 1962) but shorter than in the echidna
(Griffiths, 1968).

Herbivores

The families Vombatidae, Phascolarctidae, Phalangeridae, Petauridae, Bur-
ramyidae, and Tarsipedidae are all herbivorous, although some members of
the latter three families are known to take insects as well as plant material. It
is in the Vombatidae, Phascolarctidae, Phalangeridae, and Petauridae that
we see the greatest development of the large intestine (cecum and/or proximal colon).

The most studied nonmacropod marsupial is the brush-tailed possum

Figure 5. Digestive tract of the common wombat (*Vombatus urisinus*). Body
length = 85 cm.

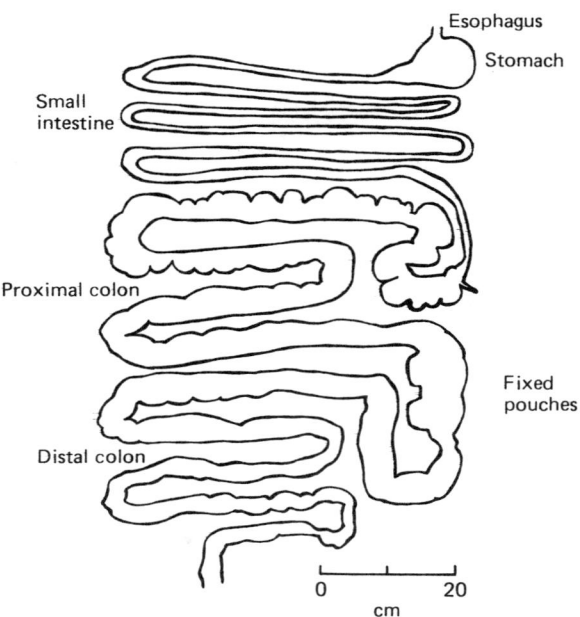

(*Trichosurus vulpecula*) in the family Phalangeridae, but recently some information has become available on other species such as the koala (*Phascolarctos cinereus*) (family Phascolarctidae) and the greater glider (*Schoinobates volans*) (family Petauridae). The other large nonmacropod herbivore is the wombat (family Vombatidae), about which little is known nutritionally except that it is a grazer with only a small cecum but a very long and wide colon (Figure 5). The capacity of the colon is approximately 70% of the total wombat digestive tract (P. N. Gowland, unpublished data). The remainder of this paper will be concerned mainly with the three foliovores *Trichosurus vulpecula*, *Phascolarctos cinereus*, and *Schoinobates volans*.

Brush-tailed possum (*Trichosurus vulpecula*). The brush-tailed possum is one of the most common and widespread marsupials in Australia, and one of the few marsupials to have adapted successfully to the presence of European man. Its success must be due in part to the wide range of food items it will accept, and it must be regarded as omnivorous. Its digestive tract, however, reflects the fact that its diet is often high in plant fiber; the small intestine is long, and both the cecum and the proximal colon are enlarged into a voluminous fermentation chamber (Figure 6).

The hindgut fermentation in the brush-tailed possum has not been quantified, but Wellard and Hume (1978) surgically removed the cecum in an

Figure 6. Digestive tract of the brush-tailed possum (*Trichosurus vulpecula*). Body length = 49 cm.

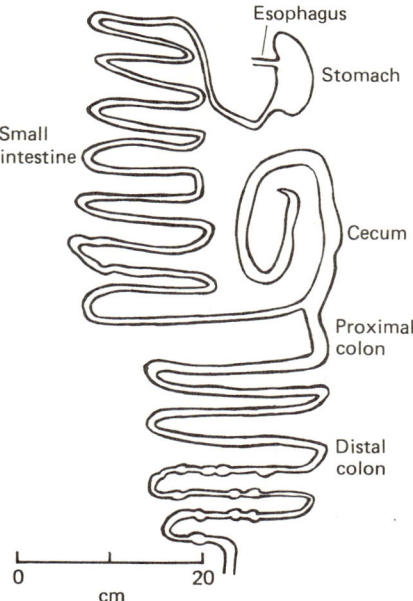

attempt to determine its significance to the animal. Cecectomy had no significant influence on dry matter intake, or on the apparent digestibility of dry matter or fiber (neutral detergent fiber; Van Soest and Wine, 1967). Rate of passage of digesta, measured by reference to the inert markers ^{51}Cr-EDTA (Downes and McDonald, 1964) and ^{103}Ru-Phenanthroline (Tan et al., 1971) was similarly unaffected by cecectomy. In the brush-tailed possum fermentation continues for some distance along the proximal colon, and perhaps this explains the absence of any overt detrimental effect on the animals of removal of the cecum; there was no evidence of hypertrophy of the proximal colon in the cecectomized possums. Thus the importance of the cecum to the nutrition of *T. vulpecula* is obscure. Certainly the high activities of the disaccharidases maltase, isomaltase, and sucrase found in the small intestine of the brush-tailed possum (Kerry, 1969) suggest that the species absorbs significant amounts of monosaccharides from the small intestine. The response in blood glucose levels in the brush-tailed possum to insulin injections (Adams and Bolliger, 1954) also supports this idea.

The success of the brush-tailed possum has led two groups to study its nitrogen economy. Cork and Harrop (1977) estimated the daily maintenance requirement of dietary nitrogen to be 311 mg kg$^{-0.75}$ body weight. Although this is higher than the estimate of 235 mg kg$^{-0.75}$ by Wellard and Hume (1978), both are low compared to eutherians, and both are in the range of values established for arid-zone, macropod marsupials (Brown and Main, 1967; Barker, 1968). They also agree with the finding of Dawson and Hulbert (1970) that the standard metabolic rate of marsupials is generally about 30% lower than the eutherian mean.

Koala (*Phascolarctos cinereus*). The koala is remarkable because of the very restricted nature of its diet, which consists almost solely of the foliage of a few *Eucalyptus* species. Koalas have also been seen to eat soil, and the terminal cecum often contains pebbles up to 1 cm in diameter (Bolliger, 1962). The nutritional significance of this is unclear.

The stomach of the koala is comparatively small, the small intestine is of intermediate length, and the cecum and proximal colon are both long and wide (Figure 7); the cecal development is perhaps the greatest of any mammal (MacKenzie, 1918). The stomach of the koala, like that of the wombat, is characterized by the presence of a gastric gland on its lesser curvature (Hingson and Milton, 1968). The gland measures about 4 cm in diameter, and contains about 25 distinct openings (Harrop and Degabriele, 1976). Similar gastric glands have also been found in the eutherian beaver (*Castor canadensis*) (Johnston, 1899). Their functional significance is unclear.

The great size of the cecum and proximal colon of the koala (Figure 7) suggests that a significant fraction of its digestible energy is absorbed as volatile fatty acids; however, the critical measurements have not yet been made. It is also possible that the hindgut microflora may detoxify components of the essential oils of Eucalyptus such as α- and β-pinene, and *p*-cymene. Eberhard et al., (1975) found that on average no more than 15% of ingested volatile oil passed through the koala gut without transformation or absorption. The ultrastructural appearance of the koala cecal epithelium suggests that it also plays an important role in water absorption (McKenzie, 1978). Eberhard (1972) found that the water content of koala fecal pellets was 44 to 48%, which is low, and characteristic of camels and kangaroo rats (*Dipodomys* sp.) maintained without water (Schmidt-Nielsen, 1964). Water turnover rates were also low (Eberhard, 1972). All of these features indicate a thrifty water economy in the koala, which is necessary because they do not drink, and thus must satisfy their needs for water from the diet.

Digestibility of grey gum (*E. punctata*) foliage by koalas has been measured by Eberhard et al. (1975) and by Harrop and Degabriele (1976). In both studies the mean apparent digestibility of dry matter was 60%, which is high for a nonruminant herbivore fed a high-fiber diet. Wellard and Hume (1976) found that juvenile leaves of *E. viminalis*, another of the koala's preferred food trees, contained 30% acid detergent fiber (Van Soest, 1963) on a dry matter basis.

The efficient digestion in the koala is probably at least partly due to the slow rate of passage through the digestive tract. Cork et al. (1977) found the mean residence time (MRT) of feed particles to be 5 to 6 days. The MRT of

Figure 7. Digestive tract of the koala (*Phascolarctos cinereus*). Body length = 51 cm.

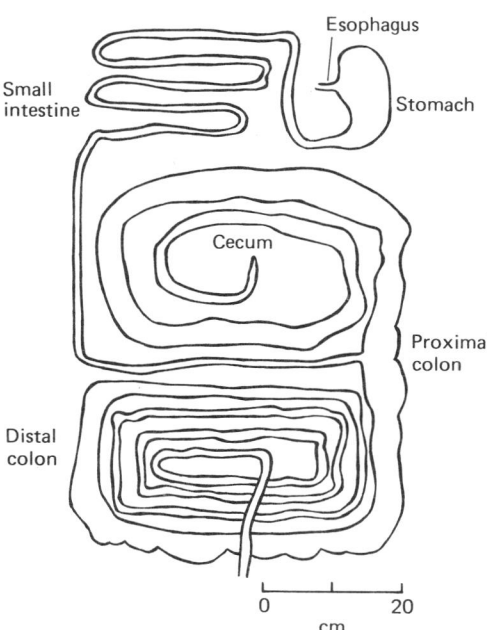

solutes was even longer, 8 to 9 days, suggesting that the cecum selectively retains solutes and fine particles, a situation known to occur in the eutherian rabbit (Pickard and Stevens, 1972). In their study with the brushtailed possum (*T. vulpecula*), Wellard and Hume (1978) also found slow rates of passage, but there was no consistent difference in the MRT of particulate and liquid digesta in either intact animals (6.3 and 5.8 days) or in cecectomized animals (5.4 and 5.6 days).

The greater glider (*Schoinobates volans*). No detailed dietary studies have been carried out with the greater glider, but Marples (1973) found the stomach contents to consist of leaf, bark, and bud debris of *Eucalyptus* spp. Thus its dietary habits appear to be very like those of the koala. Like the koala, the greater glider's digestive tract is characterized by a small stomach and a large-capacity hindgut, although unlike the koala the hindgut fermentation is confined solely to the cecum, there being no development of the proximal colon (Figure 8).

Digestibility studies have not yet been conducted with the greater glider, but Cork and Hume (1978) have measured in vitro the rate of volatile fatty acid (VFA) production in the cecal contents from wild gliders. The rate of VFA production in cecal fluid, 13 to 20 mmol liter^{-1}h^{-1}, was similar to rates measured in the hindgut of sheep fed chopped hay of lucerne (*Medicago sativa*) (Hume, 1977). The proportions in which the individual VFA were produced (mean of 60.7% acetate, 22.3% propionate, 15.1% butyrate, and 1.9% valerate in six animals) were lower in acetate and higher in propionate

Figure 8. Digestive tract of the greater glider (*Schoinobates volans*). Body length = 41 cm.

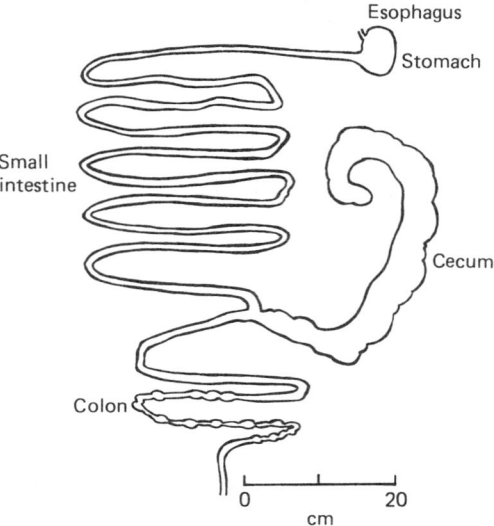

and butyrate than in the hindgut of the sheep. It can be calculated, using Marples' (1973) estimate of the daily caloric intake of wild gliders, that the total VFA production in the greater glider cecum accounts for at least 20 to 30% of the digestible energy intake of the animal. This is higher than similar estimates for the rabbit (Hoover and Heitmann, 1972) and indicates that the cecum plays an important role in the ability of the greater glider to utilize *Eucalyptus* foliage.

Rate of passage of digesta through the digestive tract of the greater glider, as in the koala and brush-tailed possum, appears to be slow. G. D. Sanson (unpublished) in a preliminary experiment found total excretion times of both particulate and liquid digesta of at least 5 days. The long excretion times recorded for this species and for the koala and the brush-tailed possum are not due entirely to the diet of *Eucalyptus* foliage because the brush-tailed possums used by Wellard and Hume (1978) were fed a semipurified diet based on bran, pollard, honey, and casein. Nevertheless, a slow rate of passage of ingesta would help to maximimize utilization of a high-fiber diet such as *Eucalyptus* leaves.

Conclusion

Our knowledge of the nutrition and metabolism of the monotremes, particularly the platypus (*Ornithorhynchus anatinus*), is embryonic. Despite numerous reports of the voracious appetite of the platypus (see Griffiths, 1978), we still have little idea of the normal range of food intake of this species, or how food intake varies throughout the year in the wild platypus. However, studies now in progress will provide much-needed information on monotreme energetics and nutrition.

Similarly, little is yet known about the nutrition of the carnivorous and omnivorous nonmacropod marsupials. However, the role of *Eucalyptus* foliage in the nutrition and ecology of the Phalangeridae, Phascolarctidae, and Petauridae is receiving attention and rapid progress can be expected in this area of marsupial nutrition.

References

Adams, D. M., and Bolliger, A. (1954). Observations on carbohydrate metabolism and alloxan diabetes in a marsupial (*Trichosurus vulpecula*). *Aust. J. Exp. Biol. Med. Sci.* 32:101–12.

Barker, S. (1968). Nitrogen balance and water intake in the Kangaroo Island wallaby, *Protemnodon eugenii* (Desmarest). *Aust. J. Exp. Biol. Med. Sci.* 46:17–32.

Bohringer, R. (In press). The bill of the platypus (*Ornithorhynchus anatinus*). *Aust. Zool.*

Bollinger, A. (1962). Gravel in the caecum of the koala (*Phascolarctos cinereus*). *Aust. J. Sci.* 24:416–17.

Brown, G. D., and Main, A. R. (1967). Studies on marsupial nutrition. V. The nitrogen requirements of the euro, *Macropus robustus*. *Aust. J. Zool.* 15:7–27.

Burrell, H. (1927). *The Platypus.* Sydney: Angus and Robertson.

Cork, S. J., and Harrop, C. J. F. (1977). Comparative nitrogen requirements and dry matter digestibilities of the bush-tail possum (*Trichosurus vulpecula*) and the rabbit (*Oryctolagus cuniculus*). *Bull. Aust. Mammal Soc. 3*:15–16, Abstr.

Cork, S. J., and Hume, I. D. (1978). Volatile fatty acid production rates in the caecum of the greater glider. *Bull. Aust. Mammal Soc. 5*:24–5, Abstr.

Cork, S. J., Warner, A. C. I., and Harrop, C. F. J. (1977). Preliminary study of the rate of passage of digesta through the gut of the koala. *Bull. Aust. Mammal Soc. 4*:24, Abstr.

Cowan, I. McT., O'Riordan, A. M. and Cowan, J. S. McT. (1974). Energy requirements of the dasyurid marsupial mouse *Antechinus swainsonii* (Waterhouse). *Can. J. Zool. 52*:269–75.

Dawson, T. J., and Hulbert, A. J. (1970). Standard metabolism, body temperature, and surface areas of Australian marsupials. *Am. J. Physiol. 218*:1233–8.

Downes, A. M., and McDonald, I. W. (1964). The chromium-51 complex of ethylene diamine tetraacetic acid as a soluble rumen marker. *Br. J. Nutr. 18*:153–62.

Eberhard, I. H. (1972). Ecology of the koala, *Phascolarctos cinereus* (Goldfuss) on Flinders Chase, Kangaroo Island. Ph.D. thesis, University of Adelaide.

Eberhard, I. H., McNamara, J., Pearse, R. J., and Southwell, I. A. (1975). Ingestion and excretion of *Eucalyptus punctata* D.C. and its essential oil by the koala, *Phascolarctos cinereus* (Goldfuss). *Aust. J. Zool. 23*:169–79.

Grant, T. R., and Carrick, F. N. (1978). Some aspects of the ecology of the platypus, *Ornithorhynchus anatinus*, in the upper Shoalhaven River, New South Wales. *Aust. Zool. 20*:181–99.

Griffiths, M. (1965). Digestion, growth and nitrogen balance in an egg-laying mammal *Tachyglossus aculeatus* (Shaw). *Comp. Biochem. Physiol. 14*:357–75.

Griffiths, M. (1968). *Echidnas*. Oxford: Pergamon Press.

Griffiths, M. (1978). *The Biology of the Monotremes*. New York: Academic Press.

Griffiths, M., and Simpson, K. G. (1966). A seasonal feeding habit of spiny ant-eaters. *CSIRO Wildl. Res. 11*:137–43.

Guiler, E. R. (1970). Observations on the Tasmanian devil, *Sarcophilus harrisii* (Marsupialia: Dasyuridae). I. Numbers, home range, movements and food in two populations. *Aust. J. Zool. 18*:49–62.

Harrop, C. J. F., and Degabriele, R. (1976). Digestion and nitrogen metabolism in the koala, *Phascolarctos cinereus*. *Aust. J. Zool. 24*:201–15.

Hingson, P. J., and Milton, G. W. (1968). The mucosa of the stomach of the wombat (*Vombatus hirsutus*) with special reference to the cardiogastric gland. *Proc. Linn. Soc. N.S.W. 93*:69–75.

Hoover, W. H., and Heitmann, R. N. (1972). Effects of dietary fiber levels on weight gain, cecal volume and volatile fatty acid production in rabbits. *J. Nutr. 102*:375–80.

Hume, I. D. (1977). Production of volatile fatty acids in two species of wallaby and in sheep. *Comp. Biochem. Physiol. 56A*:299–304.

Johnston, J. (1899). On the gastric glands of the Marsupialia. *J. Linn. Soc. London 27*:1–14.

Kerry, K. R. (1969). Intestinal disaccharidase activity in a monotreme and eight species of marsupials (with an added note on the disaccaridases of five species of sea birds). *Comp. Biochem. Physiol. 29*:1015–22.

Krause, W. J. (1972). The distribution of Brunner's glands in 55 marsupial species native to the Australian region. *Acta Anat., 82*:17–33.

Krause, W. J. (1975). Intestinal mucosa of the platypus, *Ornithorhynchus anatinus*. *Anat. Rec. 181*:251–66.

McKenzie, R. A. (1978). The caecum of the koala, *Phascolarctos cinereus:* Light, scanning and transmission electron microscope observations on its epithelium and flora. *Aust. J. Zool. 26:*249–56.

MacKenzie, W. C. (1918). *The Gastro-Intestinal Tract in Monotremes and Marsupials.* Melbourne, Australia: Critchley Parker Pty.

Marples, T. J. (1973). Studies on the marsupial glider, *Schoinobates volans* (Kerr) IV. Feeding biology. *Aust. J. Zool. 21:*213–16.

Messer, M., and Kerry, K. R. (1973). Milk carbohydrates of the echidna and the platypus. *Science 180:*201–3.

Owen, R. (1868). *On the Anatomy of Vertebrates,* vol. 3., *Mammals,* pp. 410–20. London: Longmans, Green.

Pickard, D. W., and Stevens, C. E. (1972). Digesta flow through the rabbit large intestine. *Am. J. Physiol. 22:*1161–6.

Schmidt-Nielsen, K. (1964). *Desert Animals – Physiological Problems of Heat and Water.* New York: Oxford University Press.

Sibbald, I. R., Sinclair, D. G., Evans, E. V. and Smith, D. L. T. (1962). The rate of passage of feed through the digestive tract of the mink. *Can. J. Biochem. Physiol. 40:*1391–4.

Tan, T. N., Weston, R. H., and Hogan, J. P. (1971). Use of [103] Ru-labelled tris (1,10-phenenthroline) ruthenium (II) chloride as a marker in digestion studies with sheep. *Int. J. Appl. Radiat. Isot. 22:*301–8.

Temple-Smith, P. D. (1973). Seasonal breeding biology of the platypus, *Ornithorhynchus anatinus* (Shaw, 1799), with special reference to the male. Ph.D thesis, Australian National University.

Van Soest, P. J. (1963). Use of detergents in the analysis of fibrous feeds. II. A rapid method for the determination of fiber and lignin. *J. Assoc. Off. Agric. Chem. 46:*829–35.

Van Soest, P. J., and Wine, R. H. (1967). Use of detergents in the analysis of fibrous feeds. IV. Determination of plant cell-wall constituents. *J. Assoc. Off. Agric. Chem. 50:*50–5.

Waring, H., Moir, R. J., and Tyndale-Biscoe, C. H. (1966). Comparative physiology of marsupials. *Adv. Comp. Physiol. Biochem. 2:*237–376.

Wellard, G. A. and Hume, I. D. (1976). Nitrogen metabolism in the brush-tailed possum. *Bull. Aust. Mammal Soc. 3:*50, Abstr.

Wellard, G. A., and Hume, I. D. (1978). Digestion and digesta flow in *Trichosurus vulpecula* Kerr). *Bull. Aust. Mammal Soc., 5:*26–7 Abstr.

7

Form and function of the macropod marsupial digestive tract

I. D. HUME and D. W. DELLOW

The gross anatomy of the macropod digestive tract was first described in the last century (Home, 1814, Owen, 1839–47, 1868), and in the early part of this century (Mitchell, 1916; MacKenzie, 1918). The great size and complex structure of the macropod stomach, the presence of an esophageal groove (first noted by Owen, 1839–47), and the relatively small cecum were features described as being similar to the ruminant digestive tract. The specimens examined by the early workers appear to have been mainly *Macropus giganteus* (eastern grey kangaroo), *M. parryi* (whiptail wallaby), and *Petrogale penicillata* (rock wallaby). Although Owen (1868) and MacKenzie (1918) noted some species-specific structural characteristics of the macropod stomach, the differences described were relatively minor.

The first detailed account of the digestive physiology of a macropod was given by Moir et al. (1956). Their results suggested that the quokka (*Setonix brachyurus*) was "ruminant-like" in both digestive function and metabolism. In the absence of information to the contrary, this description has subsequently been used for the entire family Macropodidae. We have recently been studying several macropod species from a wider range of environments, and this paper outlines some distinctive features in the form and function of the macropod digestive tract.

The macropod digestive tract: a comparison of eastern grey kangaroo and sheep

The stomach

The stomach of *M. giganteus* is a long, tubular organ, extensively haustrated along the greater curvature (Figure 1A). It can be divided into three regions: the sacciform forestomach (the region adjacent to, and the cul-de-sac oral to the point of entry of the esophagus), the tubiform forestomach (the main tubular body of the organ), and the hindstomach (the Gastric pouch and the adjacent region terminating at the pylorus). Although the sacciform forestomach is separated from the tubiform forestomach by a permanent ventral fold, these regions are not as distinct anatomically as are the compartments of the ruminant stomach; there are no sphincters within the macropod stomach.

The stomach contents in *M. giganteus* constitute approximately 10% of the body weight, compared with 14% for the ruminoreticulum of sheep on the same diet of chopped lucerne (*Medicago sativa*) hay (Hume, 1977a). Some 23% of the total stomach contents are found in the sacciform forestomach,

Figure 1. The stomachs of (A) *Macropus giganteus* (eastern grey kangaroo), (B) *Macropus eugenii* (tammar wallaby), and (C) *Thylogale thetis* (red-necked pademelon).

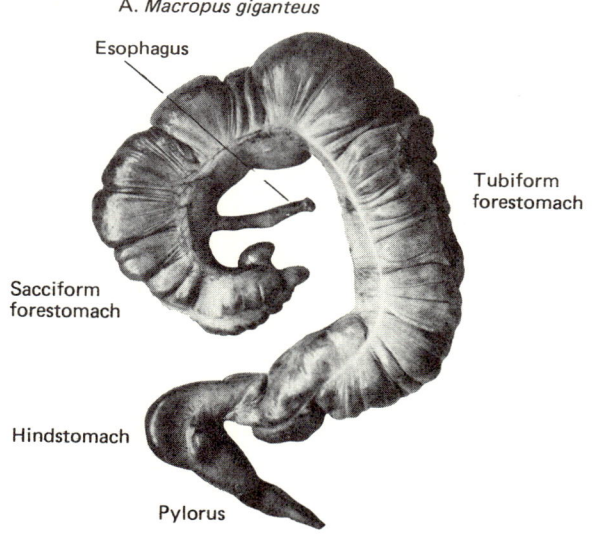

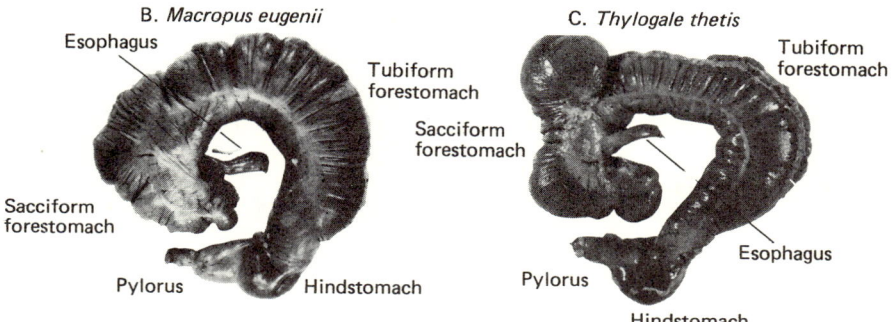

67 to 70% in the tubiform forestomach, and 7 to 10% in the hindstomach (Dellow and Hume, 1977) (Table 1). Ingested food is subjected to fermentative digestion as it passes along the length of the stomach and is only exposed to acid and peptic digestion in the hindstomach.

The abdominal segment of the esophagus is comparatively long in macropods (9 to 14 cm in adult *M. giganteus*). The esophagus enters the stomach on the lesser curvature, and is contiguous with the gastric sulcus, which extends longitudinally along the lesser curvature to the distal tubiform forestomach. Both the floor and the lips of the sulcus are lined with stratified squamous epithelium. The pouch young of *M. giganteus* exhibit a well-developed sulcus, but in the adult the sulcus is both anatomically and functionally regressive, and the lips are poorly defined. To date, regression of the gastric sulcus in the adult animal has only been observed in *M. giganteus.*

In *M. giganteus*, squamous epithelium extends throughout most of the sacciform forestomach except for the larger of the two blind sacs which is lined with an isolated thick cardiac epithelium 50 to 80 cm^2 in area. The tubiform forestomach epithelium, apart from the gastric sulcus, is extensively folded, and consists of mucin-secreting cardiac cells (Schäfer and Williams, 1876) organized into tubular glands as described for *Megaleia rufa* (red kangaroo) by Griffiths and Barton (1966). In most macropod species by far the greatest area of the stomach mucosal epithelium consists of these cardiac gland cells. Proper gastric glands are confined entirely to the gastric pouch in adult animals (Griffiths and Barton, 1966), whereas the epithelium lining the inner curvature of the hindstomach consists of glandular cells similar to those of the tubiform forestomach epithelium.

The postgastric digestive tract
The small intestine of *M. giganteus* is of considerable length, but shorter than in ruminants. The cecum and proximal colon are relatively simple in structure (Figure 2), and of similar relative size to those of the sheep. Defecation is more frequent in macropods than in sheep, partly because there is

Table 1. *Relative capacities of the three regions of the stomach in three macropod species*

Species	Contents (% of total stomach contents, wet weight basis)		
	Sacciform forestomach	Tubiform forestomach	Hindstomach
Macropus giganteus	22.8	68.0	9.3
Macropus eugenii	29.7	55.7	14.6
Thylogale thetis	50.7	39.7	9.7

Source: Dellow and Hume (1977).

no anatomical dilation of the distal colon or the rectum where fecal pellets would aggregate before being voided.

Regurgitation in the Macropodidae
Regurgitation occurs irregularly in individual animals fed either chopped lucerne hay or fresh grass, but, in contrast to that associated with rumination, it is not repetitive. Regurgitation involves several violent contractions of the abdominal muscles and the diaphragm, similar to vomiting, and in smaller species, such as *Setonix brachyurus* (quokka) (Barker et al., 1963), *Thylogale thetis* (red-necked pademelon), and *Macropus eugenii* (tammar wal-

Figure 2. Digestive tract of *Macropus giganteus* (eastern grey kangaroo).

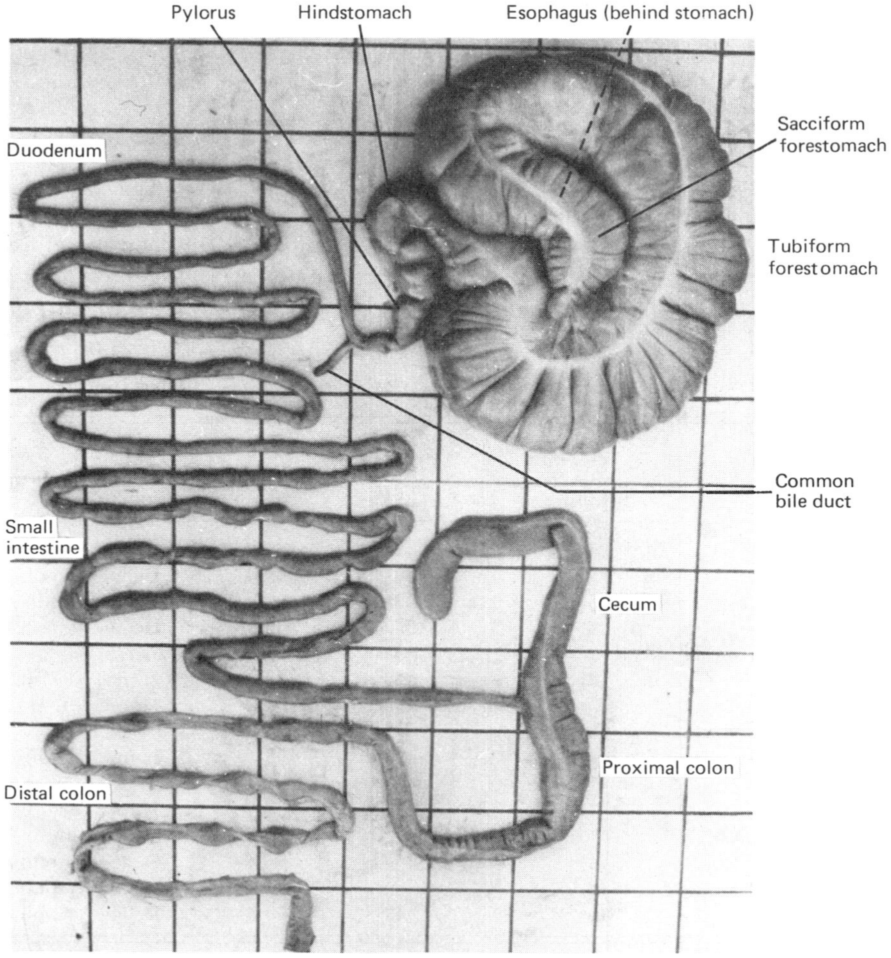

laby), this may result in ejection of a food bolus, which is immediately re-ingested and swallowed. In *M. giganteus* the food bolus is not ejected but is retained in the mouth, chewed for a short period (20 to 30 s) and then reswallowed. Regurgitation is often associated with feeding, and the frequency of occurrence can be increased by the addition of grain to the diet. Barker et al. (1963) suggested that the process be termed "merycism" rather than rumination.

Macropods may have no real requirement for rumination because they masticate their food more thoroughly than do ruminants and there is no functional barrier within the stomach to retain the larger food particles differentially. However, a number of workers have observed that macropods "chew the cud" (Moir et al., 1956). This occurs more frequently than merycism, and usually only while the animal is resting. The process does not involve regurgitation of food, but rhythmic jaw movements continue for up to 30 min, with the apparent effect of stimulating salivary secretion, as judged from the increased frequency of swallowing (Dellow, personal observation).

Intake, digestion, and digesta flow

M. giganteus and many other macropod species ususally eat less dry matter on a metabolic body weight basis (i.e., $kg^{0.75}$) than do sheep (Table 2.) This has been explained by the lower standard (basal) metabolic rate of marsupials compared with eutherians (Dawson and Hulbert, 1970), and therefore the lower maintenance energy requirement (Hume, 1974). Digestion of dry matter (and of fiber) is significantly lower in *M. giganteus* than in sheep, but

Table 2. *Intake and apparent digestibility of dry matter by macropod marsupials and sheep fed chopped lucerne hay (L) or fresh grass (G)* (Phalaris tuberosa)

Species	No. of animals	Diet	Daily intake $(g\ kg^{-0.75})$	Apparent digestibility (%)	Data source
Ovis aries	(4)	L	62.0	61.2	Hume (1977a)
	(7)	L	60.2	59.0	Dellow (unpub.)
Macropus giganteus	(3)	L	58.1	52.0	Hume (unpub.)
	(8)	L	56.7	55.5	Dellow (unpub.)
Macropus rufog-riseus	(4)	L	54.6	55.0	Hume (1977a)
Megaleia rufa	(3)	L	53.4	54.8	Hume (1974)
Macropus robustus robustus	(4)	L	61.5	53.1	Hume (unpub.)
Macropus robustus cervinus	(3)	L	52.7	61.5	Hume(1974)
Thylogale thetis	(11)	L	57.5	55.5	Dellow (unpub.)
	(4)	G	46.6	56.3	Dellow (unpub.)
Macropus eugenii	(8)	L	29.9	59.5	Dellow (unpub.)
	(5)	G	34.8	59.9	Dellow (unpub.)

there is considerable variation in digestive efficiency among macropod species (Table 2).

The lower digestibility in the macropods has been attributed to the faster rate of passage of food residues through the gut. Forbes and Tribe (1970) measured the rate of passage of stained hay particles through the digestive tract of the grey kangaroo and sheep. Mean retention time (i.e., 50% excretion time) and 90% excretion time were both shorter in the grey kangaroo (39 and 50 h) than in the sheep (52 and 89 h). Dellow (1978, and unpublished data) used the two inert markers ^{51}Cr-EDTA, which moves with the liquid phase (Downes and McDonald, 1964) and ^{103}Ru-Phenanthroline, which attaches to food particles (Tan et al., 1971) to define more clearly the pattern of digesta flow in these two herbivores. Mean retention time of particulate digesta (measured with ^{103}Ru-Phenanthroline) was shorter in the sheep than in *M. giganteus*, but there was no significant difference in 90% excretion times (Table 3). The different results may be due in part to the higher dry matter intakes of the sheep in Dellow's study, and in part to the smaller particle size of chopped lucerne hay. In addition, the stained particle technique may have an inherent bias toward larger particles; in ruminants there is differential retention of larger particles in the ruminoreticulum, and the rate of passage as measured by stained particles is influenced by the size of the particles used (Balch and Campling, 1965).

In the sheep there was minimal differentiation between the flow of particulate and liquid digesta, but in *M. giganteus* the flow rate of liquid digesta was at least twice that of the particulate phase (Table 3). In other experiments where the markers were infused via a cannula into the hindstomach, both markers were excreted in the feces at the same rate. Thus the differences in flow of the two phases of digesta must be due to the mode of flow through the stomach. The patterns of appearance of oral doses of the two markers in the feces of *M. giganteus* (Figure 3) further indicate that neither marker mixes completely with the total stomach contents as a single digesta pool, but rather each marker passes along the length of the stomach as a

Table 3. *Rate of passage (in hours) of digesta markers through digestive tract of macropod marsupials and sheep*

Species	Liquid (^{51}Cr-EDTA)			Particulate (^{103}Ru-Phe)		
	10%	50%	90%	10%	50%	90%
Ovis aries	12.6[a]	20.9[a]	38.1[a]	14.2[a]	24.5[a]	44.1[a]
Macropus giganteus	10.0[ab]	13.6[b]	19.9[b]	24.6[b]	31.2[b]	39.7[ab]
Thylogale thetis	8.5[b]	11.1[b]	17.3[b]	17.1[a]	23.9[a]	35.9[b]

Note: Means in the same column bearing different superscripts differ significantly (P < 0.05).
Data from Dellow and Hume (1977); Dellow (1978).

pulse. Preferential retention of the particulate digesta appears to be a function of the extensive haustration of the stomach wall, and contractions of these haustra presumably result in the extrusion of liquid digesta in a caudal direction. This tubular flow is in marked contrast to flow through the ovine stomach, in which ingested feed is considered to mix with the entire ruminoreticular contents as a single pool of digesta.

Production of volatile fatty acids

It has not been possible to measure volatile fatty acid (VFA) production *in vivo* in the macropod stomach because the technique depends on rapid mixing and uniform distribution of infused ^{14}C-labeled acid (Hume, 1977a). Hence comparisons between macropods and sheep have been made only in vitro. Estimates of VFA production in sheep made in vitro are invariably lower than those made in vivo. Despite this, we believe that in vitro comparisons between the species are valid.

The mean rate of VFA production in stomach contents from *M. giganteus* fed a chopped lucerne hay diet ad libitum was 33.9 mmol liter^{-1} h^{-1} equivalent to a mean of 28% of the digestible energy intake of the animal. The rate is higher than in sheep on the same diet (22.9 mmol VFA liter^{-1} h^{-1}, but because of the smaller capacity of the macropod stomach, calculated total

Figure 3. Patterns of appearance of the markers ^{51}Cr-EDTA (liquid phase of digesta) and ^{103}Ru-Phenanthroline (particulate phase of digesta) in the feces of *Ovis aries* (sheep) [top], *Macropus giganteus* (eastern grey kangaroo) [center], and *Thylogale thetis* (red-necked pademelon) [bottom].

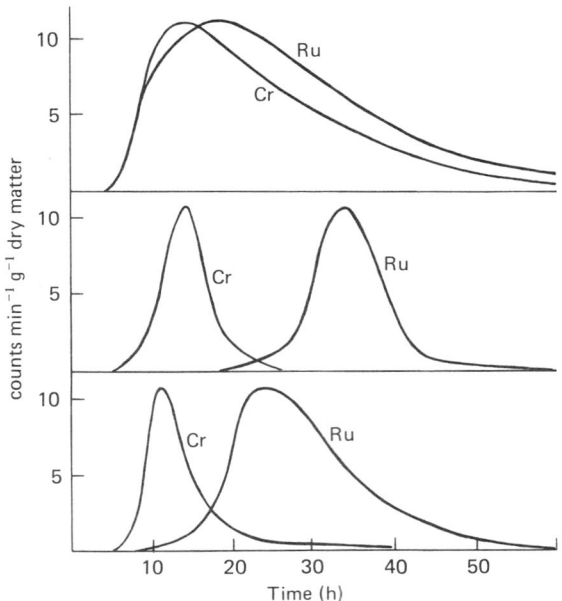

VFA production in the ruminoreticulum of sheep was similar to that of *M. giganteus*. The faster rate of VFA production in the macropod stomach can perhaps be explained by the faster flow rate of digesta through the stomach, and hence the faster turnover of the microbial population in the stomach. Rapid fermentation rates in the smaller ruminants are associated with high rumen turnover rates (Hungate et al., 1960).

Rates of VFA production in the hindgut (i.e., the cecum and proximal colon) were similar in *M. giganteus* (16.7 mmol liter^{-1} h^{-1}) and the sheep (16.0 mmol liter^{-1} h^{-1}), but because of the greater absolute capacity of the ovine hindgut, the calculated total VFA production was 6.9% of digestible energy intake in the sheep, and only 1.2% in *M. giganteus*. The individual VFA were produced in similar proportions in the two herbivores.

This comparison utilized composite samples of the mixed sacciform and tubiform forestomach contents of the grey kangaroo. In other experiments discrete samples were taken from the sacciform forestomach and from three different sites along the tubiform forestomach. The rate of production of VFA was similar in samples from the sacciform forestomach and distal tubiform forestomach in animals grazing improved pasture in spring (26.2 to 25.6 mmol liter^{-1} h^{-1} in one animal, and 19.5 to 18.5 in a second), but in animals grazing native pasture in winter there was a consistent decline in production rate from the sacciform forestomach (16.0 and 19.5 mmol liter^{-1} h^{-1} in two animals) to the distal tubiform forestomach (13.7 and 15.5 mmol liter^{-1} h^{-1}). On the poorer-quality pasture the fermentation rate in the tubiform distal forestomach appeared limited by substrate availability.

The stomach fermentation of *M. giganteus* also differs from that of the sheep in that methane does not appear to be produced in measurable quantities, at least on a chopped lucerne hay diet (Kempton et al., 1976).

Nitrogen metabolism

The maintenance nitrogen requirement of *M. giganteus* has been estimated by feeding adult animals a series of diets based on chopped oaten (*Avena* spp.) hay, starch, sucrose, and a mineral–vitamin mix, with casein added to provide a range of nitrogen concentrations. The daily maintenance requirement was 350 mg kg$^{-0.75}$ of dietary nitrogen, or 270 mg kg$^{-0.75}$ of truly digestible nitrogen (W. J. Foley and Hume, unpublished data) (Table 4). This is lower than estimated daily maintenance requirements for sheep (489 and 452 mg kg$^{-0.75}$ for dietary and truly digestible nitrogen, respectively; Moir and Williams, 1950), and can be related to the lower standard metabolic rate of marsupials compared with eutherians (Dawson and Hulbert, 1970).

Differences among macropod species
The stomach

There are significant differences among macropod species in the relative sizes of the sacciform forestomach and tubiform forestomach, and in the distribution of the different types of stomach epithelia. The stomachs of three species, *Thylogale thetis* (red-necked pademelon), *Macropus eugenii* (tammar wallaby), and *M. giganteus*, are shown in Figure 1, and the capacity of each region is expressed as a percentage of the whole stomach capacity in Table 1. Only members of the subfamily Potoroinae have a sacciform forestomach capacity relatively greater than that of *T. thetis;* in fact *Aepyprymnus rufescens* (rufous rat kangaroo) appears to have no tubiform forestomach as defined in this paper, and the sacciform forestomach constitutes most of the total stomach capacity.

Most adult macropods, especially *Macropus,* spp. (with the exception of *M. giganteus*), have a well-defined gastric sulcus running along the lesser curvature from the esophagus to the distal tubiform forestomach. Other genera in the subfamily Macropodinae (e.g. *Onychogalea, Lagorchestes, Dendrolagus,* and *Wallabia*) also have a prominent gastric sulcus, but at least two genera (*Thylogale* and *Peradorcus*) apparently have no sulcus at all. In *Petrogale* the sulcus is ill-defined. Both the presence or absence of a gastric sulcus and the position of the opening of the esophagus (i.e., whether in the sacciform or the tubiform stomach) influence the initial dispersion and subsequent movement of digesta through the stomach (Dellow, 1978).

No genera in the subfamily Potoroinae appear to have a gastric sulcus. In the Potoroinae the stomach also appears devoid of stratified squamorous epithelium, being glandular throughout. In the Macropodinae the sacciform forestomach can be either entirely squamous (e.g., *Thylogale stigmatic,* red-legged pademelon; *T. thetis,* red-necked pademelon; *Petrogale venustula,* rock wallaby), entirely glandular (e.g., *Dendrolagus lumholtzii,* tree kangaroo), or glandular except for a stratified squamous zone adjacent to the esophageal opening (e.g., *Lagorchestes conspicillatus,* spectacled hare wallaby). The nutritional significance, if any, of these variations in the pattern of mucosa distribution is unknown.

Table 4. *Maintenance nitrogen requirements of macropod marsupials and sheep*

Species	Daily requirement (mg truly digestible N kg$^{-0.75}$)	Data source
Ovis aries	452	Moir and Williams (1950)
Macropus giganteus	270	W. J. Foley and Hume (unpub.)
Macropus robustus robustus	240	W. J. Foley and Hume (unpub.)
Macropus robustus cervinus	170	Brown and Main (1967)
Macropus eugenii	250	Barker (1968)
Macropus eugenii	230	Hume (1977b)
Thylogale thetis	530	Hume (1977b)

The postgastric digestive tract

The cecum and the proximal colon of all macropods appear to be areas of microbial fermentation, but in all the species examined neither organ is large, and each has a simple gross morphology. The greatest difference among macropod species is in the relative length of the distal colon, which appears to be associated with water availability in the preferred habitat, and hence with the need for efficient absorption of water from the gut.

Intake, digestion, and digesta flow

The tammar wallaby, *Macropus eugenii*, consistently eats less dry matter on a metabolic body weight basis than do the other macropod species so far studied, suggesting that the maintenance energy requirement, and hence the basal metabolic rate of the tammar, is below the general marsupial mean established by Dawson and Hulbert (1970). The tammar and *Macropus robustus cervinus* (the Western Australian euro) are the only macropod species so far studied that appear to be similar to sheep in their ability to digest dry matter (Table 2).

The pattern of excretion of two fluid digesta markers in the feces of *Thylogale thetis* (red-necked pademelon) was intermediate between that of *M. giganteus* and the sheep (Figure 3). The less complete separation between the liquid and particulate digesta in *T. thetis* compared with *M. giganteus* can be related to the larger relative capacity of the sacciform forestomach of *T. thetis* (Table 1), which results in more complete mixing of the two markers within a single digesta pool. The longer mean retention time of particulate digesta in *M. giganteus* (Table 3) is probably attributed to the longer relative length of the tubiform forestomach in this species (Figure 2).

Production of volatile fatty acids

Hume (1977a) measured the rate of in vitro VFA production in digesta from the stomach and hindgut of *T. thetis* and *Macropus rufogriseus* (red-necked wallaby). The rate of production was higher in the stomach of *M. rufogriseus* (51.9 μmol ml^{-1} h$^-$) than in *T. thetis* (39.4 μmol ml^{-1} h^{-1}), but in the hindgut the rates were similar (26.7 and 28.7 μmol ml^{-1} h^{-1}). The contribution made by the stomach VFA production to the digestible energy intake of the animal was also greater in *M. rufogriseus* (42%) than in *T. thetis* (21%) or *M. giganteus* (28%). In all three species the contribution made by the hindgut fermentation was small (1.2 to 1.9% of digestible energy intake). Thus there appear to be significant differences among macropod species in the importance of the stomach microbial fermentation; it also appears that these differences may be related to differences in habitat preference and feeding behavior (Hume, 1977a).

Nitrogen metabolism

The maintenance nitrogen requirements of a number of macropod species are lower than those of eutherian mammals such as the sheep (Table 4).

One exception is *T. thetis*, whose maintenance requirement is similar to that of sheep, and more than twice that of other macropods such as *Macropus eugenii*, a wallaby of similar body size (Hume, 1977b). This difference is thought to be related to the more arid environment in which the tammar is found compared with the moist forest habitat of *T. thetis*. Kennedy and Hume (1978) have measured urea cycling to the gut of the tammar on high and low protein diets. Incorporation of urea nitrogen into protein in the gut was equivalent to only 34 to 53% of nitrogen intake on the high protein diet but to 103 to 112% of nitrogen intake on the low protien diet, indicating significant utilization for microbial protein synthesis of endogenous nitrogen entering the gut. Similar measurements have not yet been made with *T. thetis*.

Conclusion

Although the range of macropod species so far studied is limited, significant differences have been found in digestion, digesta flow, and microbial fermentation. Many of these differences can be related to differences in structure of the digestive tract, particularly the stomach, and to differences in habitat and feeding behavior. Extension of the work to the subfamily Potoroinae would help our understanding of interrelationships among members of the family Macropodidae.

References

Balch, C. C., and Campling, R. C. (1965). Rate of passage of digesta through the ruminant digestive tract. In *Physiology of Digestion in the Ruminant*, ed. R. W. Dougherty, pp. 108–23. Washington, D.C.: Butterworth.

Barker, S. (1968). Nitrogen balance and water intake in the Kangaroo Island wallaby, *Protemnodon eugenii* (Desmarest). *Aust. J. Exp. Biol. Med. Sci. 46*:17–32.

Barker, S., Brown, G. D., and Calaby, J. H. (1963). Food regurgitation in the Macropodidae. *Aust. J. Sci. 25*:430–2.

Brown, G. D., and Main, A. R. (1967). Studies on marsupial nutrition. V. The nitrogen requirements of the euro, *Macropus robustus. Aust. J. Zool. 15*:7–27.

Dawson, T. J., and Hulbert, A. J. (1970). Standard metabolism, body temperature, and surface areas of Australian marsupials. *Am. J. Physiol. 218*:1233–8.

Dellow, D. W. (1978). Stomach structure and flow of digesta in macropod marsupials. *Bull. Aust. Mammal Soc. 5*:25–6, Abstr.

Dellow, D. W., and Hume, I. D. (1977). Digestion, digesta flow and stomach structure in macropod marsupials and in sheep. *Bull. Aust. Mammal Soc. 4*:24, Abstr.

Downes, A. M., and McDonald, I. W. (1964). The chronium-51 complex of ethylene diamine tetraacetic acid as a soluble rumen marker. *Br. J. Nutr. 18*:153–62.

Forbes, D. K., and Tribe, D. E. (1970). The utilization of roughages by sheep and kangaroos. *Aust. J. Zool. 18*:247–56.

Griffiths, M. and Barton, A. A. (1966). The ontogeny of the stomach in the pouch young of the red kangaroo. *CSIRO Wildl. Res. 11*:169–85.

Home, E. (1814). *Lectures on Comparative Anatomy*, vol. 1. London: Bulmer.

Hume, I. D.(1974). Nitrogen and sulphur retention and fibre digestion by euros, red kangaroos and sheep. *Aust. J. Zool. 22:*13–23.

Hume, I. D. (1977a) Production of volatile fatty acids in two species of wallaby and in sheep. *Comp. Biochem. Physiol. 56A:*229–304.

Hume, I. D. (1977b). Maintenance nitrogen requirements of the macropod marsupials *Thylogale Thetis,* red-necked pademelon, and *Macropus eugenii,* tammar wallaby. *Aust. J. Zool. 25:*407–17.

Hungate, R. E., Phillips, G. D., McGregor, A., Hungate, D. P., and Buechner, H. K. (1960). Microbial fermentation in certain mammals. *Science N.Y. 130:*1192–4.

Kempton, T.J., Murray, R. M., and Leng, R. A. (1975). Rates of production of methane in the grey kangaroo and sheep. *Aust. J. Biol. Sci. 29:*209–14.

Kennedy, P. M., and Hume I. D. (1978). Recycling of urea nitrogen to the gut of the tammar wallaby (*Macropus eugenii*). *Comp. Biochem. Physiol. 61A:*117–21.

MacKenzie, W. C. (1918). *The Gastro-Intestinal Tract in Monotremes and Marsupials.* Melbourne, Australia: Critchley Parker Pty.

Mitchell, P. C. (1916). Further observations on the intestinal tract of mammals. *Proc. Zool. Soc. London,* pp. 183–251.

Moir, R. J., and Williams, V. J. (1950). Ruminal flora studies in sheep. II. The effect of the level of nitrogen intake upon the total number of free micro-organisms in the rumen. *Aust. J. Sci. Res. B3:*381–92.

Moir, R. J., Somers, M., and Waring, H. (1956). Studies on marsupial nutrition. I. Ruminant-like digestion in a herbivorous marsupial *Setonix brachyurus* (Quoy and Gaimard). *Aust. J. Biol. Sci. 9:*293–304.

Owen, R. (1839–47). Marsupialia. In *The Cyclopaedia of Anatomy and Physiology,* vol. 3, ed. R. B. Todd. London: Longman, Brown, Green, Longmans and Roberts.

Owen, R. (1868). *On the Anatomy of Vertebrates,* vol. 3., *Mammals.* pp. 410–20. London: Longmans, Green.

Schäfer, E. A., and Williams, D. J. (1876) On the structure of the mucous membrane of the stomach of the kangaroos. *Proc. Zool. Soc. London, pp.* 167–77.

Tan, T. N., Weston, R. H., and Hogan, J. P. (1971). Use of [103]Ru-labelled tris (1,10-phenanthroline) ruthenium (II) chloride as a marker in digestion studies with sheep. *Int. J. Appl. Radiat. Isot. 22:*301–8.

8

The digestive tract: insectivore, prosimian, and advanced primate

EDGAR T. CLEMENS

The principles of evolution suggest that higher forms of the order Primates were derived from the more primitive monkeylike forms, the prosimians, and that these, in turn, may have evolved from members of the order Insectivora (Szalay, 1968; Hill, 1972). The evidence is based primarily upon skeletal–structural relationships and can be tested according to physiological parameters as seen in those animals not extinct (Walker and Murray, 1975; Hiley, 1976; Müller, 1977). Presumably digestive characteristics will show structural and functional similarities or deviations in accordance with evolutionary aspects, when compared along these lines.

The observations presented in this paper describe the relationships of digesta movement and organic acid production in the gastrointestinal tract of an insectivore (the hedgehog, *Erinaceus hindei,*), a prosimian primate (the bushbaby *Galago crassicaudatus*), and an advanced primate (the vervet monkey, *Cercopithecus pygerythrus*).

Hedgehog

Some members of the order Insectivora (shrews, moles, and hedgehogs) have been in existence since the Mesozoic period 130 to 180 million years ago (Campbell, 1971). Common features of a nocturnal mode of life, insectivorous diet, limited prehension, as well as skeletal–structural relationships, provide us with an opportunity to group these animals for consideration as to possible precursors to the primates (Hill and Rewell, 1948; Hill, 1972). Structurally their digestive tract is simplified. The hedgehog's digestive tract consists of a simple stomach, a small intestine, no cecum, and a smooth noncomplex colon (Figure 1). The ileo–colonic junction is not easily defined, lacking any distinct sphincter or valve which could clearly indicate the transition from small intestine to the colon. Nor does the luminal diameter appear greatly enlarged within the region of the colon when compared to the small intestine, as seen in some other mammals lacking a cecum (Stev-

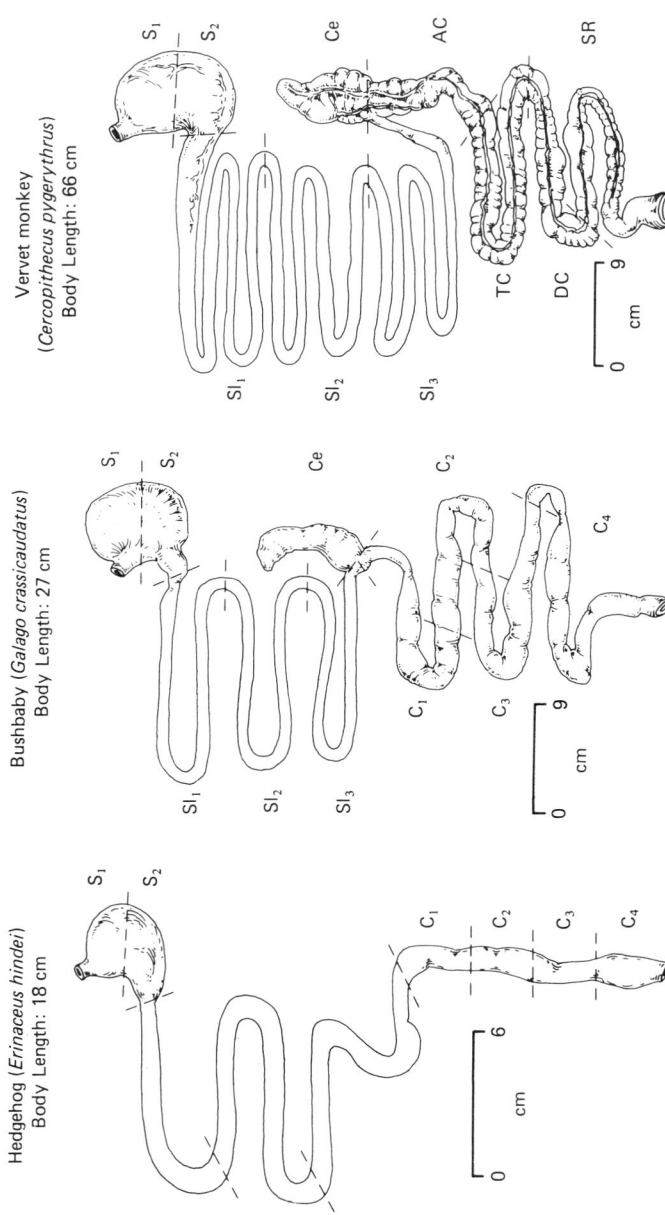

Hedgehog (*Erinaceus hindei*)
Body Length: 18 cm

Bushbaby (*Galago crassicaudatus*)
Body Length: 27 cm

Vervet monkey
(*Cercopithecus pygerythrus*)
Body Length: 66 cm

Figure 1. Gastrointestinal tracts of hedgehog, bushbaby, and vervet monkey. Sections of the tract were separated by ligatures for analysis of markers, pH, and organic acids. Symbols represent the cranial (S_1), and caudal (S_2) halves of the stomach, proximal (SI_1), middle (SI_2), and distal (SI_3) thirds of the small intestine; cecum (Ce), where present; and four equal segments of the colon (C_1–C_4). The vervet monkey's colon was divided into the ascending (AC), transverse (TC), descending (DC), and sigmoid-rectal (SR) areas. Body length refers to distance from mouth to anus in the intact animal.

ens, 1977). The colonic region is, however, easily distinguished by examining the mucosal surface.

In general, most anatomical features of the gastrointestinal tract, of these and other animals as well, function to regulate the flow of the ingesta through each of the gut segments, providing areas of retention or rapid transit in accordance with the needs for enzymatic and/or microbial digestion (Luckey, 1974).

All animals under investigation were fed a commercially prepared primate diet and were trained to consume this diet during a 1-h feeding period, twice a day at 12-h intervals. The rate of flow of ingesta through the gastrointestinal tract of the hedgehog could be approximated with markers representing the movement of fluids and particulate matter (Kotb and Luckey, 1972). The procedures used in the present study have been described by Guard (this volume).

In the hedgehog the stomach provided the only site for digesta retention, and this was limited to retaining the particulate material while allowing for the release of liquids (Figure 2). Digesta movement through the small intestine and colon was rapid, with the greater portion being excreted within 12 to 16 h after ingestion. Similarly, particulate markers, representing the solid constituents of the diet, once released from the foregut were transported quite rapidly through the remaining segments of the tract. Retention of the digesta, whether liquid or particulate matter, was never apparent in the small or large intestine. (Although the cecum is absent in the hedgehog, it is represented in Figures 2 and 3 for comparison to the data on the primates.)

The retaining of digesta within the gastrointestinal tract to enable microbial degradation is a long-accepted phenomenon (McBee, 1970; Bauchop, 1971). Microbial activity is frequently estimated by the levels of organic acids produced (Rubinstein et al., 1969; Wolin, 1974; Kay et al., 1976). The concentrations of organic acids, both lactic and volatile fatty acids (VFA), in the gastrointestinal tract of the hedgehog shows that VFA's are present within the small intestine and colon of these animals (Figure 3). Greatest variations were noted in the colon, in which concentrations occasionally reached 90 to 100 mmol liter^{-1}, similar to the lower range of levels found in the ruminoreticulum of domestic cattle (Church, 1969).

Lactic acid was nearly absent from the stomach of the hedgehog, but tended to increase progressively from the proximal small intestine to the distal colon (Figure 3), in which it reached mean levels of approximately 95 mmol liter^{-1}. The levels were somewhat higher than those observed for the volatile fatty acids in these gut segments. However, gastrointestinal lactic acid may be derived from two major sources: (1) microbial activity (primarily lactobacillus; Eyssen et al., 1965) and/or (2) the gut mucosal cells (Davenport, 1971). Therefore, we cannot with certainty attribute the increased levels of lactic acid to the activities of the microflora.

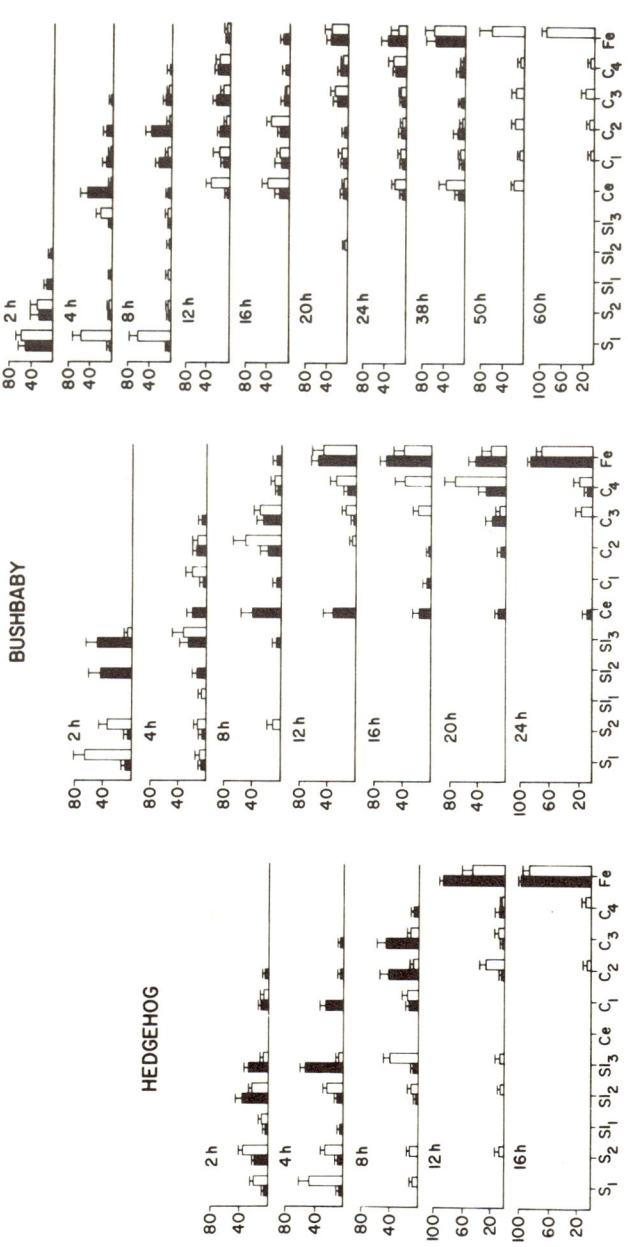

Figure 2. Percent of fluid and particle markers (± SEM) recovered from the gastrointestinal tracts of hedgehog, bushbaby, and vervet monkey at specified times after feeding. Fluid markers, polyethylene glycol (MW 4000) and phenol red, were used for only the 2 through 38 h postfeeding periods. Particulate markers, polyethylene tubing (2 mm OD, 1.3 sp. gr.), were cut into lengths of 2 mm for the hedgehog and 5 mm for the bushbaby and the vervet monkey. Symbols along the abscissa represent the sections of tract as given in Figure 1 legend. Fe refers to the fecal excretion (N = 3).

Bushbaby

Bushbabies are believed to have been in existence for approximately 60 million years (Charles-Dominique, 1977). The bushbaby and hedgehog are similar in their nocturnal mode of life and insectivorous diet, but because of its prehensile advantage, the bushbaby is taxonomically considered as a lower form of primate (Hill, 1972).

The digestive tract of the bushbaby is similar to that of the hedgehog. It has a balloonlike stomach, a small intestine, and a smooth, noncomplex colon (Figure 1). It differs from that of the hedgehog in having a well-defined cecum (Hill, 1966; Hill, 1972). In addition, the colon has a larger diameter than the small intestine, which would allow for a larger relative volume.

In order to determine functional similarities or differences between the bushbaby and hedgehog not attributable to dietary differences, the bushbaby was fed a diet identical to that of the hedgehog and on the same 12-h feeding schedule.

In the bushbaby the flow of digesta through the gastrointestinal tract was

Figure 3. Mean values (± SEM) obtained 2, 4, 8 and 12 h after feeding for volatile fatty acids (VFA), lactic acid, and pH along the gastrointestinal tracts of the hedgehog (●), bushbaby (▲), and vervet monkey (o). Symbols along the abscissa represent the sections of tract as given in Figure 1 (N = 12).

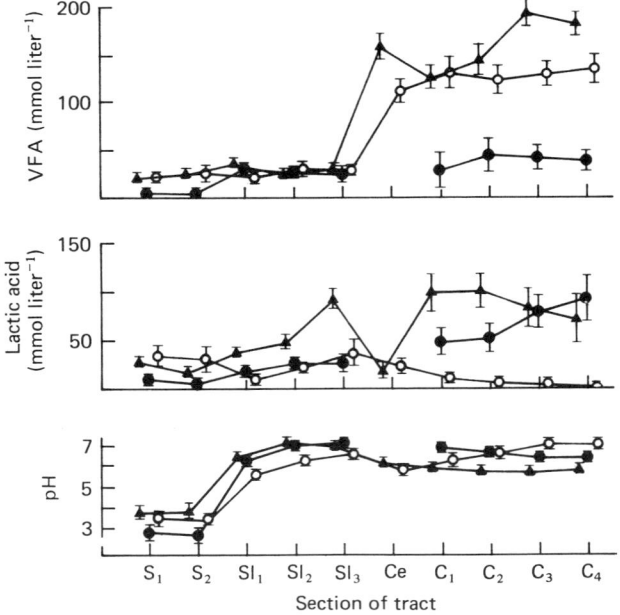

rapid, as it was in the hedgehog (Figure 2). The liquid and water-soluble portion of the meal left the foregut rapidly and tended to accumulate in the distal small intestine and cecum. Cecal retention of liquid was obvious throughout most of the day, whereas liquid appeared to move rapidly through the colon. Particulate matter, in contrast, was retained in the stomach and passed rapidly through the small intestine. Particulate markers in no case entered the cecum of these animals, suggesting some form of control that permits the entrance of liquid and perhaps particles of very minute size. Particulate matter moved rapidly through the colon of the bushbaby, and retention of such material in the more distal regions of the colon may have been due to either colonic retention for enhanced digestive processes, or simply rectal control for periodic fecal evacuation.

The levels of organic acids within the gastrointestinal tract of the bushbaby indicated that those microbes that tend to produce VFA's exist primarily in the cecum and colon (Figure 3). Highest levels were found in the cecum and distal colon where quantities of liquid and/or particles accumulated. The variations in colonic VFA levels were directly associated with the flow of the meal through these gut segments, being higher when larger quantities of the meal were present and decreasing as the meal was evacuated from these areas (Clemens, unpublished data). An interesting feature is that even though quantities of liquid and particulate marker accumulated in the distal small intestine, the levels of VFA in this area were similar to those in more proximal segments, but significantly lower than in the cecum and colon where digesta retention was also apparent. This suggests that the retention of the digesta was not the primary requirement for the presence of VFA-producing microorganisms. Since the pH in the distal small intestine, cecum, and colon was not grossly different (see Figure 3), luminal pH was not the ruling factor either.

When compared to the hedgehog, concentrations of VFA were significantly higher within the bushbaby's colon, whereas they did not differ within the stomach and small intestine. Concentrations of VFA in the cecum and colon of the bushbaby equaled, and in many cases exceeded, levels observed in the foregut of ruminants (in which the advantage of extensive microbial fermentation and VFA production is well known) (Hungate, 1966; Church, 1969).

The other organic acid of major interest is lactic acid (Figure 3). Concentrations of lactic acid were highest in the distal small intestine and colon of the bushbaby. Note that lactic acid was nearly absent from the cecum of these animals; this indicates that lactic acid was not produced within nor did it enter the cecum, suggesting the presence of a group of microorganisms whose fermentative process produce VFA's but not lactic acid. The other possibility, of course, is that lactic acid entered the cecum of the bushbaby but was readily absorbed or selectively utilized. This, however, seems unlikely because the VFA concentration was not reduced.

In general, concentrations of lactic acid within the gastrointestinal tract of the hedgehog and the bushbaby were similar, differing slightly at only a few sites along the tract.

Vervet monkey

The vervet monkey represents a form of advanced primate believed to be at an evolutionary level equal to that of the domestic cat and dog (Hill, 1972). Certainly its digestive tract (Figure 1) ranks at the level of the more advanced primates, including man. Anatomically, and perhaps physiologically, it is very similar to that of man (Hill, 1966). Only the cecum of the vervet monkey shows major anatomical differences. Although it is sacculated like man's cecum, it is proportionally larger and lacks the appendix found in this area of the human gut (Hill and Rowell, 1948).

When compared to the insectivore and the prosimian, the digestive tract of the vervet monkey differs mainly in the appearance of the large bowel, which exhibits a higher degree of complexity in both cecum and colon.

When considering the vervet monkey, again on the same feeding program and diet as the hedgehog and bushbaby, we see that the ingested material was retained within the digestive tract for an extended period of time (Figure 2). The time required for the fecal appearance of one-half of the marker $(T_{1/2})$ was approximately 30 h, about three times longer than in the bushbaby $(T_{1/2} \simeq 12$ h), and three to four times longer than in the hedgehog $(T_{1/2} \simeq 9$ to 10 h). It is interesting to note that the mean transit time of fluid and small particulate markers in man is similar, 37 ± 3 h (Hinton et al., 1969).

In the vervet monkey, gastric emptying of liquid and particles was slower than in the hedgehog or bushbaby, and there were no signs of retention in the distal small intestine as was evident in the bushbaby. The cecum of the vervet monkey allowed entrance of particulate material as well as liquid, with perhaps a slight preference for the former. The distribution and retention of both liquid and particulate matter in the colon of this animal were especially interesting. While it took only 4 h for the markers to reach the large bowel, most of the material remained in these gut segments for the next 20 h, with a partial retention up to at least 38 h after initial administration. Retention appeared to be uniform throughout the entire cecum and colon. The slow movement and the distribution of markers from cecum to rectum would allow for the mixing of many meals. While the limits of this experiment only allowed for the measurement of fluid marker distribution up to 38 h, a more complete study indicated that both fluid and particulate markers may be retained in the gastrointestinal tract of the vervet monkey for up to 5 days (Clemens, unpublished data).

In spite of the extended colonic retention in the vervet monkey, VFA levels were no greater than those observed in the bushbaby with its less complex colon and more rapid transit (Figure 3). This is not to say that the greater length of time available for absorption of these acids provided no advantage. The concentrations present may represent a tissue threshold, a level above which the excess is readily absorbed (Crane, 1968). In any event, volatile fatty acids were found primarily in the hindgut of all species (hedge-hog, bushbaby, and vervet monkey), and were not a major constituent in the foregut or midgut contents.

The levels of lactic acid were considerably lower throughout the greater portions of the gastrointestinal tract of the vervet monkey than in the hedge-hog and bushbaby, and particularly within the hindgut (Figure 3). Only within the foregut were lactic acid concentrations similar in the three species.

Conclusions

In the past we have assumed certain requirements for fermentative diges-tion to occur within the digestive tract of an animal, the conditions being fluid volume, digesta retention, appropriate pH, and absorption or removal of the fermentation end products (primarily organic acids) (Church, 1969). Such conditions exist within the foregut of ruminants, and it is here that the greatest investigative efforts have been directed. More recent investigations, however, demonstrate comparable activity within the digestive tract of many nonruminant herbivores (Alexander and Davies, 1963; Argenzio et al., 1974; Clemens, 1977), omnivores (Alexander and Davies, 1963; Argenzio and Southworth, 1975; Clemens et al., 1975a), and carnivores (Stevens, 1977), as well as in birds (Clemens et al., 1975b).

The data presented in this paper demonstrate fermentative digestion in primates and give evidence of this process in a primitive mammal, the hedge-hog. More important, the data cast doubts upon the relevance of the condi-tions assumed necessary for microbial activity to occur.

The requirement for retention of digesta appears limited. Microbial activ-ity, as is evident by the levels of organic acids produced, occurs within the digestive tract of the hedgehog and bushbaby where the retention is mini-mal. The life cycle of the intestinal microbes is sufficiently short to allow a 10-fold increase in population within a 24-h period (Gibbons and Kapsima-lis, 1967). In addition these microbes may exist at many sites within each segment of the tract, even in the absence of the food constituents (Luckey, 1974). It therefore seems plausible that the microbial activity can occur within a limited time as the digesta pass through the animal's tract. On the other hand, the retention of the ingesta at a particular site within the tract does not necessarily imply that fermentative digestion will occur, as is evi-dent in the distal small intestine of the bushbaby.

The comparisons suggest evolution of the digestive tract from a structur-ally simple tract as in the hedgehog to tracts of more complexity, with the development of the cecum in the bushbaby and the sacculation of the cecum and colon in the vervet monkey. Whereas there appears to be an association of primitive–simple and advanced–complex, primitive digestive processes do not necessarily mean simple, nor do advanced mean complex. This is most evident in the complexity of the hyrax, believed to be one of the oldest living ungulates (Clemens, 1977). Although in most cases gut complexity tends to slow the rate of digesta passage, gastrointestinal microbial activity and organic acid production are features in primitive as well as advanced species, regardless of the rate at which the digesta move through the tract.

The nutritional value that an animal may derive from the fermentative digestive process and organic acid production cannot readily be assessed. There is good evidence, however, that, as in the ruminant stomach, VFA's are readily absorbed from the colon of herbivores (Argenzio et al., 1974) and omnivores (Argenzio and Southworth, 1975). It is reasonable to assume that colonic absorption also occurs in the insectivore and primates discussed above.

These studies were supported by the University of Nairobi, Research and Publications Grants.

References

Alexander, F., and Davies, M. E. (1963). Production and fermentation of lactate by bacteria in the alimentary canal of the horse and pig. *J. Comp. Pathol. 73*:1–8.

Argenzio, R. A., and Southworth, M. (1975). Sites of organic acid production and absorption in gastrointestinal tract of the pig. *Am. J. Physiol. 228:454–60.*

Argenzio, R. A., Southworth, M., and Stevens, C. E. (1974). Sites of organic acid production and absorption in the equine gastrointestinal tract. *Am. J. Physiol. 226:*1043–50.

Bauchop, T. (1971). Stomach microbiology of primates. *Annu. Rev. Microbiol. 25:*429–36.

Campbell, B. G. (1971). *Human Evolution*, pp. 34–39. London: Heinemann Educational Books.

Charles-Dominique, P. (1977). *Ecology and Behavior of Nocturnal Primates*, p. 3. London: Gerald Duckworth.

Church, D. C. (1969). *Digestive Physiology and Nutrition of Ruminants*, pp. 151–85. Corvallis, Oregon: D. C. Church.

Clemens, E. T. (1977). Sites of organic acid production and patterns of digesta movement in the gastrointestinal tract of the rock hyrax. *J. Nutr. 107*:1954–61.

Clemens, E. T., Stevens, C. E., and Southworth, M. (1975a). Sites of organic acid production and pattern of digesta movement in the gastrointestinal tract of swine. *J. Nutr. 105*:759–68.

Clemens, E. T., Stevens, C. E., and Southworth, M. (1975b). Sites of organic acid production and pattern of digesta movement in the gastrointestinal tract of geese. *J. Nutr. 105*:1341–50.

Crane, R. K. (1968). A concept of the digestive-absorptive surface of the small intestine. In *Handbook of Physiology*, vol. 5, eds. C. F. Code and W. Heidel, Sec. 6, pp. 2535–42. Washington, D.C.: American Physiological Society.

Davenport, H. W. (1971). *Physiology of the Digestive Tract*, 3rd ed. Chicago: Year Book Medical Publishers.

Eyssen, H., Swaelen, E., K-Gindifer, Z., and Parmentier, G. (1965). Nucleotide requirements of *Lactobacillus acidophilus* variants isolated from the crops of chicks. *Antonie van Leeuwenhoek 31*:241–8.

Gibbons, R. J., and Kapsimalis, B. (1967). Estimates of the overall rate of growth of the intestinal microflora of hamsters, guinea pigs, and mice. *J. Bacteriol. 93*:510–12.

Hiley, P. G. (1976). The thermoregulatory responses of the galago (*Galago crassicaudatus*), the baboon (*Papio cynocephalus*) and the chimpanzee (*Pan satyrus*) to heat stress. *J. Physiol. 254*:657–71.

Hill, W. C. O. (1966). *Primates: Comparative Anatomy and Taxonomy*, vol. 6. Edinburgh: Edinburgh University Press.

Hill, W. C. O. (1972). *Evolutionary Biology of the Primates*. London and New York: Academic Press.

Hill, W. C. O., and Rewell, R. E. (1948). The caecum of Primates. *Trans. Zool. Soc. London 26*:199–256.

Hinton, J. M., Lennard-Jones, J. E., and Young, A. C. (1969). A new method for studying gut transit times using radioopaque markers. *Gut 10*:842–7.

Hungate, R. E. (1966). *The Rumen and Its Microbes*. New York: Academic Press.

Kay, R. N. B., Hoppe, P., and Maloiy, G. M. O. (1976). Fermentative digestion of food in the colobus monkey, *Colobus polykomos*. *Experientia 32*:485–7.

Kotb, A. R., and Luckey, T. D. (1972). Markers in nutrition. *Nutr. Abstr. Rev. 42*:813–45.

Luckey, T. D. (1974). Introduction: The villus in chemostat man. *Am. J. Clin. Nutr. 27*:1266–76.

McBee, R. H. (1970). Metabolic contributions of the cecal flora. *Am. J. Clin. Nutr. 23*:1514–18.

Müller, E. F. (1977). Energiestoffwechsel, Temperaturregulation und Wasserhaushalt beim Plumplori (*Nycticebus coucang*, Boddaert 1785). Ph.D. thesis, Eberhard-Karls Universität zu Tübingen, West Germany.

Rubinstein, R., Howard, A. V., and Wrong, O. M. (1969). In vivo dialysis of faeces as a method of stool analysis. *Clin. Sci. 37*:549–64.

Stevens, C. E. (1977). Comparative physiology of the digestive system. In *Dukes' Physiology of Domestic Animals*, 9th ed., ed. M. J. Swenson, pp. 216–32. Ithaca, New York: Comstock Publishing Associates.

Szalay, F. (1968). The beginnings of primates. *Evolution 22*:19–36.

Walker, D., and Murray, P. (1975). An assessment of masticatory efficiency in a series of anthropoid primates with special reference to the Colobinae and Cercopithecinae, pp. 135–50. In *Primate Functional Morphology and Evolution*, ed. R. H. Tuttle. Paris: Mouton Publishers.

Wolin, M. J. (1974). Metabolic interactions among intestinal microorganisms. *Am. J. Clin. Nutr. 27*:1320–8.

9

Evolution of mammalian homeothermy: a two-step process?

C. RICHARD TAYLOR

In this paper I address the simple question: How did mammalian homeothermy evolve? To answer this question, we need to know which mechanisms of the thermoregulatory system were possessed by the first mammals, and to trace the physiological modifications of this system as mammals evolved. To the extent that we are successful, we may be able to identify accurately "primitive" and "advanced" thermoregulatory mechanisms.

Unfortunately, we cannot make physiological measurements on fossils. To trace the evolutionary history of physiological systems, we are forced to rely on measurements from the living descendants of the first mammals and their reptilian ancestors.

Many measurements have been made on the temperature regulatory system of reptiles (Templeton, 1970), primitive mammals (Dawson, 1973), and advanced eutherian mammals (Bligh, 1973; Whittow, 1971; Robertshaw, 1977). We know the extent to which monotremes, marsupials, and insectivores (three groups of mammals considered on historical and anatomical grounds to retain many conservative mammalian characteristics) can regulate their body temperature in response to heat and cold; we know the relative roles of variable insulation and metabolic heat production in maintaining a constant body temperature in the cold; and we know the capabilities of these animals for increasing the rate of evaporative cooling in hot environments. We even know a little about temperature regulation of some of these animals in response to exercise heat loads (Baudinette, 1977; Baudinette et al., 1976; Dawson and Taylor, 1973; Taylor, 1977b). Yet, the crucial question remains: How useful is this information in reconstructing the evolution of mammalian homeothermy?

We must be careful not to jump to the conclusion that some aspect of the thermoregulatory system is primitive, simply because it is possessed by a primitive mammal. We must remember that those mammals we now call primitive have survived to the twentieth century. We have no guarantees that they haven't adapted to changing situations during the vast period of time over which mammalian evolution occurred. For example, because we observe that monotremes possess a low body temperature, it does not necessarily follow that the first mammals also had low body temperatures; or because primitive insectivores estivate or hibernate, we cannot conclude that the first mammals estivated or hibernated. It is tempting to view evolution of a physiological system as a continuous sequence of refinement, always leading to a better system. A consequence of this logic is that those "primitive mechanisms" retained by primitive mammals are inferior to "advanced mechanisms" possessed by advanced mammals. The fallacy of this logic is evident when we reflect on the fact that these "primitive mechanisms" have survived for tens to hundreds of million years in competition with "advanced systems". There must be something about these mechanisms that is rather good, for them to have persisted to the present.

There are some wonderful road signs in East Africa which sometimes appear 20 yards beyond a sign warning of an extreme danger. These signs read "You have been warned!" The department of public works is absolved of any further responsibility if you choose to ignore their warning. The analogy is obvious. Only insofar as we are cautious and careful in coordinating physiological data from living animals with information from the fossil record, do we have a chance of surviving with a reasonable hypothesis for the evolution of a physiological system. Although a particular hypothesis may be reasonable, and even may make some testable predictions, we must always remember that it still may not be "true." There may be other ways to interpret the same data that are equally good or better. Nevertheless, a working hypothesis for the evolution of physiological systems can be quite productive; it can pull together a great deal of information into a simple logical scheme; it can point out where information is missing; and, most important, it can be exciting, just as maneuvering a vehicle down an escarpment on slippery mud can be exciting.

Formulating a scheme for the evolution of mammalian homeothermy

Reptilian homeotherms

The advantages of a constant body temperature seem obvious. At any instant hundreds of complex chemical reactions are taking place within a vertebrate. All of these reactions are finely tuned to operate at an optimal rate at some particular temperature. Changes in temperature will alter the rates of the various reactions differently and disturb this fine tuning. Fish, amphibians, and reptiles all possess thermoregulatory mechanisms which enable them to maintain body temperature close to "optimum" during activity (Whittow, 1970). It seems likely, therefore, that the reptilian ancestors of the first mammals possessed some of these same mechanisms. Before we can begin to formulate a scheme for the evolution of mammalian homeothermy, we must ask: What are the physiological mechanisms that distinguish mammalian from reptilian homeotherms?

Many reptiles regulate their body temperature very precisely under changing environmental conditions (Templeton, 1970). They utilize many of the same mechanisms used by mammals to control the rate of heat gain from and heat loss to the environment: behavior, posture, and changes in blood flow. An area of their hypothalamus (as in mammals) integrates temperature information and regulates heat gain and heat loss mechanisms. Some reptiles even increase evaporative cooling to prevent overheating. The magnitude of this increase is not sufficient to allow them to maintain their body temperature more than a degree or two below ambient temperature, if at all (Schmidt-Nielsen, 1964; Templeton, 1970). However, it may give them more time to escape to a cooler environment. Indeed this seems to be the function of increased evaporative cooling in small mammals (Taylor, 1974). Reptiles appear to have gone small mammals "one better" in maximizing the effects of their limited evaporative cooling: they have developed a vascular arrangement that selectively cools the brain (Crawford, 1972). They also use this arrangement to selectively warm the brain in cool environments (Heath, 1964). It seems reasonable to postulate that the reptilian ancestors of mammals, the cynodonts, also possessed many of these same mechanisms for regulating heat gain and heat loss. Furthermore, many of the cynodonts were large, and, like the present-day Komodo dragons (McNab and Auffenberg, 1976), they probably had a nearly constant body temperature as a result of the thermal inertia of their large body mass. In an excellent paper, Spotila and his colleagues (Spotila et al., 1973) have developed a physical model for quantifying the thermal inertia in large reptiles. Their model predicts that these large reptiles would have had a nearly constant body temperature.

Mammalian innovations

If living reptiles possess such sophisticated thermoregulatory mechanisms, what "new" mechanisms do mammals possess? There appear to be three major "innovations" in mammalian homeothermy: (1) insulation via fur and subcutaneous fat; (2) about a fivefold jump in resting metabolism (when normalized for differences in body temperature and size), and (3) the ability to increase metabolic heat production to maintain a constant body temperature in cold environments.

These innovations are not unique to mammals. Birds have adopted essentially the same three mechanisms for regulating their body temperatures (Dawson and Hudson, 1970). It is tempting to conclude that these mechanisms evolved in some common reptilian ancestor. The appeal of this idea is that one need postulate that homeothermy evolved only once. A consequence is that mammal-like reptiles and dinosaurs would have been homeotherms. Bakker (1975) has presented many arguments to support this view.

The idea of "hot blooded dinosaurs" appeals to the imagination and has resulted in a popular book (Desmond, 1976) and a "Nova" TV program bearing this catchy title.

Some basic differences in the thermoregulatory systems of mammals and birds suggest that homeothermy evolved independently in these two vertebrate classes. The same parts of the brain are not used in the same way for regulating body temperature. Cooling the preoptic area of the hypothalamus increases heat production in mammals, but not in birds (Snapp et al., 1977). Fur and feathers certainly appear to have evolved independently as mechanisms for increasing insulation. Also, birds apparently do not possess subcutaneous fat (Templeton, 1970). On the basis of these differences, it seems reasonable to conclude that homeothermy in birds and mammals evolved independently.

A two-step hypothesis for evolution of mammalian homeothermy
The next obvious question is: Were these innovations acquired independently or simultaneously by mammals? Crompton et al. (1978) proposed that the evolution of mammalian homeothermy involved two steps. In the first step the acquisition of fur and the ability to regulate metabolic heat production would have enabled the first mammals to expand into a temporally wide nocturnal niche. As long as these first mammals regulated their body temperature at a low level of 25 to 30 °C, they could have accomplished this with a reptilian type of energetics. We propose that a second step occurred when these nocturnal mammals invaded a diurnal niche. This step necessitated a higher body temperature (about 35 to 40 °C) and a mammalian type of energetics.

What is the evidence to support this two-step hypothesis? The fossil record, as reviewed by Crompton (in this volume), indicates that the first mammals were small nocturnal insectivores. They were an order of magnitude smaller than their closest reptilian ancestors (Crompton and Jenkins, 1978), and possessed relatively larger brains (Jerison, 1973). The increase in the relative size of their brain appears to be attributable mainly to improved senses of smell and hearing. This led Jerison to the conclusion that they were nocturnal, the improvement in these senses having compensated for a loss of visual information. Their skeleton indicates that they operated in much the same habitat as many small living insectivores (Jenkins and Parrington, 1976).

What was an optimal body temperature for the first mammals?
Most biologists probably consider that a body temperature of 35 to 40 °C, common to most living birds and mammals, represents some sort of "optimal" temperature for homeotherms. It is often suggested that mammals evolved slowly toward this optimum. Low body temperatures then would be "primitive," reflecting an inferior thermoregulatory system. An example of

this argument is presented in C. J. Martin's much-quoted study of the development of homeothermy in monotremes and marsupials (Martin, 1902). He proposed a simple rank-order for the development of homeothermy. He states that the echidna is the "... lowest in the scale of warm blooded animals. Its attempts at homeothermism fail to the extent of 10 °C when the environment varies from 5–35 °C. During cold weather the echidna abandons all attempts at homeothermism and hibernates for four months." Martin's "ascending scale" of physiological superiority proceeds sequentially through the platypus and marsupials to eutherian mammals. Many physiologists hold to a similar conception of the evolution of homeothermy. Indeed the generalization that emerges most clearly from T. J. Dawson's comprehensive and thoughtful review of the temperature regulation of primitive mammals (Dawson, 1973) is that lower body temperatures appear to be a characteristic of living primitive mammals.

Another way to interpret the significance of a low body temperature is as an adaptation for conserving energy. If one accepts this view, an optimal body temperature is the lowest one that can be maintained by a mammalian thermoregulatory system. The thermoregulatory system of primitive mammals then becomes energetically "superior" to that of advanced mammals. Studies on the echidna (Schmidt-Nielson et al., 1966), platypus (Grant and

Figure 1. Effects of low body temperature and reptilian energetics on width of thermal neutral zone (3000-g mammal). Oxygen consumption is plotted as a function of ambient temperature for an echidna having a body temperature of 28 °C (after Schmidt-Nielsen et al., 1966); that predicted for a typical mammal of the same size having the same insulation and a body temperature of 39 °C; and that predicted for an early mammal of the same size having the same insulation and a body temperature of 28 °C. (See text for discussion.)

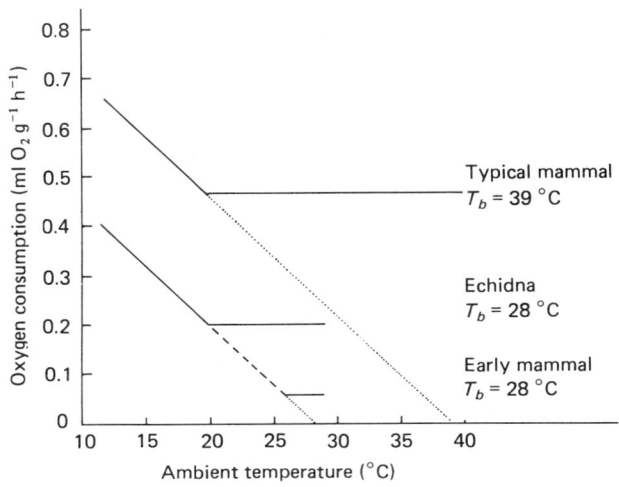

Dawson, 1978), and numerous marsupials (Dawson, 1973) show that these animals regulate their body temperatures in the cold about as precisely as similar-sized "advanced" mammals but at a lower energetic cost. This can be seen from the measurements of Schmidt-Nielsen et al. (1966) on the "primitive" echidna and comparing them to a typical "advanced" eutherian mammal (assuming that both possess the same insulation) (Figure 1). The echidna regulates its body temperature about 10 °C lower than the eutherian mammal (between 28 and 30 °C). Its metabolic rate shows a Q_{10} effect of about 2.1 and is only about 40% that of the eutherian mammal. Therefore the echidna maintains its body temperature within its thermoneutral zone at 40% the energetic cost.

An apparent disadvantage of the echidna's low body temperature and low metabolism is a narrower thermoneutral zone (Figure 1). Both the echidna and the eutherian mammal have the same lower critical temperature (20 °C). However, other factors being equal, the thermoneutral zone of the eutherian mammal will extend 10 °C further on the upper end of the zone because of this animal's 10 °C higher body temperature. It should be noted that mammals normally do not choose to rest in ambient temperatures above their thermoneutral zone and probably encounter these conditions only for brief periods until they can find a cooler microclimate. One interpretation of the increased metabolism in response to cold could be an "emergency mechanism" enabling animals to maintain their body temperature while they escape the cold (in much the way we have considered increased evaporative cooling an emergency mechanism for escaping heat by small animals).

To a certain extent the concept of a thermoneutral zone is a physiologist's artifact. Being enamored with this concept (probably because it is so measurable), physiologists tend to forget that it really has very little to do with animals in nature. Lower critical temperatures apply only to resting animals with maximal insulation. Once an animal begins to exercise, its effective lower critical temperature may fall. The heat produced by the exercise could help keep the animal warm. For example, if an echidna could maintain the same maximum insulation while walking at 2 km h^{-1} as at rest, it would not require any additional heat to keep warm until air temperatures below -11 °C (Table 1).

Crompton et al. (1978) have argued that an optimal body temperature for the first mammals was the lowest one that they could maintain without prolonged evaporative cooling, and that because of their nocturnal niche this would have been about 10 °C lower than the 40 °C common to most living birds and mammals. This lower body temperature still would have been 2 to 3 °C above the highest ambient temperatures they were likely to encounter. Therefore, the animals would not have had to resort to prolonged evaporative cooling. They could have maintained this lower body temperature with a reptilian type of energetics (one-third to one-fifth that predicted for a mammal), although this would have raised their lower critical temperature from 20 to 26 °C (Figure 1). Walking at a speed of 2 km h^{-1}, they could still have encountered ambient temperatures of −4 °C without having to increase heat production to keep warm (provided their insulation remained the same).

For small mammals to invade the day, they would have needed a higher body temperature and a mammalian type of energetics. It is impossible for small vertebrates (less than 1 kg body mass) to sustain the high rates of evaporative cooling necessary to maintain their body temperature even a few degrees below ambient temperature for more than a few hours. The problem becomes more acute the smaller the animal. The rate at which heat flows into an animal from a hot environment will be approximately proportional to its surface area; thus the smaller the animal is, the larger the relative amount of water that will have to be evaporated to keep cool (Schmidt-Nielsen, 1964). Small 30-g animals would have to drink almost continuously to avoid lethal levels of dehydration while maintaining their body temperature only a few degrees below ambient temperature. A nocturnal fossorial homeotherm could avoid this problem with a body temperature of 25 to 30 °C, whereas a diurnal animal that encountered solar radiation would need its temperature closer to 40 °C. A higher metabolic rate would have been necessary to sustain the additional 10 °C gradient during the night, insulation and other factors being equal. Thus the fivefold jump in metabolic rate would be needed in order for mammals to invade a diurnal niche.

Our hypothesis that a low body temperature was optimal for the first mammals runs counter to the conventional wisdom that an optimal body temperature should be as high as possible. One assumed advantage of a higher temperature is that it enables animals to do things faster. A little reflection reveals some problems with this idea. On an evolutionary scale, contraction velocity of muscles appears to have been limited by structural constraints (McMahon, 1975) rather than temperature (Crompton et al.,

Table 1. *Effect of increased metabolism during exercise on lower critical temperature of typical eutherian mammal, echidna, and early mammal*

	Lower critical temperature [d]	
	At rest	Walking (2 km h^{-1})
Typical eutherian mammal[a]	20	−22
Echidna[b]	20	−11
Early mammal[c]	26	− 4

[a]$T_{body,}$ 39 °C; mammalian energetics.
[b]$T_{body,}$ 28 °C; mammalian energetics.
[c]$T_{body,}$ 28 °C; reptilian energetics.
[d]Calculated from the equations given by Taylor (1977a) for 3-kg body mass and 2 km h^{-1} walking speed, and assuming that insulation was the same at rest and while walking.

1978). For example, the intrinsic velocity of a muscle of a 100-kg antelope or ostrich is only one-third as fast as that of an equivalent muscle in a 30-g mouse or quail, despite the fact that both animals have the same body temperature. If temperature were limiting an animal's speed, the antelope could have its body temperature 10 °C lower than the mouse and still run at the same speed.

Testing an energetic prediction of the two-step hypothesis

Have any insectivores remained in the nocturnal-fossorial niche during the evolution of mammals? If so, our hypothesis would predict that these animals would have retained the low body temperature and reptilian type of energetics. The Tenrecidae and Erinaceidae are two families of insectivores descended from forms that appear to have remained in the nocturnal insectivorous niche. The Tenrecidae, which have been isolated from the mainstream of mammalian evolution for about 75 million years, retain many primitive morphological characteristics, and their ecological niche and body temperature on Madagascar are similar to those we have proposed for the first mammals (Eisenberg and Gould, 1970). Body temperature of *Setifer setosus* averaged about 29.5 °C while they foraged, and ambient temperature ranged between 21 and 24 °C. Other species of Tenrecidae also have low body temperatures, a few degrees above the highest temperatures they encounter in nature. Our hypothesis would predict that these animals would have completed only the first step in the evolution of mammalian homeothermy and would have retained the reptilian type of energetics.

Do the energetic measurements on these insectivores confirm our prediction? Dawson and Hulbert (1970) have shown that when the standard metabolic rates of marsupials, monotremes, and other mammals with a low body temperature are normalized to a body temperature of 38 °C using a Q_{10} of 2.5, their metabolic rates are close to those predicted for eutherian mammals of the same size by Kleiber's equation (Kleiber, 1961). However, the resting metabolic rate of reptiles treated in the same manner is one-third to one-fifth that of mammals. We should, therefore, by using this normalization, be able to distinguish a reptilian from a mammalian type of energetics. If the lowest values for resting metabolic rate reported by Hildwein (1970) for *Tenrec ecaudatus* are handled in this way, we obtain reptilelike resting metabolic rates less than one-third those predicted for a mammal. However, if the highest reported values are used, we obtain nearly three-fourths the predicted value. A similar situation exists for hedgehogs (Shkolnik and Schmidt-Nielsen, 1976). Although resting metabolic measurements have proved extremely valuable in broad comparative studies, it is more difficult to use them to decide whether a particular species possesses a reptilian or a mammalian type of energetics.

Crompton et al. (1978) were able to avoid much of the variability inherent in resting metabolic measurements by comparing the energetics of animals

over a range of exercise intensities. They found that the metabolic rate of
either a reptile or a mammal exercising at a given intensity was much more
reproducible than its resting metabolic rate. Bakker (1972) has compared
the energetics of lizards and mammals over a range of running speeds. He
found the rates of oxygen consumption of a lizard and a mammal of the
same size increased at about the same rate with increasing speed, but the
metabolic rate of the lizard was lower by a nearly constant amount at any
speed (Figure 2). This difference was approximately the same as the difference that Dawson and Hulbert (1970) would have predicted between a resting lizard and resting mammals after normalizing the rates to the same body
temperature. Therefore, these authors decided to use measurements of the
relationships between rates of oxygen consumption and running speed,
rather than resting rates, to distinguish between reptilian and mammalian
energetics. Data from three species of insectivores treated in this way show
that they all possess more nearly a reptilian than a mammalian type of

Figure 2. (A) Relationship between speed and predicted rate of oxygen
consumption for a mammal (dashed line) and lizards (dotted line) for a
hypothetical mammal and lizard each weighing 675 g and with a body
temperature of 38 °C (Taylor, 1977a). Similar predicted relationships are
also plotted in the other graphs for a mammal and a lizard of the same
weight and body temperature as our experimental animals. (B) Tenrec, 675
g, 35.7 °C; (C) setifer, 120 g, 28 °C; (D) hedgehog, 1.05 kg, 37.5 °C; (E)
opossum, 2.7 kg, 36.2 °C; (F) echidna, 5.04 kg, 31 °C. A comparison of our
experimental data (solid line in each graph) with predicted relationships
shows that the tenrec, setifer and hedgehog possess lizard-like energetics,
whereas the opossum and echidna possess mammal-like energetics. (Based
on data in Crompton et al., 1978.)

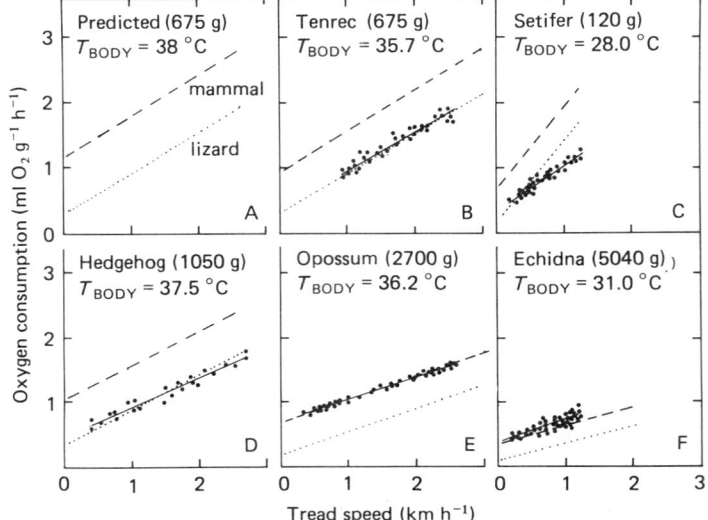

energetics (Figure 2). In contrast, a marsupial (opossum) and a monotreme (echidna) seem to have acquired a mammalian type of energetics.

Conclusions

Our hypothesis was that mammalian homeothermy was acquired in two steps: (1) the acquisition of fur together with the ability to increase heat production in the cold; and (2) a high body temperature together with a three- to fivefold increase in resting metabolism above that of their reptilian ancestors. We proposed that the first step enabled mammals to invade a temporally wide nocturnal niche previously unexploited by reptiles, and the second step enabled them to invade a diurnal niche. This hypothesis predicts that groups of mammals that have remained in the nocturnal niche during the evolution of mammals would have taken only the first step, and would retain a low body temperature and a reptilian type of metabolism. The insectivores appear to have remained in this nocturnal niche and conform to this prediction of our hypothesis. According to our hypothesis the fact that most living nocturnal mammals (including monotremes and marsupials) do not possess a reptilian type of energetics suggests that some ancestor was diurnal, and that this trait was difficult to reacquire, or that it imposed limits on ambient temperature that were too restrictive to compensate for the energetic savings.

References

Bakker, R. T. (1975). Experimental and fossil evidence of the evolution of tetrapod bioenergetics. In *Perspectives in Biophysical Ecology*, eds. D. Gates and R. Schmerl, pp. 365–97. New York: Springer Verlag.

Bakker, R. T. (1972). Locomotory energetics of lizards and mammals compared. *Physiologist 15*:278.

Baudinette, R. V. (1977). Locomotory energetics in a marsupial, *Setonix brachyurus*. *Aust. J. Zool. 25*:423–28.

Baudinette, R. V., Nagle, K. A., and Scott, R. A. D. (1976). Locomotory energetics in dasyurid marsupials. *J. Comp. Physiol. 109*:159–68.

Bligh, John. (1973). *Temperature Regulation in Mammals and Other Vertebrates*, eds. A. Neuberger and E. L. Tatum. New York: North-Holland.

Crawford, E. C., Jr. (1972). Brain and body temperatures in a panting lizard. *Science 177*:431–33.

Crompton, A. W., and Jenkins, F. A., Jr. (1978). African mesozoic mammals. In *The Evolution of African Mammals*, eds. V. J. Maglio and H. B. S. Cooke. Cambridge, Mass.: Harvard University Press.

Crompton, A. W., Taylor, C. R., and Jagger, J. A. (1978). Evolution of homeothermy in mammals. *Nature 272*:333–6.

Dawson, T. J. (1973). "Primitive" mammals. In *Comparative Physiology of Thermoregulation*, vol. 3, *Special Aspects of Thermoregulation*, ed. G. C. Whittow, pp. 1–46. New York: Academic Press.

Dawson, T. J., and Hulbert, A. J. (1970). Standard metabolism, body temperature, and surface areas of Australian marsupials. *Am. J. Physiol. 218*: 1233–8.

Dawson, T. J., and Taylor, C. R. (1973). Energetic cost of locomotion in kangaroos. *Nature* 246:313–4.

Dawson, W. R., and Hudson, J. W. (1970). Birds. In *Comparative Physiology of Thermoregulation;* vol. I, *Invertebrates and Nonmammalian Vertebrates*, ed. G. C. Whittow, pp. 224–302. New York: Academic Press.

Desmond, A. J. (1976). *The Hot Blooded Dinosaurs.* New York: Dial Press/James Wade.

Eisenberg, J. F., and Gould, E. (1970). The Tenrecs: A study in mammalian behavior and evolution. *Smithson. Contrib. Zool. 27.* Washington, D.C.: Smithsonian Institution Press.

Grant, T. R., and Dawson, T. J. (1978). Temperature regulation in the platypus, *Ornithorhynchus anatinus*: Maintenance of body temperature in air and water. *Physiol. Zool. 51*:1–6.

Heath, J. E. (1964). Head–body temperature differences in horned lizards. *Physiol. Zool. 37*:273–9.

Hildwein, G. (1970). Capacités thermorégulatrices d'un mammifère insectivore primitif, le tenrec; leurs variations saisonnières. *Arch. Sci. Physiol. 25*:55–71.

Jenkins, F. A., Jr. and Parrington, F. R. (1976). The post-cranial skeletons of the Triassic mammals *Eozostrodon, Megazostrodon* and *Erythrotherium. Phil. Trans. B(13) 273*:387–431.

Jerison, H. J. (1973). *Evolution of the Brain and Intelligence.* New York: Academic Press.

Kleiber, M. (1961). *The Fire of Life: An introduction to Animal Energetics.* New York: Wiley.

Martin, C. J. (1902). Thermal adjustment and respiratory exchange in monotremes and marsupials. A study in the development of homeothermism. *Phil. Trans. R. Soc. London Ser. B 195*:1–37.

McMahon, T. A. (1975). Using body size to understand the structural design of animals: Quadrupedal locomotion. *J. Appl. Physiol. 39*:619–27.

McNab, B. K., and Auffenberg, W. (1976). The effect of large body size on the temperature regulation of the Komodo dragon, *Varanus komodoensis. Comp. Biochem. Physiol. 55A*:345–50.

Robertshaw D., ed. (1977). *Environmental Physiology II. International Review of Physiology*, vol. 15. Baltimore: University Park Press.

Schmidt-Nielsen, K. (1964). *Desert Animals: Physiological Problems of Heat and Water.* New York: Oxford University Press.

Schmidt-Nielsen, K., Dawson, T. J., and Crawford, E. C., Jr. (1966). Temperature regulation in the echidna (*Tachyglossus aculeatus*). *J. Cell. Physiol. 67*:63–71.

Shkolnik, A., and Schmidt-Nielsen, K. (1976). Temperature regulation in hedgehogs from temperate and desert environments. *Physiol. Zool. 49*:56–64.

Snapp, B. D., Heller, H. C., and Gospe, S. M., Jr. (1977). Hypothalamic thermosensitivity in California quail. *J. Comp. Physiol. 117*:345–57.

Spotila, J. R., Lommen, P. W., Bakken, G. S., and Gates, D. M. (1973). A mathematical model for body temperatures of large reptiles: Implications for dinosaur ecology. *Am. Nat. 107(955)*:391–404.

Taylor, C. R. (1974). Exercise and thermoregulation. In *MTP International Review of Science*, Series One, *Environmental Physiology*, ed. D. Robertshaw, pp. 163–84. London: Butterworth.

Taylor, C. R. (1977a). The energetics of terrestrial locomotion and body size in vertebrates. In *Scale Effects in Animal Locomotion*, ed. T. J. Pedley, pp. 127–141. London: Academic Press.

Taylor, C. R. (1977b). Exercise and environmental heat loads: different mechanisms for solving different problems? In *MTP International Review of Science, Physiology*,

Series Two, *Environmental Physiology*, ed. D. Robertshaw, pp. 119–46. Baltimore: University Park Press.

Templeton, J. R. (1970). Reptiles. In *Comparative Physiology of Thermoregulation*. vol. 1, *Invertebrates and Nonmammalian Vertebrates*, ed. G. C. Whittow, pp. 167–221. New York: Academic Press.

Whittow, G. C., ed. (1970). *Comparative Physiology of Thermoregulation*, vol. 1, *Invertebrates and Nonmammalian Vertebrates*. New York: Academic Press.

Whittow, G. C., ed. (1971). *Comparative Physiology of Thermoregulation*, vol. 2, *Mammals*. New York: Academic Press.

10

Have some mammals remained primitive thermoregulators?
Or is all thermoregulation based on equally primitive brain functions?

JOHN BLIGH

"Primitive," when used to describe a contemporary species, means that it possesses conservative anatomical or physiological features characteristic of an early stage in the evolution of other species of the same common stock. The reason for such unchanged form and function could lie in the stability of the environment or in the adequacy of the conservative traits; for had the organism been unable to adapt to environmental changes, it would have failed, and if the organism had been able to adapt and had needed to, it would have done so. Thus "primitive," either generally or in some functional particularity, does not imply any failure to come to terms with the environment, but rather the absence of the need to do so. Every species here now is evidently sufficiently well adapted to its natural environment to continue to exist, whether or not we regard it as primitive.

The term "primitive homeotherm" has been used to describe mammals that are not permanently homeothermic, or which are thermally relatively labile. These terms seem to imply an assumption that mammals were still progressing toward homeothermy at the time of their radiation, which occurred about 70 million years ago, and that some lines have since moved further in this direction than some others have.

However, it may not be correct to assume that evolving mammals, since their divergence from a reptilian line, have been moving progressively from poikilothermy through heterothermy or homeothermy, or to consider those contemporary mammals that are not truly homeothermic as inferior in the thermoregulatory sense.

Evidence of premammalian evolution of homeothermy
There is plenty of evidence that the various forms of mammalian heterothermy do not indicate only partial progression toward homeothermy, but, rather, that such species have changed from homeothermy to heterothermy in the course of further evolution, or can do so intermittently in response to

special environmental challenges such as severe seasonal or nychthemeral temperature changes.

I can mention a few important pieces of this evidence:

1. Many reptiles are quite good regulators of their body temperatures (see Bligh, 1973): by behavioral means they control the uptake of heat from or loss to the environment. This implies that a capacity to thermoregulate already existed in the reptilian ancestors of both birds and mammals, and may explain the similarities between avian and mammalian thermoregulation (Bligh, 1973). The distinction between mammals and their reptilian ancestors may thus relate more to the change from ectothermy to endothermy than to a change from poikilothermy to homeothermy. The well-developed thermoregulatory capacity of the monotremata—as revealed by the studies of Schmidt-Nielsen et al. (1966) and Grant and Dawson (1978)—supports the view that heterothermy in the mammals may have nothing to do with primitiveness versus specialization.

2. In every mammalian line in which seasonal hetero- or poikilothermic hibernators occur there are other species that qualify as permanent homeotherms. Indeed the hibernators themselves, when not hibernating, may be finely regulating homeotherms. Thus the heterothermy of the hibernant seems to be an additional acquisition by an accomplished homeotherm.

3. The nychthemeral heterothermy of the mammals in arid and semiarid environments – the camel in particular, but also the donkey, giraffe, and other species – is another aspect of mammalian thermoregulation that appears to be more an adaptation to the environment than evidence of a primitive function. The South American camelidae do not share the heterothermy of the camel, and Bligh et al. (1975) have considered it more likely that the relatively heterothermic dromedary camel was derived from homeothermic stock than vice versa. The further increase in thermolability in the camel when dehydrated (Schmidt-Nielsen et al., 1957) is presumably effected through the central nervous influence on thermoregulation of other central nervous functions, and comparable with those temporary changes in thermoregulation induced in a man by fever and exercise.

4. During fever, the thermoregulatory set-point is said to be reset at a higher level. However, Kluger and others (see Kluger, 1978) have shown that pyrogens also affect the behavioral thermoregulation of fishes, amphibia, and reptiles. Thus, a thermoregulatory set-point, if such a thing exists, is obviously not limited to endothermic and homeothermic animals, or even just to mammals irrespective of their thermostatic prowess. On whatever central nervous structure or function a pyrogen acts as if to "elevate the set-point," it must be something much older than the mammals.

The thesis that the neurology of homeothermy is primitive

These pieces of information suggest that mammalian thermoregulation does not depend on a postreptilian fundamental change in the nature of the central nervous machinery concerned with thermoregulation, but rather on a conservative neuronal arrangement that has remained capable of considerable plasticity in its performance. This would allow the acquisition of many different patterns of thermoregulation without a need for fundamental changes in the neuronal processes.

Is it possible that the basic neurology of mammalian thermoregulation is really quite unchanged and thus still capable of diversification, not only between species, but also within a species in different physiological and environmental situations? There is at least an arguable case in support of this proposition.

Homeothermy as a component of homeostasis
Homeostasis must be nearly as ancient as life itself; and further control of the internal environment became necessary once the layers of cells in a clone were too thick for adequate diffusion between the external environment and the innermost cells of the clone. Thus it is likely that even the simplest central nervous system contains structures concerned with the control of the internal environment. In the vertebrates the hypothalamus, which is much

Figure 1. Diagrammatic representation of fixed (reflex) relations between sensors (S) and response effectors (E). (A) A series of direct connections between sensors and effectors with central nervous interactions between these pathways. (B) A single sensor-to-effector pathway in which the influences of other pathways are represented as summed excitatory and inhibitory actions at a single junctional synapse between the afferent and efferent fibers. (C) Two pathways as in B linked by reciprocal inhibitory connections.

A

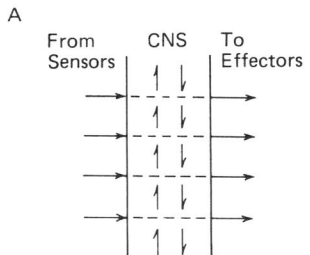

B

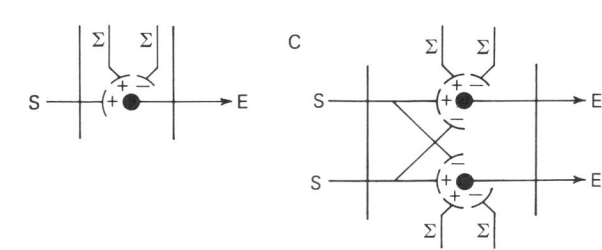

concerned with the autonomic processes by which homeostasis is effected, is indeed an ancient structure of the brain.

For each homeostatic function there must be some kind of set-point process; there is nothing particular about homeothermy in this respect. Thus it is noteworthy that the brain structure particularly necessary for homeothermy, the hypothalamus, is also essential for other more ancient homeostatic functions. There is, therefore, the distinct possibility that the neurology of homeothermy is based on that of homeostasis generally.

The central neurology of homeostasis: a possibility

Considered in the simplest possible terms, a central nervous system is concerned with two primary functions: (1) the effecting of appropriate responses to the external environment, and (2) the effecting of the stabilization of the internal environment.

At the autonomic (or automatic) level these functions can be considered as dependent on fixed neuronal connections between sensors and effectors, with interconnections between these pathways by which signals from the various sensors can give rise to coordinated patterns of responses.

This general principle of fixed sensor–effector pathways with interconnections between them is illustrated by Figure 1A. Any single function can be represented as in Figure 1B, in which excitatory and inhibitory influences from other pathways impinge on the particular pathway, and modify the relations between sensor and effector.

Mutually inhibitory relations between two reciprocal functions (e.g., extension and flexion of a locomotory limb) are based on cross inhibition between two sensor–effector pathways as represented in Figure 1C. Bligh (in press as of July 1979) showed that a homeostatic function could be achieved by a no more complex neuronal arrangement than this provided there were two populations of sensors of the controlled variable with reciprocal disturbance/activity characteristics. Such sensors, linked to reciprocally acting effectors, could alone achieve the stability of the sensed modality, as was suggested for thermoregulation by Vendrik (1959).

The central neurology of thermostasis: a theoretical consideration

Two populations of thermosensors with activity/temperature profiles that are reciprocal over an "operational" range of temperature are known to exist, and are known to be linked, somehow, to two categories of thermoregulatory effector functions: controlled heat production and controlled heat loss. Figure 2A shows the theoretical relations between sensor activities and effector activities under steady state conditions if one assumes direct sensor-to-effector pathways with crossing inhibitory links between them. These sensor and effector patterns are essentially those that are now known to exist. Figure 2B shows that excitatory and inhibitory influences acting on both thermosensor-to-thermoregulatory-effector pathways from elsewhere in the CNS would create shifts, in one direction or the other, of the "null-point," "threshold point," or "set-point" body temperature at which thermo-regulatory heat production and heat loss are minimal.

Variations in these four categories of influences from elsewhere in the CNS (excitation and inhibition of heat production and excitation and inhibition of heat loss) could simulate, and may actually explain, virtually every variation in thermoregulatory patterns known to occur between species, and within a species seasonally and in other circumstances.

The central neurology of thermostasis: experimental evidence

That central nervous functions are immensely complex and that homeo-thermy depends on an endogenous set-point reference signal generator within the CNS are well entrenched concepts. This probably means that this much simpler concept of thermostasis (and of homeostasis in general) will

Figure 2. (A) Cold (C) and warm (W) sensors with reciprocal activity/tem-perature relations, linked to heat production (HP) and heat loss (HL) effectors with crossing inhibition between the sensor-to-effector pathways. This model will effect a null-point or threshold temperature below which thermoregulatory heat production will occur, and above which thermo-regulatory heat loss will occur.(B) Excitatory and inhibitory influences from elsewhere in the CNS shown here as summed effects, modify the relations between thermosensors and thermoregulatory effectors, such that a com-mon threshold temperature for HP and HL may be shifted upward (as in fever) or downward (as in exercise), or the threshold for HP and HL may separate, as may occur in hibernation and other kinds of temporary heterothermy.

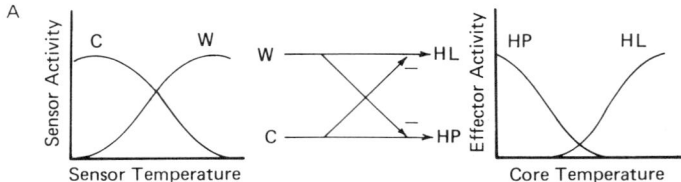

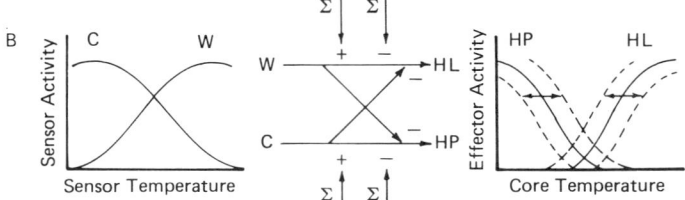

lack the respectability that comes from popular appeal. There is, however, a body of evidence obtained over a number of years at the Institute of Animal Physiology at Babraham, Cambridge, England that appears to be wholly consistent with this concept.

When synaptically active substances are injected through a permanently implanted cannula into a lateral cerebral ventricle of an unanesthetized sheep, the thermoregulatory effects indicate that these substances act at populations of synapses concerned with specific thermoregulatory functions. The original study by Bligh et al. (1971) was with the putative transmitter substances 5-hydroxytryptamine (5-HT) and noradrenaline (NA) and the cholinomimetic substance carbamylcholine (CCh). The results were expressed in terms of the simple neuronal format indicated in Figure 3. Many studies have since been done, either to test the effects of other putative transmitter substances or to test the validity of the model by seeing whether specific synaptic interferences by drugs had the effects on thermoregulation predicted from the model. The predictions have been extraordinarily accurate, and although these further studies have clarified some issues and required some changes in the model, the principal features have been strengthened. The model that is consistent with all our results is given in Figure 4. I have previously pointed out how very similar this neuronal concept based entirely on synaptic interference studies in the sheep, is to the simplest neuronal connections between thermosensors and thermoregulatory effectors that are needed to account for the relations between thermal disturbances and thermoregulatory responses in mammals generally, and in man in particular (Bligh, in press).

Conclusions

Here I argue, in support of Vendrik's (1959) hypothesis, (1) that the very existence of "cold" and "warm" sensors with opposing activity/temperature profiles renders a separate endogenous set-point mechanism for thermoregulation unnecessary; (2) that reciprocal inhibition, which is a characteristic of many reflex functions at the level of the spinal cord, would have the effect of creating a null-point temperature at which both heat production and heat loss would be minimal; and (3) that the infinity of possible variations in the excitatory and inhibitory influences on the sensor-to-effector pathways from elsewhere in the CNS, or by neural or hormonal feedback, could be the basis of the many different body temperature patterns, both between species and within species at different seasons or in different circumstances.

It is implied that other homeostatic functions might likewise depend on the reciprocal activities of two populations of the sensors of one modality, and that this basic means of achieving homeostasis is essentially a primitive one in principle, being based on little more than the simplest sensor–effector relations of spinal reflexes.

In some circumstances there is a distinct virtue in remaining physiologically conservative – as opposed to becoming highly specialized – because the former often serves to preserve the organism's capacity to effect variations in a function without irreversibly changing the basic process. One of the reasons why mammals have been able to establish themselves in so wide a range of environments may be that the evolution of homeothermy has not involved irreversible specialization, so that various forms of heterothermy are still selectable options, at least in some mammalian lines.

Figure 3. The neuronal model used by Bligh et al. (1971) to express the thermoregulatory effects in sheep, goat and rabbit of intracerebroventricular injections of 5-hydroxytryptamine (5-HT), noradrenaline (NA), and carbamylcholine (CCh). W = warm sensors; C = cold sensors; EHL = evaporative heat loss; PVMT = peripheral vasomotor tone; HP = heat production.

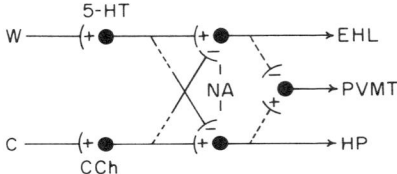

Figure 4. A simplified version of the neuronal model used by Bligh (in press) to summarize the thermoregulatory effects in the sheep of intracerebroventricular injections of putative transmitter substances where: H = histamine; 5-HT = 5-hydroxytryptamine; ACh = acetylcholine; CCh = carbamylcholine; NA = noradrenaline; DA = dopamine; EHL = evaporative heat loss; PVMT = peripheral vasomotor tone; HP = heat production.

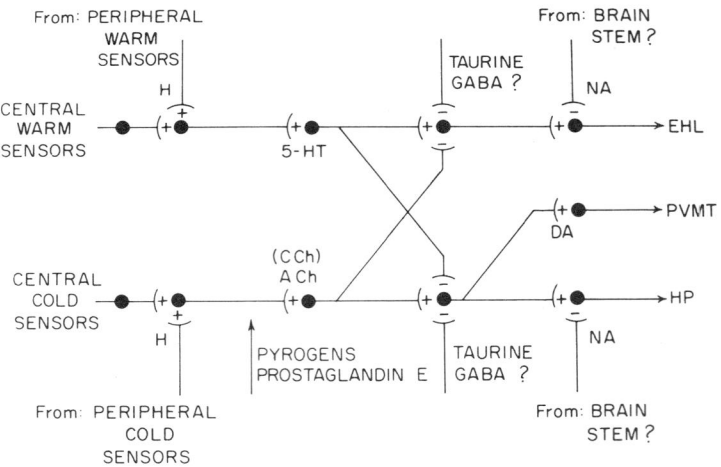

References

Bligh, J. (1973). *Temperature Regulation in Mammals and other Vertebrates.* Amsterdam: North Holland.

Bligh, J. (In press.) The central neurology of mammalian thermoregulation. *Neuroscience.*

Bligh, J., Baumann, I., Sumar, J., and Pocco, F. (1975). Studies of body temperature patterns in South American Camelidae. *Comp. Biochem. Physiol. 50A:*701–8.

Bligh, J., Cottle, W. H., and Maskrey, M. (1971). Influence of ambient temperature on the thermoregulatory responses to 5-hydroxytryptamine, noradrenaline and acetylcholine injected into the lateral cerebral ventricles of sheep, goats and rabbits. *J. Physiol. London. 212:*377–92.

Grant, T. R. and Dawson, T. J. (1978). Temperature regulation in the platypus, *Ornithorhynchus anatinus:* Maintenance of body temperature in air and water. *Physiol. Zool. 51:*1–6.

Kluger, J. J. (1978). The evolution and adaptive value of fever. *Am. Sci. 66:*38–43.

Schmidt-Nielsen, K., Schmidt-Nielsen, B., Jarnum, S. A. and Houpt, T. R. (1957). Body temperature of the camel and its relation to water economy. *Am. J. Physiol. 188:*103–12.

Schmidt-Nielsen, K., Dawson, T. J., and Crawford, E. C. (1966). Temperature regulation in the echidna (*Tachyglossus aculeatus*). *J. Cell. Comp. Physiol. 67:*63–71.

Vendrik, A. J. H. (1959). The regulation of body temperature in man. *Ned. Tijdschr. Geneesk. 103:* 240–4.

11

What is a primitive thermoregulatory system?

H. CRAIG HELLER

A variable body temperature (T_b) has frequently been cited as a primitive character for mammals. Variability, however, is not necessarily an indication of a simple regulatory system. A sophisticated regulator can operate with a broad or a narrow band width, and its settings may be altered as occasion demands. I shall argue in this paper that there are situations in which it is adaptive for the mammalian thermoregulatory system to operate with a broad dead band between thresholds for active heat loss and heat production responses, and there are also situations in which it is adaptive to regulate T_b at levels other than the usual 37 °C.

Variability of T_b may be a manifestation of specialized, advanced adaptations. Investigations of the central nervous thermoregulatory system in question are necessary before conclusions can be reached as to whether observed variations in T_b are regulated changes, are due to broad band regulation, or are truly due to limited regulatory abilities. Little information is available on the central nervous thermoregulatory systems of monotremes, marsupials, and insectivores, groups generally considered to be phylogenetically primitive, and it is therefore not possible for me to discuss their thermoregulatory systems. However, by examining eutherian species that show adaptive variability in T_b regulation, I can discuss what is not necessarily primitive, and perhaps indicate comparative studies that should be made on monotremes, marsupials, and insectivores to reveal overall evolutionary trends in the mammalian thermoregulatory system.

Background

Information coming exclusively from studies of eutherians tells us that integrative functions in the mammalian thermoregulatory system are largely localized in the hypothalamus. This part of the brain is itself temperature-sensitive, providing a major feedback loop in the system. Therein lies the rationale of the studies I shall describe.

The characteristics of the thermoregulatory system can be measured under given conditions by manipulating hypothalamic temperature (T_{hy}) and measuring one or more thermoregulatory responses. For example, an awake, quiet mammal at a thermoneutral ambient temperature (T_a) has a basal rate of metabolic heat production (MHP) that is not influenced by variations of T_{hy} over a certain range. If T_{hy} is lowered below this range, the rate of MHP increases proportionately to the level of cooling. This thermoregulatory response can therefore be described by a threshold T_{hy} (or T_{set}) and a proportionality constant (or α) that relates the response rate to T_{hy} values below T_{set}. Another thermoregulatory response such as evaporative water loss can similarly described, but of course it will have a different T_{set} and α.

The influences of other variables on the thermoregulatory system can be measured in terms of the changes they induce in the values for T_{set} and α for the various thermoregulatory responses. For example, a decrease in T_a, and therefore skin temperature, generally induces an increase in the T_{set} and/or α values for thermoregulatory responses, and an increase in T_a induces the opposite shifts. Because of these shifts in characteristics of hypothalamic thermosensitivity, the activation of heat conservation/production responses is not dependent upon a fall in core temperature in a cold environment, nor is the activation of heat loss responses dependent upon a rise of core temperature in a warm environment. Variations in core T_b are thereby minimized in a changing thermal environment. Conversely, if natural variations in T_b are regulated and are due to specialized adaptations of the thermoregulatory system, they should be associated with appropriate shifts in characteristics of hypothalamic thermosensitivity in a stable thermal environment.

Body temperature during sleep

Daily fluctuations of T_b with a peak during the active/waking phase and a low during the inactive/sleeping phase have long been recognized in mammals. These daily fluctuations have at least two components, a circadian one (for a review see Aschoff et al., 1974) and a sleep-related one (for a review see Heller and Glotzbach, 1977). The sleep-related influences on T_b are for two reasons of special relevance to this paper. First, even the earlier literature made it fairly clear that the fall in T_b at the onset of sleep was regulated in that it was accompanied by coordinated changes in thermoregulatory effector mechanisms. Second, sleep is a conservative mammalian character in that it and the distinct electrophysiologically defined stages of rapid eye movement (REM) sleep and non-rapid eye movement or slow wave sleep (nREM or SWS) are present in all species investigated.

Recent investigations of hypothalamic thermosensitivity in several rodent species during electrophysiologically defined sleep have shown that there is a decline in the regulated T_b during nREM sleep (which normally constitutes about 80% of total sleep time in mammals) (Glotzbach and Heller, 1976;

Florant et al., 1978). Typical results from two species are shown in Figure 1. There is clearly a decline in the T_{set} as well as the α for the MHP response during nREM sleep as compared to wakefulness. Such results could reflect either a resetting of the regulator to a lower level (a regulated decline in T_b) or a general relaxation of the regulatory mechanism (a greater band width for regulated T_b). This latter possibility is unlikely in light of the observation that in a warm environment the transition from wakefulness to nREM sleep is accompanied by an increase in active heat loss (see Heller and Glotzbach, 1977). Although good data comparing the T_{set} and α values for a heat loss response during wakefulness and nREM sleep do not yet exist, it seems that the conclusion will be that the primitive mammalian adaptation of nREM sleep involves a mechanism for lowering the regulated T_b.

Studies of thermoregulation and hypothalamic thermosensitivity during REM sleep have yielded a very different story. Thermoregulatory responses measured in different thermal environments during transitions from nREM to REM sleep have shown no indication of a consistent shift in regulated T_b. In fact, such studies have suggested that there is a lack of T_b regulation during REM sleep (Parmeggiani and Rabini, 1967; Shapiro et al., 1974). Measurements of hypothalamic thermosensitivity have confirmed that interpretation (Glotzbach and Heller, 1976; Parmeggiani et al., 1973; Florant et al., 1978). As can be seen in Figure 1, there is virtually no hypothalamic

Figure 1. Characteristics of hypothalamic thermosensitivity as a function of arousal state in a kangaroo rat (*Dipodomys heermanni*) and a marmot (*Marmota flaviventris*). Data points obtained during wakefulness are solid circles, those obtained during nREM sleep are open circles, and those obtained during REM sleep are open triangles. The kangaroo rat measurements were made at a T_a of 30 °C, and the α_{MHP} during wakefulness was -3.9 W kg^{-1} °C^{-1} and during nREM sleep was -1.9 W kg^{-1} °C^{-1}. The marmot measurements were made at a T_a of 15 °C, and the α_{MHP} during wakefulness was -0.9 W kg^{-1} °C^{-1} and during nREM sleep was -0.4 W kg^{-1} °C^{-1}. (The marmot graph is redrawn from Florant et al., 1978.)

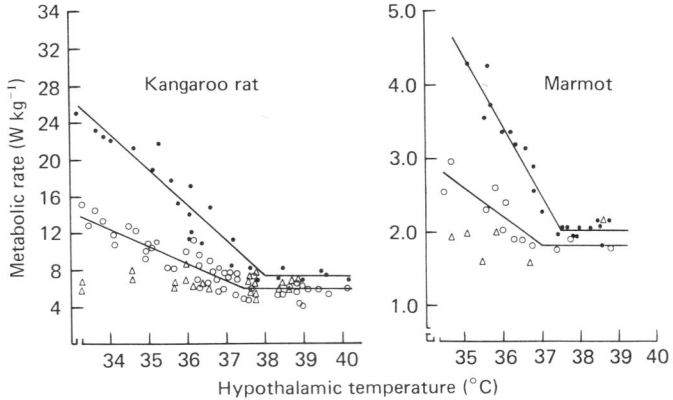

thermosensitivity during REM sleep. The thermoregulatory deficiencies associated with REM raise the interesting question of whether this sleep stage is more primitive than nREM sleep. It had been argued on other grounds that nREM sleep evolved together with endothermy in birds and mammals as an energy-conserving adaptation and is therefore of more recent origin than REM sleep (Berger, 1975). It will be very interesting to find out whether or not nREM and REM sleep in monotremes and marsupials involve similar thermoregulatory phenomena as they do in eutherians.

Hibernation and torpor

Hibernation is the most extreme example of T_b variability in mammals. In the older literature it was often proposed that hibernators are imperfect homeotherms possessing a primitive thermoregulatory system and compensatory thermal resistance to cold that have largely been lost by the more advanced mammals. Now it is quite clear that hibernators do not have deficient thermoregulatory systems. When euthermic they display thermoregulatory abilities and characteristics of hypothalamic thermosensitivity equivalent to those of nonhibernators (Heller et al., 1974; for a review see Heller and Glotzbach, 1977). The torpid hibernator remains sensitive to temperature and can respond to dangerously low values of T_a either by arousing to euthermia or by elevating MHP to maintain a greater gradient between T_b and T_a. The major site of temperature sensitivity during torpor in some species of hibernators is in the head region and probably in the hypothalamus (Lyman and O'Brien, 1974; Heller and Colliver, 1974). Manipulations of T_{hy} in hibernating ground squirrels and marmots have shown that the characteristics of hypothalamic thermosensitivity are qualitatively similar in hibernation and in euthermia, but the values of T_{set} and α are much lower (Figure 2) (Heller and Colliver, 1974; Florant and Heller, 1977). Additional experiments on ground squirrels and marmots entering hibernation have revealed the continuity of the thermoregulatory system over the range of T_b experienced by the hibernator (Heller et al., 1977; Florant et al., 1978).

Hibernation definitely seems to be an extension of the normal range for mammalian temperature regulation. Therefore, it is more likely to be a specialized or advanced rather than a primitive character, and must have required the evolution of many specialized adaptations.

One specialized adaptation of the thermoregulatory system of hibernators is found in the temperature sensitivities of individual hypothalamic neurons. The proposal of continuity of the thermoregulatory system was difficult to accept because most studies of the temperature sensitivities of hypothalamic units in mammals reveal rather narrow ranges of sensitivity with most units becoming silent at temperatures not much below 30 °C. A comparison of thermosensitivities of hypothalamic units of a hibernator (the hamster) and a nonhibernator (the guinea pig) showed that in contrast to the normal narrow thermosensitive range of the units in the nonhibernator, units in the hibernator's hypothalamus had continuous response curves over a much broader range of temperatures. The units in the hibernator had a higher mean firing rate at normal T_b and were active to much lower temperatures (Wünnenberg et al., 1976). Cooling the skin of the hibernator flattened the temperature response curves of its hypothalamic units and extended their lower limits into the range of T_b experienced during hibernation (Speulda and Wünnenberg, 1977).

Hibernation involves an extreme downward resetting of the T_b regulator and appears to be a specialized adaptation. Could this adaptation have evolved as an extension of the more generalized and hence primitive mammalian characteristic of a downward resetting of the T_b regulator during nREM sleep? Electrophysiological studies of hibernation suggest that this is the case. The cortical electroencephalogram of ground squirrels and marmots entering hibernation was scorable by usual criteria down to a brain temperature (T_{br}) of about 25 °C. Clearly, hibernation is entered through sleep (South et al., 1969; Walker et al., 1977). Moreover, the ground squirrel studies of Walker et al. (1977) showed that the increase in total sleep time (TST) at the onset of hibernation is due to an increase in nREM sleep only. REM sleep steadily declined during entrance and was rarely seen below a T_{br} of 28 °C. Similar EEG recordings from round-tailed ground squirrels esti-

Figure 2. Characteristics of hypothalamic thermosensitivity of a golden-mantled ground squirrel (*Citellus lateralis*) during euthermia and hibernation as determined by responses in metabolic rate to changes of hypothalamic temperature. The data from the hibernating animal were obtained at a T_a of 3.5 °C when T_b was 3.7 °C. The data from the euthermic animal were obtained at a T_a of 10 °C when T_b was 38 °C. The calculated α_{MHP} for hibernation was 0.35 W kg^{-1} °C^{-1} and the calculated α_{MHP} for euthermia was 7.44 W kg^{-1} °C^{-1}.

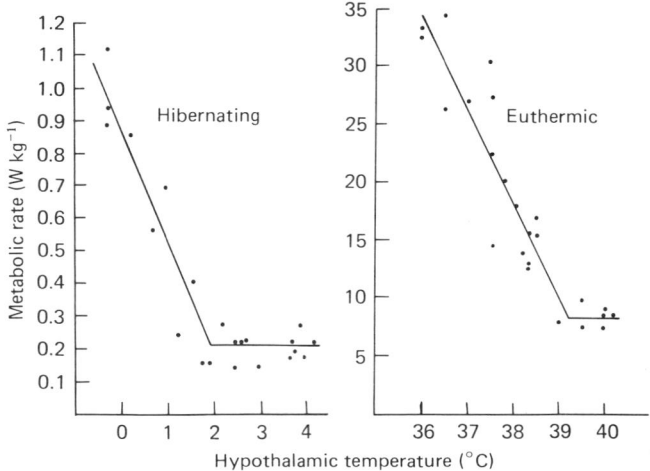

vating at rather high T_a's, so that T_{br} never fell below 25 °C during torpor, showed that the torpid animal spent over 90% of the time asleep, and REM was rarely observed (Walker et al., 1979). It seems very plausible that mammalian torpor evolved as an energy-conserving strategy out of basic sleep-related mechanisms, especially those subserving nREM sleep.

Response to warm environments

Adaptations to very warm environments may also involve variations in the regulated T_b. The example I shall discuss exploits the sleep-related resetting of the regulator. It comes from studies of the thermoregulatory systems of several species of small ground-dwelling squirrels that are day-active in extremely hot, arid environments in which the scarcity of free water precludes the use of evaporative water loss as a major thermoregulatory strategy. Instead, their strategy is to employ broad band regulation permitting T_b to rise during short periods of above-ground activity alternating with retreats to a cooler microhabitat where excess body heat can be lost passively. The scope for heat storage during periods of exposure is enhanced by hypothermia during the periods of heat loss.

In a study of three species of chipmunk (genus *Eutamias*), Chappell et al. (1978) observed responses to heat stress in the laboratory. The experiment consisted of gradually heating the whole animal with a photoflood lamp positioned over the metabolism chamber or gradually heating the animal's hypothalamus by means of water-perfused thermodes until the animal showed what was interpreted as active heat loss behavior. The heating was then terminated, and the animal was left undisturbed while its T_b declined. By this procedure maximum and minimum T_b values were obtained that

Figure 3. Maximum and minimum body temperatures tolerated by three *Eutamias* species. Lines show the ranges; bars signify two standard deviations to each side of the mean. Open bars are data from awake animals, diagonally striped bars are data from sleeping animals, stippled bars are data from whole-body heatings, and solid bars are data from hypothalamic heatings (redrawn from Chappell et al. 1978).

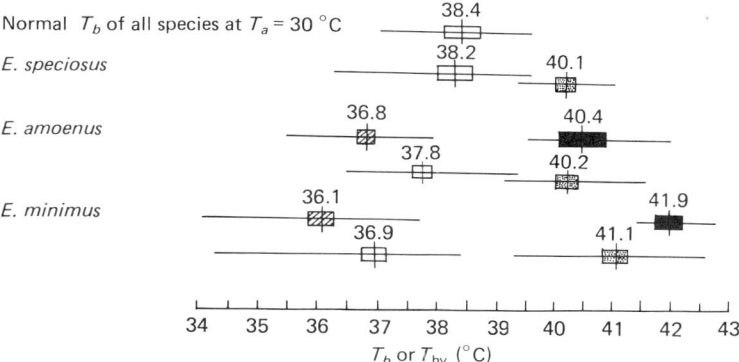

were believed to reflect the band width for euthermic T_b regulation. Definite interspecific differences in the behavioral responses to hyperthermia were seen. *Eutamias amoenus*, the species from the sagebrush–pinon pine habitat, and *E. minimus*, the species from the sagebrush habitat, almost always assumed a prone, sprawling posture as a sign of heat stress. The species from the mesic pine forest, *E. speciosus*, behaved quite differently; it never sprawled in response to hyperthermia. Whole-body heatings of *E. speciosus* invariably resulted in frantic activity as T_b approached 40 to 40.5 °C. The response of *E. speciosus* to hypothalamic heating was not distinct. It assumed a crouched, alarm posture with ears flattened back and remained immobile even as T_{hy} reached extremely high levels (above 42.5 °C). The behavior of *E. speciosus* following hyperthermia was also qualitatively different from that of *E. amoenus* and *E. minimus*. These latter two species slept almost immediately after about half of the heating episodes, whereas *E. speciosus* invariably remained awake. None of the animals was observed to sleep in the experimental apparatus except after induction of hyperthermia. The results of these experiments are summarized in Figure 3.

Comparisons of the upper T_b limits with the lower T_b limits reveal a greater band width for regulation of T_b in these three species as we go from a more mesic habitat to increasingly arid and hot habitats. Such broad band regulation with a wide dead band between thresholds for active thermoregulatory responses has obvious adaptive significance for species using the heat-storage strategy. The greater the difference between minimum and maximum tolerated T_b, the greater is the scope for heat storage and the longer is the time the animal can remain active in a hot environment before it must seek a heat sink (Chappell et al., 1978).

Broad band width is not a fixed characteristic of the thermoregulatory system of *E. minimus* and *E. amoenus*; rather it is a specific adaptive response

Figure 4. Measurements of hypothalamic thermosensitivity in eutherian mammals plotted as a function of body weight. (Redrawn from Heller, 1978.)

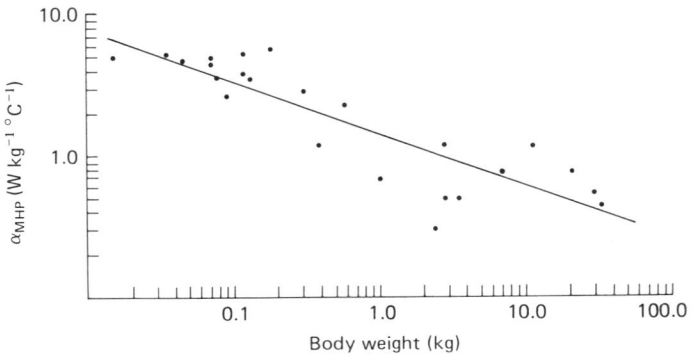

of the thermoregulatory system to thermal stress. In the absence of a high heat load, both of these species maintain their T_b within a narrow range; and at a thermoneutral T_a of 30 °C the values of T_{set} for the MHP responses of these three species lie between 38 and 39 °C. There must, therefore, be mechanisms whereby the normal characteristics of the thermoregulatory system are altered upon exposure to heat. In the case of *E. minimus* and *E. amoenus*, one of those mechanisms can be the lowering of the T_{set} which occurs during nREM sleep. There must be another mechanism, however, since hypothermia also occurred in the waking animals.

Conclusions

The purpose of this paper was to point out that fluctuations in T_b may not indicate a primitive, in the sense of a deficient, thermoregulatory system, but instead may be the manifestation of advanced, specialized adaptations. We may return to the title, "What is a primitive thermoregulatory system?" but I would suggest that at the present time this is a nonproductive question. We can do more by a broad comparative approach that seeks to discover which features of the thermoregulatory system are most general and which are highly specialized or derived. Because very little information exists on the characteristics of the central nervous regulator of T_b in marsupials, monotremes, and insectivores, much work lies ahead. For example, in Figure 4 the hypothalamic thermosensitivities of a broad size range of eutherians are plotted versus body size, and a general relationship seems to emerge. Where will the monotremes, marsupials, and insectivores fit into this pattern? Is the very general mammalian phenomenon of sleep accompanied, in monotremes and marsupials, by similar thermoregulatory adjustments to those we have described in eutherians? Is torpor a regulated phenomenon in noneutherians, and if so is it an extension of sleep-related mechanisms as it appears to be in eutherians? It is hoped that the documentation of evolutionary trends and relationships in the vertebrate thermoregulatory system will lead to conclusions about what the primitive mammalian system was like.

This paper was written while I was a guest in the laboratory of Dr.Werner Rautenberg at the Ruhr Universität, Bochum, West Germany. Much of the work discussed in this paper has been supported by grants from the U. S. National Institutes of Health.

References

Aschoff, J., Biebach, H., Heise, A., and Schmidt, T. (1974). Day–night variation in heat balance. In *Heat Loss from Animals and Man,* eds. J. L. Monteith and E. F. Mount, pp. 147–72. London: Butterworth.

Berger, R. J. (1975). Bioenergetic functions of sleep and activity rhythms and their possible relevance to aging. *Fed. Proc. 34*:97–102.

Chappell, M. A., Calvo, A. V., and Heller, H. C. (1978). Hypothalamic thermosensitivity and adaptations for heat-storage behavior in three species of chipmunks (*Eutamias*) from different thermal environments. *J. Comp. Physiol. 125*:175–83.

Florant, G. L., and Heller, H. C. (1977). CNS regulation of body temperature in euthermic and hibernating marmots (*Marmota flaviventris*). *Am. J. Physiol.* 232:R203–8.

Florant, G. L., Turner, B. M., and Heller, H. C. (1978). Temperature regulation during wakefulness, sleep, and hibernation in marmots (*Marmota flaviventris*). *Am. J. Physiol.* 235:R82–8.

Glotzbach, S. F., and Heller, H. C. (1976). Central nervous regulation of body temperature during sleep. *Science* 194:537–9.

Heller, H. C. (1978). Hypothalamic thermosensitivity in mammals. In *Effectors of Thermogenesis*. eds. L. Girardier and J. Seydoux, *Experientia* suppl. 32, 348 pp.

Heller, H. C., and Colliver, G. W. (1974). CNS regulation of body temperature during hibernation. *Am. J. Physiol.* 227: 583–9.

Heller, H. C., and Glotzbach, S. F. (1977). Thermoregulation during sleep and hibernation. In *Environmental Physiology II, MTP International Review of Physiology*, vol. 15, ed. D. Robertshaw, pp. 147–72. London: Butterworth.

Heller, H. C., Colliver, G. W., and Anand, P. (1974). CNS regulation of body temperature in euthermic hibernators. *Am. J. Physiol.* 227:576–82.

Heller, H. C., Colliver, G. W., and Beard, J. (1977). Thermoregulation during entrance into hibernation. *Pflügers Arch.* 369:55–9.

Lyman, C. P. and O'Brien, R. C. (1974). A comparison of temperature regulation in hibernating rodents. *Am. J. Physiol.* 227:218–23.

Parmeggiani, P. L., and Rabini, C. (1967). Shivering and panting during sleep. *Brain Res.* 6:789–91.

Parmeggiani, P. L., Franzini, C., Lenzi, P., and Zamboni, G. (1973). Threshold of respiratory responses to preoptic heating during sleep in freely moving cats. *Brain Res.* 52:189–201.

Shapiro, C. M., Moore, A. T., Mitchell, D., and Yodaiken, M. L. (1974). How well does man thermoregulate during sleep? *Experientia* 30:1279–81.

South, F. E., Breazile, J. E., Dellman, H. D., and Epperly, A. D. (1969). Sleep, hibernation, and hypothermia in the yellow-bellied marmot (*M. flaviventris*). In *Depressed Metabolism*, ed. X. J. Musacchia and J. F. Saunders, pp. 277–312. New York: Elsevier.

Speulda, E., and Wünnenberg, W. (1977). Thermosensitivity of preoptic neurons in a hibernator at high and low ambient temperatures. *Pflügers Arch.* 370:107–9.

Walker, J. M., Glotzbach, S. F., Berger, R. J., and Heller, H. C. (1977). Sleep and hibernation in ground squirrels (*Citellus* spp.): Electrophysiological observations. *Am. J. Physiol.* 233:R213–21.

Walker, J. M., Garber, A., Berger, R. J., and Heller, H. C. (1979). Sleep and estivation (shallow torpor): Continuous process of energy conservation. *Science* 204:1098–1100.

Wünnenberg, W., Merker, G., and Speulda, E. (1976). Thermosensitivity of preoptic neurons in a hibernator (golden hamster) and a non-hibernator (guinea pig). *Pflügers Arch.* 363:119–23.

12

The evolution of energy metabolism in mammals

A. J. HULBERT

Niggle was a painter . . . . He was the sort of painter who can paint leaves better than trees.
Tolkien, 1964

Yet Niggle wanted to paint a tree. In this contribution I want to paint a small aspect of an immense tree, the phylogenetic tree, and specifically those branches concerned with the mammals and their evolution. Most of the evidence as to the functional development of these branches must come from the detailed study of leaves, the individual species, that we know, largely from structural evidence, to be on a particular branch. It is only possible to study a few leaves, and therefore we must continually be aware of our choice of leaves lest they be too unusual.

The mammals can be divided into three major groups – the monotremes (more properly the Prototherians), the marsupials (Metatherians), and the placentals (Eutherians) – and it has been common to regard them as decreasing in "primitiveness" in this sequence. Even an elementary student of evolution realizes that all extant animals are mosaics, to varying degrees, of "primitive" and "advanced" (or more recently adaptive) features. But which features are which? To avoid too many value judgements I will describe these mammals as members of different "phylogenetic" groups and will comment on which features I consider to be "primitive" (i.e., conservative) or "advanced" (i.e., derived) at the end of this contribution.

I intend to examine the evolution of energy metabolism in mammals under three broad sections: first, the reptilian–mammalian difference in metabolism, then the various "phylogenetic" levels of metabolism, and last the various "ecological" levels of metabolism. In each section I will be concerned with the role of the thyroid hormones in regulating the level of metabolism and also with the role of organ and cell metabolism in determining the metabolic level of the whole organism.

The reptilian-mammalian difference in metabolism

Krogh (1916), in a comparison of the metabolism of a curarized dog and several poikilotherms, stated that his results indicated that "the oxidative energy of the tissues is greater in the warm-blooded than in a cold-blooded organism." In 1932, Benedict extended this comparison by pointing out a quantitative difference in the levels of energy metabolism in "cold"- and "warm"-blooded animals when both were measured at 37 °C. He found that the energy metabolism of his cold-blooded animals (predominantly snakes) was about one-fifth of that for a man of the same size. Hemmingsen (1960) pointed out that within the animal kingdom there appear to be three general levels of energy metabolism and that these are for unicellular organisms, multicellular poikilotherms, and homeotherms in ascending order. He found that, when expressed at the same body temperature, the level of standard energy metabolism in the homeotherms (birds and mammals) was approximately four to five times that for the poikilotherms. The values of standard metabolism for reptiles and mammals are presented in Table 1.

Table 1. *A comparison of the standard metabolism of reptiles and mammals*

	Standard metabolism (W kg$^{-0.75}$)	Body temperature (°C)[a]	Normalized (38 °C) standard metabolism (W kg$^{-0.75}$)[b]	Data source
Reptiles				
Tuatara	0.19	30	0.39	Wilson and Lee (1970)
Lizards	0.40	30	1.02	Bartholomew and Tucker (1964)
Snakes	0.48	30	1.02	Dmi'el (1972)
Crocodiles	0.29	23	1.06	Huggins et al. (1971)
Turtles	0.15	20	0.58	Hutton et al. (1960)
Mean			0.81	
Mammals				
Monotremes, Echidnas	0.92	31	1.55	Dawson et al. (in press)
Platypus	2.21	32	3.45	Grant and Dawson (1978)
Marsupials	2.37	35	3.00	Dawson and Hulbert (1970)
Eutherians				
Advanced	3.34	38	3.34	Kleiber (1961)
Insectivores	2.76	35	3.63	Dawson (1973)
Edentates	1.69	33	2.66	Dawson (1973)
Mean			2.94	

[a]In the reptiles this is the temperature at which the standard metabolism was measured.
[b]The Q_{10} values used in these "normalizations" were generally obtained from the source of original data. They varied from 2.1 to 3.3; where a measured Q_{10} was not available from the literature, a value of 2.5 was used.

Hemmingsen (1960) has shown the energy metabolism to be proportional to the three-fourths power of body weight throughout most of the animal kingdom. Thus metabolism values in this table are expressed on a kg $^{3/4}$ basis, as this negates the effect of body size. What can be seen in Table 1 is that when the metabolism is compared at the same body temperature in all these groups, the mammals form an obvious group, all having a metabolism about four times the value for the poikilotherms.

Heath (1968) suggested that the shift from the reptilian sprawling gait to the erect, limb-supported posture of modern mammals and the resultant tonus of muscles in mammals is a major factor in their greater heat production. This is not supported by the evidence, for in the resting dog (a mammal with erect posture) the tonus of skeletal muscle does not contribute significantly to the total energy metabolism. Galvão et al. (1963) found a decrease of only 17% in the metabolism of dogs that had their CNS totally destroyed, and this is not much more than the contribution of the brain itself to the dog's metabolism (Jansky, 1965). Also, many mammals have both a sprawling gait and a typical mammalian level of energy metabolism (e.g., monotremes); and the energy cost of standing above lying in man, cattle, and sheep is about 9%, and in the horse there is no extra cost (Brody, 1945). Hence it would appear that increased muscle tonus is a negligible contributor to this "quantum" increase in metabolism.

One major difference between modern homeothermic mammals and poikilotherms is the size of their brain. Jerison (1973) has shown that the average size of the brain in a mammal is approximately 10-fold greater than the average brain in a reptile of the same body size. Because the brain is a metabolically active organ, it is obvious that an increase in brain size will lead to an increase in the animal's total energy metabolism. However, a little reflection will show it to be only a minor contributor to the reptilian–mammalian difference in metabolism. In man, the brain contributes approximately 20% of the total heat production (Jansky, 1965); but man has a particularly large brain, and in other mammals the brain's contribution is considerably less. It has been estimated to contribute approximately 3% in the rat and about 12% in the dog (Jansky, 1965). Hence we may say that about 5 to 10% of a mammal's metabolism is probably due to the evolutionary increase in its brain size.

We can further analyze the reptilian–mammalian difference if we examine types of energy metabolism other than the standard metabolism. Benedict (1932), besides comparing homeotherm and poikilotherm metabolism at 37 °C, also compared them at other temperatures. He found that curarized and pithed mammals whose body temperature had dropped considerably (down to 16 to 20 °C) still had a metabolism that was several times greater than that of similar-sized, cold-blooded animals at the same temperature. This large difference was also present when the metabolism of a natural hibernator, the marmot, was compared with similar-sized poikilotherms at the same body temperature. More recent studies of metabolism during mammalian hibernation (or torpor) have reported levels that are typically "mammalian" rather than "reptilian" (Hammel et al., 1968).

There appears to be a fundamental difference between reptiles and mammals in regard to energy metabolism. This idea is further supported when one examines the cellular evidence available. Jansky (1962) showed that the whole animal cytochrome oxidase activity was very similar to the maximal metabolism in several species of rodents. Because cytochrome oxidase is the last link in the chain of respiratory enzymes, it was considered that its activity would give an indication of the maximal oxidative possibilities of an organism. Robin and Simon (1970) found a strong correlation between tissue cytochrome oxidase activity and tissue oxygen consumption within a single species. They also found a correlation between myocardial cytochrome oxidase activity and whole animal oxygen consumption in a turtle, a bird, and four mammals. It is of interest to note that, although the turtle and rat studied by these authors had similar weights, the rat had a myocardial cytochrome oxidase activity more than six times that of the turtle.

More enlightening is the recent evidence of Bennett (1972), who examined the activities of seven enzymes in the liver and skeletal muscle of three lizard species and the laboratory rat. He found that although there was no difference in the activities of the "soluble" (cytoplasmic) enzymes between the lizards and the rat, the activities of the mitochondrial enzymes were all very much lower in the reptiles. Bennett suggested that it is "quite probable that the hiatus in organismal metabolic rates between reptiles and homeotherms is a reflection of a greater number of mitochondria per unit of tissue weight in the latter." A small amount of data concerning the relative amounts of mitochondria in poikilotherms and homeotherms is available (Weibel et al., 1969; Eisenberg and Kuda, 1975; Mobley and Eisenberg 1975; Hulbert and Popham, unpublished results). It suggests that mammals have a greater amount of mitochondria per unit cell volume (at least in liver and muscle tissue), but the difference in volume of mitochondria is not of the same magnitude as the difference in organismal levels of metabolism. Because the most important structural parameter, with respect to the metabolic capacity of a tissue, is not the volume of mitochondria but rather the total surface area of their inner membranes, it is obvious that a well-controlled morphometric study of the mitochondria of tissues in representative reptiles and mammals would give us more information about this difference between reptiles and mammals.

A most important aspect of the reptilian–mammalian difference relates to the role of the thyroid gland. Removal of the thyroid gland in a mammal results in a considerable decrease in its standard metabolism, whereas in the lizard at 20 °C there is no change. Thyroidectomy in reptiles kept for some time at higher temperatures (30 °C) does show a decreased level of metabolism (Maher, 1965). However, since the reptilian–mammalian comparison is made with reptiles generally kept at room temperature (even though their oxygen consumption may be measured at 30 °C), we can expect that in these reptiles there is no calorigenic effect due to the thyroid hormones. It is probably not just that the thyroid hormones cannot exert their calorigenic effect at 20 °C, but, when kept at 20 °C, the reptilian thyroid gland shows no significant secretion of thyroid hormones (Hulbert and Augee, unpublished).

The cellular differences between reptiles and mammals may, at least in part, be due to the actions of the thyroid hormones. Among its many effects, triiodothyronine increases the amount of several mitochondrial enzymes as well as the surface area of mitochondrial membranes from both heart and liver tissue (Reith et al., 1973). In liver it also results in an increase in the fraction of the cell volume occupied by mitochondria. A fundamental difference between poikilotherms and homeotherms is related to the fluidity and functional behavior of their respective cell membranes, and this may also be due to their different thyroid gland activities (Hulbert, 1978).

In summary, if we examine the level of metabolism in mammals, we can account for the following: approximately one-fourth may be regarded as some form of "reptilian-level" heritage, another 5 to 10% may be the contribution of the evolutionary increase in brain size, and possibly a half can be accounted for as the portion due to the "calorigenic" action of the thyroid hormones. If these approximations are correct (and it must be remembered they are very approximate), then there is still some "unaccounted for" portion.

"Phylogenetic" levels of metabolism in mammals

Although Martin (1902) first compared the metabolism of monotreme, marsupial, and eutherian mammals, it has only been in the last decade and a half that a comprehensive comparison has been made. Dawson and Hulbert (1970) showed that marsupials from a wide range of ecological niches had essentially the same level of metabolism, and that this level was about 70% of that for eutherians. This low level of metabolism was also associated with a lower body temperature. Recently it has been found that all three extant monotreme species (Grant and Dawson, 1978; Dawson et al., in press) have low levels of metabolism. The metabolism of the two genera of echidnas is the same, but the platypus has a higher level. The platypus will be considered in the third section of this paper; in this section we will consider only the echidnas.

I will briefly examine the marsupial–eutherian difference in metabolism. If marsupials had a lower intrinsic level of metabolism unrelated to their lower body temperature, then we might expect either the size or the metabolic intensity of the main body organs (or both) to be lower in marsupials. The relative contribution of five major organs to body weight is about the same in the marsupial mouse (*Antechinus stuartii*) and in a similar-sized eutherian mouse (*Mus musculus*), although individual organs may vary between the species (Hulbert, unpublished results). These organs (liver, heart, kidneys, lungs, and brain) are responsible for approximately 33% and 43% of the standard metabolism in the rat and dog respectively (Jansky, 1965) and are thus significant contributors to the level of standard metabolism. It may be that the tissues of a marsupial are less "metabolically intense" than those of a similar-sized eutherian. If cytochrome oxidase activity can give an indication of cellular metabolic intensity (as it appears in the reptilian–mammalian comparison), then the liver of the marsupial mouse is no less "metabolically intense" than the liver of the similar-sized eutherian mouse when both are measured at the same temperature (Hulbert, unpublished results). Thus the admittedly scarce evidence (i.e., the organ contribution to body composition and the metabolic intensity of a major organ, the liver) suggests that there is no intrinsically lower level of tissue metabolism in a marsupial when compared to a similar-sized eutherian mammal. This conclusion is also supported by a ratio of "summit" to "standard" metabolism for some marsupials that is similar to that for eutherian mammals (Hulbert and Dawson, 1974a).

Although the thyroid plays a significant role in eutherian metabolism, little is known of its role in the metabolism of marsupials and monotremes. A recent study of thyroid function in similar-sized monotreme, marsupial, and eutherian mammals (Figure 1) has shown that removal of the thyroid results in a significant and similar reduction in the standard metabolism of the bandicoot (*Isoodon macrourus*) and the rabbit (*Oryctolagus cuniculus*). How-

Figure 1. The role of the thyroid hormones in the standard metabolism of a monotreme (*Tachyglossus aculeatus*), a marsupial (*Isoodon macrourus*), and a eutherian mammal (*Oryctolagus cuniculus*). C = control measurement before thyroidectomy. T_x = metabolism measured about 4 weeks after thyroidectomy. N = 5 in each case. (From unpublished results of Hulbert and Augee.)

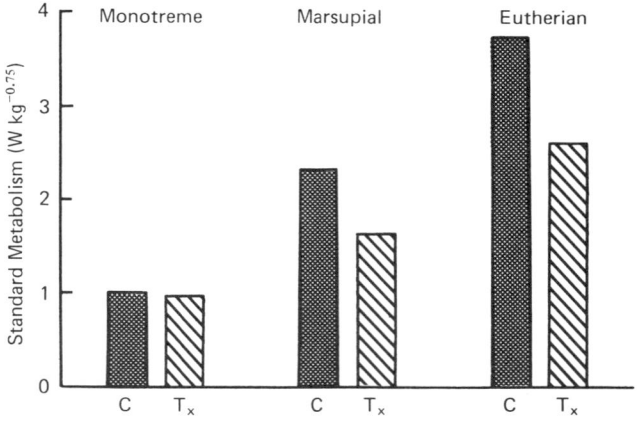

ever, in the echidna (*Tachyglossus aculeatus*) virtually no reduction in the level of standard metabolism was observed. This suggests that the thyroid hormones are responsible for the same relative proportion of metabolism in marsupials and eutherians, whereas the lower level of metabolism in echidnas is, in part, due to a reduced role of thyroid hormones.

Because the different levels of metabolism in mammals are often associated with different body temperatures, it is helpful, when comparing the different groups, to make some correction (Table 1). When expressed at the same body temperature, the simple increase in metabolic level from "primitive" monotremes to "advanced" eutherians disappears. In the monotremes, the echidnas retain a low metabolism after temperature normalization, whereas the platypus, after normalization, shows a high level of metabolism. Such normalizations must be treated with caution, but the basic assumption that metabolism increases with temperature in monotremes and marsupials has been verified many times (Aleksiuk and Baldwin, 1973; Schmidt-Nielsen et al., 1966; Hulbert and Dawson, 1974a). Thus it can be concluded that the main reason for the different levels of metabolism in these mammalian groups is the different level of regulated body temperature in each group. However, as proposed above, the low metabolic level in echidnas may also be related to a low thyroid activity. Possibly the low metabolism in the edentates may also be associated with a low output of thyroid hormone (Goffart, 1971).

Although different body temperatures may be the primary explanation of the different levels of metabolism in monotremes, marsupials, and other phylogenetic groups of mammals, I have not considered why the level of body temperature should tend to be higher in the more "advanced" mammals. Because a higher body temperature results in a higher level of metabolism, and hence a higher requirement for food, we could expect that in most situations it might be selectively disadvantageous. We must expect the tendency for higher body temperatures in the more advanced groups of mammals to be the result of an even greater selective advantage. An increase in body temperature affects the rate of many biochemical and biophysical processes. For example, an increase in temperature increases conduction velocity and decreases the refractory period of nerves (Kehl and Morrison, 1960). It also increases the rate of contraction and decreases the latent period of muscle contraction (Alpert et al., 1972). These effects imply that, other things being equal, an increase in body temperature would result in faster response both to a given situation and to any change in that situation. It is not difficult to appreciate how such an effect of increased body temperature on the neuromuscular system would be of great evolutionary advantage. Presumably, it is the same selective advantage that leads to the evolution of the "giant" axons in some invertebrates (Roeder, 1967).

An intriguing possible advantage of a high body temperature involves one of the basic building stones of evolution, that is, spontaneous mutation. It has been shown in *Drosophila melanogaster* that an increase in temperature results in an increased spontaneous mutation rate (Serra, 1968). Whether temperature would affect the mutation rate in mammalian cells and whether such an increase would be of evolutionary advantage is, of course, not known. It is, however, worthy of consideration.

"Ecological" levels of metabolism in mammals

Not all examples of reduced metabolism are found in whole phylogenetic groups, nor are they predominantly due to a low body temperature. Within many mammalian groups there are examples of a low metabolism that can be regarded as an adaptation to the environment inhabited, hence "ecological" levels of metabolism. Although the levels of metabolism in many arctic, temperate, and tropical mammals are similar (Scholander et al., 1950), there are many examples where the desert species of a particular group of mammals possess lower metabolic rates than their nondesert relatives. The camel has a metabolism (Schmidt-Nielsen et al., 1967) that is below that of other ungulates, and a large number of desert rodents have metabolic rates that are 80 to 85% of those of nondesert rodents (MacMillen and Lee, 1970; Hart, 1971). It must be remembered that not all desert members of a group need have a reduced metabolism. Shkolnik (this volume) has documented a large number of ecological reductions of metabolism in a variety of mammalian groups.

"Ecological" and "phylogenetic" levels of metabolism should not be regarded as mutually exclusive. Some desert marsupials have a lower level of metabolism than their nondesert relatives. An example is the rabbit-eared bandicoot (*Macrotis lagotis*), which has a level of metabolism that is about 80% of that of marsupials in general (Hulbert and Dawson, 1974b). Temperate and tropical species of bandicoots have the normal marsupial level of metabolism; thus the rabbit-eared bandicoot represents an ecological reduction in metabolism superimposed on a low phylogenetic level. A similar situation exists with desert and nondesert hedgehogs (Shkolnik, this volume).

In some of the desert rodents the low level of metabolism is associated with a reduced secretion rate of thyroid hormone (Yousef and Johnson, 1975). Many desert ground squirrels also have both a low metabolism (Hudson and Deavers, 1973) and a low level of thyroid activity (Hudson and Deavers, 1976). There is some indication that the level of thyroid activity in the rabbit-eared bandicoot is lower than that in other marsupials (Hulbert, unpublished results). Perhaps many of these ecological reductions in metabolism are controlled by a reduced activity of the thyroid gland.

Not all ecological levels of metabolism are reductions. Many marine mammals have higher metabolic rates than terrestrial mammals of similar size (Scholander, 1940; Ridgway, 1972). These ecological levels of metabolism may also be, at least partly, controlled by the thyroid. Porpoises have large thyroid glands and a high output of thyroid hormone (Ridgway, 1972). The high level of metabolism in the platypus compared to the other monotremes has been suggested as an example of an aquatic adaptation (Grant and Dawson, 1978). It would be of interest of know something about the activity of its thyroid gland.

Conclusions

The level of metabolism in modern mammals is much greater than that found in modern reptiles. Only a small part of this difference is due to the evolutionary increase in brain size, and most of it is reflected in differences at the cellular level. A large part of this evolutionary increase is due to the calorigenic action of the thyroid hormones.

The levels of metabolism in mammals vary, but they can loosely be grouped in two categories, either phylogenetic or ecological levels of metabolism. Phylogenetic levels of metabolism are largely the result of different levels of regulated body temperature in the various phylogenetic groups of mammals and should be looked at in a macroevolutionary perspective. Ecological levels of metabolism are more related to individual species and are found in some of the mammals that occupy specific habitats such as desert and aquatic environments. Some of the ecological levels of metabolism may be controlled by the activity of the thyroid gland. Ecological levels should be considered in a more microevolutionary perspective. Phylogenetic and ecological levels of metabolism should not be considered mutually exclusive.

Returning to the question of "primitive" and "advanced" features, I think we should consider a low body temperature in mammals to be, in general, a "primitive" feature (or more properly a conservative feature) and thus a high body temperature to be, in general, an "advanced" (or derived) feature. We should avoid thinking of a low metabolism as a "primitive" feature because many individual species within a group of mammals (and indeed some breeds within a species) have a low level of metabolism that is adaptive for their environment (e.g., many desert species). These could possibly be considered as "advanced" low levels of metabolism.

This work was supported by a grant from the University of Wollongong Research Fund and in part by a National Health and Medical Research Council grant.

References

Aleksiuk, M., and Baldwin, J. (1973). Temperature dependence of tissue metabolism in monotremes. *Can. J. Zool. 51:*17–19.

Alpert, N. R., Hamrell, B. B., and Halpern, W. (1972). The mechanical properties of isolated papillary muscle from the thirteen-lined ground squirrel. In *Hibernation and Hypothermia, Challenges and Perspectives*, ed. F. South et al., pp. 421–55. Amsterdam: Elsevier.

Bartholomew, G. A., and Tucker, V. A. (1964). Size, body temperature, thermal conductance, oxygen consumption, and heart rate in Australian varanid lizards. *Physiol. Zool. 37*:341–54.

Benedict, F. G. (1932). The physiology of large reptiles. *Carnegie Inst. Washington Publ.* 425.

Bennett, A. W. (1972). A comparison of activities of metabolic enzymes in lizards and rats. *Comp. Biochem. Physiol. 42B*:637–47.

Brody, S. (1945). *Bioenergetics and Growth*, with special reference to the efficiency complex in domestic animals. New York: Reinhold.

Dawson, T. J. (1973). "Primitive" mammals. In *Comparative Physiology of Thermoregulation*, vol. 3, ed. G. C. Whittow, pp. 1–46. New York: Academic Press.

Dawson, T. J., and Hulbert, A. J. (1970). Standard metabolism, body temperature, and surface areas of Australian marsupials. *Am. J. Physiol. 218*:1233–8.

Dawson, T. J. Grant, T. R., and Fanning, D. (In press). Standard metabolism of monotremes and the evolution of homeothermy. *Aust. J. Zool.*

Dmi'el, R. (1972). Relation of metabolism to body weight in snakes. *Copeia 1972*:179–81.

Eisenberg, B. R., and Kuda, A. M. (1975). Stereological analysis of mammalian skeletal muscle. *J. Ultrastruct. Res. 51*:176–87.

Galvão, P. E., Tarasantchi, J., and Magalhaes, C. A. E. (1963). Energy metabolism of dogs after total destruction of the central nervous system. *Am. J. Physiol. 204*:330–1.

Goffart, M. (1971). *Function and Form in the Sloth.* Oxford: Pergamon Press.

Grant, T. R., and Dawson, T. J. (1978). Temperature regulation in the platypus, *Ornithorhynchus anatinus:* Production and loss of metabolic heat in air and water. *Physiol. Zool. 51*:315–32.

Hammel, H. T., Dawson, T. J., Abrams, R. M., and Andersen, H. T. (1968). Total calorimetric measurements on *Citellus lateralis* in hibernation. *Physiol. Zool. 41*:341–57.

Hart, J. S. (1971). Rodents. In *Comparative Physiology of Thermoregulation*, vol. 2, ed. G. C. Whittow, pp. 1–149. New York: Academic Press.

Heath, J. E. (1968). The origins of thermoregulation. In *Evolution and Environment*, ed. E. T. Drake, pp. 259–78. New Haven: Yale University Press.

Hemmingsen, A. M. (1960). Energy metabolism as related to body size and respiratory surfaces and its evolution. *Rep. Steno Mem. Hosp. 9*:1–110.

Hudson, J. W., and Deavers, D. R. (1973). Metabolism, pulmocutaneous water loss and respiration of eight species of ground squirrels from different environments. *Comp. Biochem. Physiol. 45A*:69–100.

Hudson, J. W., and Deavers, D. R. (1976). Thyroid function and basal metabolism in the ground squirrels *Ammospermophilus leucurus* and *Spermophilus* spp. *Physiol. Zool. 49*:425–44.

Huggins, S. E., Hoff, H. W., and Valentinuzzi, M. E. (1971). Oxygen consumption of small caimans under basal conditions. *Physiol. Zool. 44*:40–7.

Hulbert, A. J. (1978). The thyroid hormones: A thesis concerning their action *J. Theor. Biol. 73*:81–100.

Hulbert, A. J., and Dawson, T. J. (1974a). Thermoregulation in perameloid marsupials from different environments. *Comp. Biochem. Physiol. 47A*:591–616.

Hulbert, A. J., and Dawson, T. J. (1974b). Standard metabolism and body temperature of perameloid marsupials from different environments. *Comp. Biochem. Physiol. 47A*:583–90.

Hutton, K. E., Boyer, D. R., Williams, J. C., and Campbell, P. M. (1960). Effect of temperature and body size upon heart rate and oxygen consumption in turtles. *J. Cell. Comp. Physiol. 55*:87–93.

Jansky, L. (1962). Maximal steady state metabolism and organ thermogenesis in mammals. In *Comparative Physiology of Temperature Regulation*, eds. J. P. Hannon and E. Viereck, pp. 175–91, Fort Wainwright, Alaska: Arctic Aeromedical Laboratory.

Jansky, L. (1965). Adaptability of heat production mechanisms in homeotherms. *Acta Univ. Carol. Biol. 1:*1–91.

Jerison, H. J. (1973). *Evolution of the Brain and Intelligence*. New York: Academic Press.

Kehl, T. H., and Morrison, P. R. (1960). Peripheral nerve function and hibernation in the thirteen-lined ground squirrel, *Spermophilus tridecemlineatus. Bull. Mus. Comp. Zool. Harv. Univ. 124:*387–400.

Kleiber, M. (1961). *The Fire of Life: An Introduction to Animal Energetics*. New York: John Wiley.

Krogh, A. (1916). *The Respiratory Exchange of Animals and Man*. London: Longmans, Green.

MacMillen, R. E., and Lee, A. K. (1970). Energy metabolism and pulmocutaneous water loss of Australian hopping mice. *Comp. Biochem. Physiol. 35:*355–69.

Maher, M. J. (1965). The role of the thyroid gland in the oxygen consumption of lizards. *Gen. Comp. Endocrinol. 5:*320–5.

Martin, C. J. (1902). Thermal adjustment and respiratory exchange in monotremes and marsupials. *Phil. Trans. Roy. Soc. Lond. Ser. B 195:*1–37.

Mobley, B. A., and Eisenberg, B. R. (1975). Sizes of components in frog skeletal muscle measured by methods of stereology. *J. Gen. Physiol. 66:*31–45.

Reith, A., Brdiczka, D., Nolte, J., and Staudte, H. W. (1973). The inner membrane of mitochondria under influence of triiodothyronine and riboflavin deficiency in rat heart muscle and liver. *Exp. Cell Res. 77:*1–14.

Ridgway, S. H. (1972). Homeostasis in the aquatic environment. In *Mammals of the Sea: Biology and Medicine*, ed. S. H. Ridgway, pp. 590–747, Springfield, Illinois: Charles C Thomas.

Robin, E. D., and Simon, L. M. (1970). How to weigh an elephant: Cytochrome oxidase as a rate-governing step in mitochondrial oxygen consumption. *Trans. Assoc. Am. Physicians 83:*288–97.

Roeder, K. D. (1967). *Nerve Cells and Insect Behavior*. Cambridge, Massachusetts: Harvard University Press.

Schmidt-Nielsen, K., Dawson, T. J., and Crawford, E. C. (1966). Temperature regulation in the echidna (*Tachyglossus aculeatus*). *J. Cell. Physiol. 67:*63–72.

Schmidt-Nielsen, K., Crawford, E. C., Newsome, A. E., Rawson, K. S., and Hammel, H. T. (1967). Metabolic rate of camels: Effect of body temperature and dehydration. *Am. J. Physiol. 212:*341–6.

Scholander, P. F. (1940). Experimental investigation on the respiratory function in diving mammals and birds. *Hvalradets Skr. 22:*1–131.

Scholander, P. F., Hock, R., Walters, V., and Irving, L. (1950). Adaptation to cold in arctic and tropical mammals and birds in relation to body temperature, insulation and basal metabolic rate. *Biol. Bull. Mar. Biol. Lab. Woods Hole 99:*259–71.

Serra, J. A. (1968). *Modern Genetics*, vol. 3. London: Academic Press.

Tolkien, J. R. R. (1964). *Tree and Leaf*. London: George Allen and Unwin.

Weibel, E. R., Stäubli, W., Gnägi, H. R., and Hess, F. A. (1969). Correlated morphometric and biochemical studies on the liver cell. *J. Cell Biol. 42:*68–91.

Wilson, K. J., and Lee, A. K. (1970). Changes in oxygen consumption and heart-rate with activity and body temperature in the tuatara, *Sphenodon punctatum. Comp. Biochem. Physiol. 33:*311–22.

Yousef, M. K., and Johnson, H. D. (1975). Thyroid activity in desert rodents: A mechanism for lowered metabolic rate. *Am. J. Physiol. 229:*427–31.

13

Metabolic capabilities of monotremes and the evolution of homeothermy

TERENCE J. DAWSON and T. R. GRANT

The sequence of evolutionary changes that led to the development of ho-meothermy in mammals has long been debated. Martin (1902) examined the body temperatures, metabolism, and thermoregulation of various reptiles, monotremes, marsupials, and eutherians and agreed with the suggestion made by Sutherland (1897) that, on the basis of body temperature, mono-tremes and marsupials represent a stage of physiological development inter-mediate between the relatively accurate homeothermism of higher mammals and the rudimentary regulation of reptiles. For many years Martin's studies were considered as indicating that in mammals homeothermic competence and metabolic capability evolved gradually together. Although more recent work has not substantiated the actual values obtained by Martin, it has been confirmed that monotremes (Schmidt-Nielsen et al., 1966, Augee and Ealey, 1968) and marsupials (Dawson and Hulbert, 1970) have low body tempera-tures as well as resting metabolic rates. These results and the wide range of body temperatures and metabolic levels reported for various species belong-ing to primitive or conservative placental groups (Dawson, 1973), could be considered as supporting the suggestion that homeothermic and metabolic capabilities evolved in a relatively gradual manner.

These long-standing ideas now have been challenged. The assumed direct association between an increasing level of metabolism and body temperature and homeothermic competence does not necessarily hold. This is particu-larly indicated by the marsupials because many now have been shown to be excellent homeotherms (Dawson, 1973). The acquisition of higher metabolic levels also has been suggested to occur in major steps rather than gradually. It has been proposed that all homeotherms or endotherms belong to a specific metabolic group that is distinct from the poikilothermic or ectother-mic reptiles, and that the development of homeothermy was associated with a major jump in metabolic capability (Dawson and Hulbert, 1970).

A further hypothesis on the evolution of homeothermy in mammals has come from Crompton et al. (1978), who proposed that mammals initially invaded a nocturnal niche without their metabolism being increased significantly above reptilian levels. A second step enabled mammals to invade a diurnal niche, and this involved the acquisition of higher body temperatures and metabolic rates. The hypothesis was claimed to be supported by data that indicated that among mammals considered primitive, the Insectivora were still at the initial reptilian stage, whereas monotremes and marsupials had evolved a distinct mammalian level of metabolism.

Fundamental in the ideas of Dawson and Hulbert (1970) and Crompton et al. (1978) has been the procedure of adjusting or "normalizing" metabolism to a common body temperature of 38 °C, using a Q_{10} of between 2 and 3. Dawson and Hulbert expressed their data for marsupials and those for the metabolism of the monotreme *Tachyglossus aculeatus*, given by Schmidt-Nielsen et al. (1966), in the weight-independent terms ($kg^{0.75}$) of Kleiber (1961) and adjusted the values to a "placental body temperature" of 38 °C. These normalized minimal metabolic rates were then found to be similar to those of eutherian mammals. Crompton et al. (1978) examined the metabolism of a range of primitive species during exercise. Metabolic costs of running on a treadmill were compared with normalized predicted responses to similar exercise for lizards and "advanced" mammals. The normalization involved an adjustment of the predicted responses from an assumed body temperature of 38 °C to the average body temperatures measured for each of the species examined. After this procedure the mammals they examined fell into either a reptilian or a mammalian energetic category.

Behind the idea of normalization is the concept that there may be certain intrinsic levels of metabolic capability, and that body temperature variations lead to the considerable scatter in reported data. The difficulty with this approach resides in the variability it seeks to overcome. Unfortunately, depending on the data used, different conclusions can be drawn. Dawson and Hulbert (1970) used results for the monotreme, *T. aculeatus* (Schmidt-Nielsen et al. 1966), but other information is now available for monotremes, and its normalization could lead to interpretations different from those originally derived (Augee and Ealey, 1968; Augee, 1976; Grant and Dawson, 1978b).

Because monotremes are perhaps the most conservative of mammals and retain many archaic mammalian characteristics, knowledge of their metabolic characteristics is central to a discussion of the evolution of homeothermy and higher metabolic levels. We have been working on the thermal relations of the platypus *Ornithorhynchus anatinus* (Grant and Dawson 1978a, b) and the other genera of monotremes, and are in a position to clarify their metabolic relationships. To do this we endeavored to obtain the metabolic rates of the monotremes under standard conditions similar to those used in the study of marsupials by Dawson and Hulbert (1970). Crompton et al. (1978) claimed that the use of metabolic responses to locomotion could avoid variability in resting measurements, suggesting that the metabolic rate of a mammal exercising at a given intensity was more reproducible than its resting metabolic rate. Unfortunately, they did not provide experimental evidence in support of this claim.

The metabolic characteristics of all three genera of monotremes could be determined because the Taronga Zoological Park, Sydney, allowed the examination of their long-beaked echidnas (*Zaglossus bruijni*), the largest of extant monotremes. Two *Z. bruijni* were obtained from the highlands of New Guinea and were kept in the zoo's reverse daylight house. Their diet was similar to that described by Collins (1973). Their sex was not determined. The echidnas *T. aculeatus* and the platypuses *O. anatinus* were obtained locally. The conditions under which the *T. aculeatus* were kept are similar to those described by Schmidt-Nielsen et al. (1966), and those for the platypus by Grant et al. (1977).

The equipment and techniques used were as described by Dawson and Hulbert (1970) for Australian marsupials, so that the results would be directly comparable. Oxygen consumption was measured in an open circuit system, and the factor used to convert oxygen consumption in liters per hour to watts was 5.59. In the studies on the platypus, body temperature (deep colonic) was measured by radio telemetry (Grant and Dawson, 1978a) whereas copper–constantan thermocouples were used for *Z. bruijni* and *T. aculeatus*. The animals were fasted for 24 h prior to measurement in a controlled temperature chamber under thermoneutral conditions. The assessed thermoneutral temperatures were for *O. anatinus* 25 °C, *T. aculeatus* 25 °C, and *Z. bruijni* 20 °C. Measurements were made only when an animal was resting quietly and its oxygen consumption and body temperature were stable and had been so for at least 30 min. Further details of the techniques are given in Dawson et al. (1978) and Grant and Dawson (1978b).

The results in Table 1 show that while resting body temperature was similar for all species, standard oxygen consumption was markedly different. Although the limited number of individuals studied precludes a proper analysis, it seems that the SMR of the three species may vary in relation to body weight in a manner similar to that of both marsupials and eutherians (Figure 1). Consequently metabolism was expressed in the weight-independent terms of Kleiber (1961), $kg^{0.75}$, to obtain appropriate interspecific comparisons. The values of the weight-independent SMR of *Z. bruijni* and *T. aculeatus* were similar, but that of *Z. bruijni* was significantly lower than that of *T. aculeatus* ($P < 0.01$). The metabolic rate of the platypus was over twice that of the other two monotremes and only slightly lower than that generally reported for marsupials. The values for *Z. bruijni* and *T. aculeatus* are very low when compared with most other mammalian values. Our values for *T. aculeatus* are below values generally reported for this animal (Dawson, 1973), and this concerned us somewhat. However, the results were consistently reproducible, and other recent measurements on resting echidnas by other

Table 1. *Body temperatures and metabolic relationships of three genera of monotremes, and of marsupials and eutherian mammals*

Animal	No. of animals	Weight range (kg)	T_{body} (°C)	O_2 Consumption (liter kg^{-1} h^{-1})	Heat production (W $kg^{-0.75}$)	Heat production normalized to 38 °C (W $kg^{-0.75}$)
Z. bruijni	2	10.73–16.53	31.7±0.07	0.081±0.009	0.86±0.02	1.53[b]
T. aculeatus	6	2.64–4.22	31.3±0.62	0.132±0.006	0.98±0.05	1.81[b]
O. anatinus	5	1.01–1.62	32.1±0.16	0.363±0.034	2.21±0.22	3.80[b]
Marsupials[a]		–	35.5	–	2.35	3.00[b]
Eutherians[a]		–	38.0	–	3.34	3.34

Values given are group means ± standard deviation.

[a] Derived from Dawson and Hulbert (1970).
[b] A Q_{10} of 2.5 was assumed in this adjustment.

workers (Augee, 1976; A. J. Hulbert, personal communication) support our current values. Any slight activity can markedly affect the metabolic rate when the basal rate is so low.

Our results show that while there are considerable differences in the metabolic characteristics of monotremes, their body temperatures are at similar levels. The platypus is of particular interest because its SMR differs so markedly from that of the other two monotremes. From an evolutionary point of view it appears that the two monotreme families have long been distinct. The oldest and only known Tertiary monotreme fossil, *Obdurodon insignis* (Woodburne and Tedford, 1975; Archer et al., 1978) is from the Miocene of South Australia and is referable to the family Ornithorhynchidae. Additionally, biochemical evidence based on changes in the amino acid sequences in the globins of myoglobin and hemoglobin suggests a divergence of the Tachyglossidae and Ornithorhynchidae commencing between 28 and 52 million years ago (Whittaker et al., 1978). This long divergence of the two monotreme lines does not in itself explain the large difference in their metabolic levels. Other mammalian groups have been distinct much longer without such differences resulting. It is possible to infer therefore that there has been marked pressure for adaptive changes. Aquatic and semiaquatic mammals often have acquired elevated metabolic rates, apparently to cope with their thermally demanding environment (Irving, 1973), and it is probable that this is also true for the platypus (Grant and Dawson, 1978a,b).

Over much of its present distribution in Eastern Australia the platypus is found in tableland areas where, in winter, it may feed for extended periods in water with temperatures approaching 0 °C. How does the platypus man-

Figure 1. The relationships between metabolic heat production and body weight for monotremes, marsupials, and eutherians. The dashed lines represent suggested relationships between SMR and body weight for monotremes and the solid lines for marsupials and eutherians (Table 1).

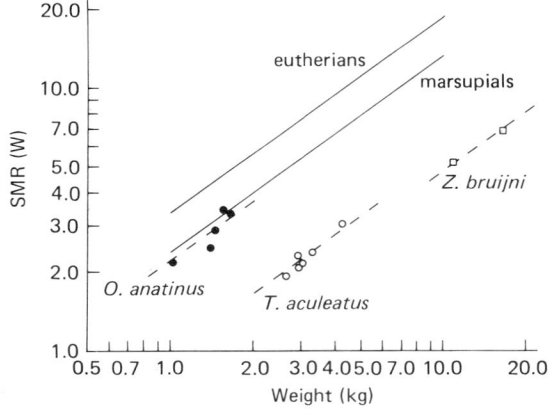

age this? The metabolic capability of the platypus, although high compared with other monotremes, is still considerably lower than many eutherian aquatic and semiaquatic species such as the muskrat, *Ondatra zibethica* (Figure 2). However, the platypus has adaptive characteristics concerned with retarding heat loss (Grant and Dawson, 1978b), notably the fur and the cardiovascular specializations.

Although not particularly deep, the fur of the platypus is more dense than that of any other comparable species except the sea otter, *Enhydra lutris* (Tarasoff, 1972). Associated with the high density are kinked underfur fibers and spatulate guard hairs which facilitate the holding of a layer of still air next to the body when the platypus is in water. The possibility of further waterproofing of the pelage is presented by the occurrence of sebaceous glands associated with hair follicles. Possibly the most important feature of the platypus fur is its ability to maintain insulative integrity in water. Platypus fur decreased in insulation by only 60 to 70% in water, whereas that of other semiaquatic species such as the beaver, *Castor canadensis*, decreased by more than 90%. The cardiovascular specializations are several, of which the complex vascular bundles in the hind limbs are strongly suggestive of a sophisticated counter-current heat exchange system (Grant and Dawson, 1978b). The overall effectiveness of temperature regulation of the platypus is seen in its ability to maintain body temperature for several hours in near-freezing water.

The monotremes appear to be a metabolically diverse group of competent

Figure 2. Metabolic responses of platypus (O.a.) and muskrat (O.z.) in air and water. Steady state values (except for the muskrat in water below 25 °C; for which T_{body} dropped steadily). T_a is ambient temperature in either air or water.

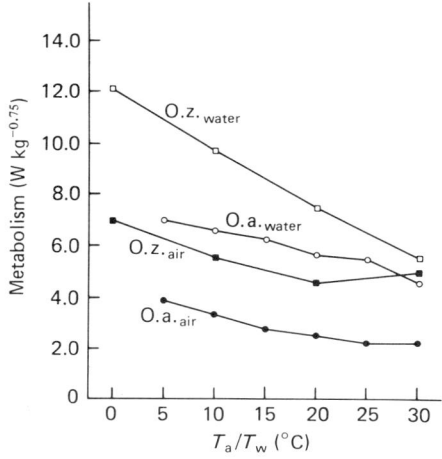

homeotherms, although they maintain a low body temperature (see Dawson et al., 1978, for data on the thermoregulatory abilities of *Z. bruijni*). What then do our current results imply about the evolution of homeothermy? First, they do not support the suggestion of Dawson and Hulbert (1970) that there is a distinct intrinsic mammalian metabolic level. If the results presented in this paper are "normalized" to a eutherian level of 38 °C, considerable variation remains (Table 1). It is now evident that there is more variability in metabolic levels of primitive mammals than was considered by Dawson and Hulbert (1970). This is further supported by the results of Shkolnik and Schmidt-Nielsen (1976) for hedgehogs of the insectivore family Erinaceidae, which also show a variable metabolic pattern. We consider that this evidence weighs against the hypothesis that the evolution of homeothermy was associated with a specific jump in metabolic capability. Although some of the variability observed may reflect adaptive adjustments to life in different habitats (Grant and Dawson, 1978b; Shkolnik and Schmidt-Nielsen, 1976), this does not alter the case. Primitive mammals can have low body temperatures that are well regulated, associated with a considerable range of metabolic levels. In some, such as *Z. bruijni*, the SMR is only one-fourth of the normal eutherian level, yet they appear to be competent homeotherms (Dawson et al., 1978). The normalization of metabolic rates to a common body temperature does not obscure these metabolic differences, and we see no physiological justification for the procedure.

The new data we present on the standard metabolic rates of monotremes do not support or necessarily contradict, the hypothesis of Crompton et al. (1978) that the evolution of homeothermy in mammals took place in two distinct steps. But because this hypothesis rests, in part, on the use of a normalization procedure, we urge caution in its acceptance. Several of the species examined by Crompton et al. attained body temperatures during exercise considerably above those previously reported under resting conditions (Dawson, 1973), whereas others did not. Perhaps these differences in the response of body temperature to exercise reflect differences in thermoregulatory capabilities during exercise, or such things as the greater thermal inertia of larger animals, differing thermoneutral zones, intensity of exercise, and different locomotory characteristics, rather than metabolic capabilities. In the data of Crompton et al. it can be noted that the species with the lowest average body temperatures, and therefore those for which the normalization procedure would have the greatest effect, are species (*Setifer setosus* and *T. aculeatus*) that exercised at the lowest speeds. Taylor et al. (1970) have reported that the rectal temperature of all running animals increased with increasing speed. In the light of these difficulties, it is perhaps also premature to use locomotory energetics and to neglect characteristic resting metabolic levels in consideration of the evolution of homeothermy.

The work was supported by a grant from the Australian Research Grants Committee. The technical assistance of D. Fanning in much of this work is acknowledged.

References

Archer, M., Plane, M., and Pledge, N. (1978). A review of the middle Miocene mono-treme *Obdurodon insignis* based on new cranial and postcranial remains. *Aust. Zool.* 20:9–27.

Augee, M. L. (1976). Heat tolerance of monotremes. *J. Thermal. Biol.* 1:181–4.

Augee, M. L., and Ealey, E. H. M. (1968). Torpor in the echidna, *Tachyglossus acule-atus. J. Mammal.* 49:446–54.

Collins, L. R. (1973). *Monotremes and Marsupials. A Reference for Zoological Institutions.* Washington, D. C.: Smithsonian Institution Press.

Crompton, A. W., Taylor, C. R., and Jagger, J. A. (1978). Evolution of homeothermy in mammals. *Nature (London)* 272:333–6.

Dawson, T. J. (1973). "Primitive" mammals. In *Comparative Physiology of Thermoregula-tion*, vol. 3, ed. G. C. Whittow, pp. 1–46. New York: Academic Press.

Dawson, T. J., and Hulbert, A. J. (1970). Standard metabolism, body temperature, and surface areas of Australian marsupials. *Am. J. Physiol.* 218: 1233–8.

Dawson, T. J., Fanning, D., and Bergin, T. J. (1978). Metabolism and temperature regulation in the New Guinea monotreme, *Zaglossus bruijni. Aust. Zool.* 20:99–103.

Grant, T. R., and Dawson, T. J. (1978a). Temperature regulation in the platypus, *Ornithorhynchus anatinus*: Maintenance of body temperature in air and water. *Phys-iol. Zool.* 51:1–6.

Grant, T. R., and Dawson, T. J. (1978b). Temperature regulation in the platypus, *Ornithorhynchus anatinus*: Production and loss of metabolic heat in air and water. *Physiol. Zool.* 51:315–32.

Grant, T. R., Williams, R., and Carrick, F. N. (1977). Maintenance of the platypus, *Ornithorhynchus anatinus*, in captivity under laboratory conditions. *Aust. Zool.* 19:117–23.

Irving, L. (1973). Aquatic mammals. In *Comparative Physiology of Thermoregulation*, ed. G. C. Whittow, vol. 3, pp. 47–96. New York: Academic Press.

Kleiber, M. (1961). *The Fire of Life: An Introduction to Animal Energetics.* New York: Wiley.

Martin, C. J. (1902). Thermal adjustment and respiratory exchange in monotremes and marsupials. *Phil. Trans. R. Soc. London Ser. B.* 195:1–37.

Shkolnik, A., and Schmidt-Nielsen, K. (1976). Temperature regulation in hedgehogs from temperate and desert environments. *Physiol. Zool.* 49:56–64.

Schmidt-Nielsen, K. Dawson, T. J., and Crawford, E. G., Jr. (1966). Temperature regulation in the echidna (*Tachyglossus aculeatus*) *J. Cell. Phsyiol.* 67:63–72.

Sutherland, A. (1897). The temperature of reptiles, monotremes and marsupials. *Proc. R. Soc. Victoria* 9:57–67.

Tarasoff, F. J. (1972). Anatomical observations on the river otter, sea otter and harp seal with reference to thermal regulation and diving. Ph. D. thesis, McGill Univer-sity, Montreal.

Taylor, C. R., Schmidt-Nielsen, K., and Raab, J. L. (1970). Scaling of energetic cost of running to body size in mammals. *Am. J. Physiol.* 219:1104–7.

Whittaker, R. G., Fisher, W. K., and Thompson, E. O. P. (1978). A review of protein sequence studies of monotreme globins. *Aust. Zool.* 20:57–68.

Woodburne, M. O. and Tedford, R. H. (1975). The first tertiary monotreme from Australia. *Am. Mus. Novit.* No. 2588:1–11.

14

Energy metabolism in hedgehogs: "primitive" strategies?

AMIRAM SHKOLNIK

In the mid-1960s several studies convincingly put an end to the long-standing speculation that a purported thermoregulatory inadequacy was related to an archaic phylogenetic status.

A study of temperature regulation in the echidna (Schmidt-Nielsen et al., 1966) concluded that "the echidna in cold surroundings maintains a well regulated body temperature, and this is consistent with the definition of homeothermy – although the level (about 30 °C) is below the usual level of mammals." It concluded that body temperature in the echidna results from a balance between heat loss and heat gain regulated in ways similar in nature and efficiency to those known in the "higher" placental mammals. Also, in a review of marsupial thermoregulation, Schmidt-Nielsen (1964) had stated: "The fact is that the marsupials, after a separate evolutionary history of perhaps 100 million years, represent the same functional level as placental mammals, and with many striking similarities."

Extensive studies by Dawson and his co-workers have reached the same conclusion and made it clear that even though the body temperature and the standard metabolic rates (SMR) of marsupials are lower than those usually found in eutherian placental mammals, some marsupials are excellent homeotherms, utilizing mechanisms to maintain body temperature similar to those of advanced eutherians. In his comprehensive review of thermoregulation in primitive mammals, Dawson (1973) has convincingly redeemed both monotremes and marsupials from the stigma of imperfection and inferiority.

Data to support the application of this concept to hedgehogs (Insectivora) were, however, meager at that time. Their low, sometimes fluctuating body temperature and their much lower than expected standard metabolic rates according to Kleiber's "Mouse to Elephant" curve, were widely held against these insectivores, and even in the late sixties they were considered "primitive" in their thermoregulatory capacities. Being considered inferior and "primitive" among advanced kindred has a humiliating connotation that a closer examination can hardly support.

Hildwein and Malan (1970), in their study of the European hedgehogs, and Shkolnik and Schmidt-Nielsen (1976), studying three species of Erinaceid hedgehogs, finally provided the evidence that hedgehogs possess thermoregulatory capacities similar to those of other eutherians.

Over a wide range of low ambient temperatures hedgehogs are capable of maintaining their body temperature constant, although at a level 3 to 4 °C lower than typical of other eutherians. At air temperatures close to or exceeding body temperature, hedgehogs tend to develop hyperthermia; but so do most other small nocturnal mammals (Hart, 1971), and hedgehogs are strictly nocturnal. Hyperthermia is by no means always a disadvantage. Under certain circumstances it may be considered an effective strategy for coping economically with environmental heat loads.

The mechanisms that hedgehogs utilize to maintain their body temperature are similar to those known in other homeotherms. At low temperatures their metabolic rate may increase up to four- to fivefold, and this capacity to increase heat generation in the cold is similar to that of other eutherians. The mechanisms they apply in hot environments to dissipate heat by evaporation are also similar to those of other eutherians.

Hedgehogs are distributed over a wide range of climatic conditions. From northern areas in Eurasia to extreme tropical desert, different species inhabit different areas. Nevertheless the body temperature in the various species is the same (Table 1). This characteristic demonstrates again their competence as thermoregulators. However, there is another aspect to the wide distribution of hedgehogs that also defies the concept of primitive – if the connotation is inadequate. Hedgehog species that differ in their geo-

Table 1. *Standard resting metabolic rates of three species of hedgehogs*

Species	No. of animals	Body mass (g)[a]	Body temperature (°C)[a]	Metabolic rate ($\dot{V}O_2$)	
				Measured (ml h^{-1})	Percent of predicted[b]
Erinaceus europaeus	5	749 ±183	34.00 ±0.74	337.05	98
Hemiechinus auritus	8	397 ±60	33.75 ±0.61	150.86	74
Paraechinus aethiopicus	5	453 ±59	34.19 ±0.64	108.75	51

[a] Mean ± standard deviation.
[b] $\dot{V}O_2$ (ml h^{-1}) = 2.3 M(g)$^{0.75}$ (Dawson and Hulbert, 1970).
Source: Shkolnik and Schmidt-Nielsen (1976).

graphical distribution possess physiological adaptations to their particular natural environments similar to those described in other eutherians. In this respect adaptation to the desert environment, a habitat most hostile to life, is of interest. This seems particularly significant in view of frequent statements that although "primitive" mammals are excellent thermoregulators in the cold, they are still deficient in hot environments.

In the study mentioned above the three species of hedgehogs were selected to represent three geographic distribution patterns (Harrison, 1964) (in Israel, where various zoogeographic units meet, these species may be found in close proximity):

1. The European hedgehog (*Erinaceus europaeus*)–abundant in cold to temperate areas from Northern Europe to the Mediterranean Sea.

2. The long-eared desert hedgehog (*Hemiechinus auritus aegyptius*) – (commonly found in North Africa and the Middle East where it inhabits mainly semiarid terrain, particularly coastal sand dunes and loessy inland plains.

3. The Ethiopian hedgehog (*Paraechinus aethiopicus*) – a tropical species that lives exclusively in extreme desert and is abundant in oases in the central Sahara, Arabia, the Sinai, and Southern Negev.

Although all three species were found to be effective thermoregulators, quantitative differences exist in the mechanisms they utilize to maintain body temperature constant. Many of these differences can be interpreted as specific adaptations to their particular natural environments.

Such an interpretation may be applied to the conspicuous differences in the level of the standard metabolic rate among the three species–differences that do not agree with the differences in their body mass.

Let me remind you that the standard metabolic rate (SMR) of eutherians may be allometrically predicted from their body mass according to Kleiber's (1961) widely accepted equation. Dawson and Hulbert (1970) found that the SMR of marsupials with body temperatures similar to that found in the hedgehogs amounts to about 70% of that of eutherians, and suggested an equation to relate SMR to body mass in this group.

From Table 1 we see that the SMR measured in the *Erinaceus* indeed agrees with the value expected for a marsupial of a similar body weight. The SMR of the desert species, however, deviates greatly from the predicted values, and that of the *Paraechinus* barely exceeds 50% of this level. This characteristic of the desert-dwelling hedgehogs needs further clarification.

In order to facilitate a comparison of the metabolic rates of hedgehogs with those of rodents with corresponding distribution, the SMR of the hedgehogs were normalized to a body temperature of 38 °C (assuming a Q_{10} of 2.5). Once this correction was accomplished, the SMR of *Erinaceus* agreed with the value predicted for eutherians in general, according to the "Mouse to Elephant" curve (Kleiber, 1961). The other two species, however, in accordance with the extent of their penetration of the desert, show SMR values lower than that predicted – the SMR of the *Paraechinus* being 50% of that value.

Table 2. *Standard resting metabolic rates in mammals from various ecological habitats*

Animal	Distribution	Body weight (g)	Metabolic rate Measured ml O_2 h^{-1}	Percent of predicted[a]	Data source
Murid mice					
Apodemus mustacinus	Mesic	36	44.2	103	Yahav (unpub.)
Acomys cahirinus	Desert to mesic	48	68.3	76	Shkolnik and Borut (1969)
Acomys russatus	Extreme desert	51	72.5	62	Shkolnik and Borut (1969)
Canids					
English pointer	Mesic	24 000	8400	108	Banhölzer (1976)
Cananite dog	Semidesert	17 000	4760	85	Meir and Shkolnik (unpub.)
Saluki dog	Desert	19 000	5130	82	Meir and Shkolnik (unpub.)
Fennec fox	Extreme desert	1 125	432	60	Banhölzer (1976)
Goats					
Mountain goat	Mediterranean	32 500	9046.5	98	Meir and Shkolnik (unpub.)
Bedouin goat	Desert	16 500	3651.0	66	Meir and Shkolnik (unpub.)

[a] $\dot{V}O_2$ (ml h^{-1}) $= 3.8$ M(g)$^{0.75}$.

The similarities between the SMR values in hedgehog and rodent species are indeed striking as SMR declines from mesic to desert habitats in ecologically equivalent species. Data from various eutherian groups point out that a low SMR emerges as a general characteristic of desert mammals (Tables 1 and 2).

The significance of a low metabolism in the desert may seem vague. Table 3 shows that in terms of water economy it is highly significant (in addition to its obvious significance in saving energy in a habitat where food is scarce). Above the lower critical temperature a *Paraechinus* evaporates at only a half to a third the rate of *Erinaceus* in order to stabilize its body temperature (Table 3). Considering that evaporative water loss at such temperatures greatly exceeds the losses through the other avenues (urine and feces), any saving significantly improves the entire water balance.

The thermal conductance of these animals provides a surprising contrast to what might be expected. Above the lower critical temperature, "advanced" mammals tend to increase their overall (core to environment) thermal conductance. The significance of this mechanism was convincingly discussed in a study of the desert jackrabbit (Dawson and Schmidt-Nielsen, 1966). Following their way of calculating conductance, we find that this mechanism is as equally well developed in *Erinaceus* and *Hemiechinus* as in the jackrabbit (Table 4). However, in *Paraechinus*, the desert-dwelling species, thermal conductance is lower than in the other hedgehogs, and, surprisingly enough, does not increase even when air temperature approaches body temperature. The monotreme echidna (*Tachyglossus aculeatus*), on the other hand, falls within the same pattern as advanced eutherians.

Does this lack of reaction to an environmental thermal stress cast a shadow of "primitiveness" on the Ethiopian hedgehog? A substantial change in thermal conductance can be attributed mainly to vasomotor mechanisms. Dawson has suggested that a high tissue insulation (low conductance) may be due to a low peripheral blood flow which is related to a low level of metabolic activity. There are good indications that this is the case in the Ethiopian hedgehog. An adaptive strategy for living in the desert that is based on a low metabolism requires a low thermal conductance if body temperature is to be maintained, but changes in thermal conductance are unessential in a nocturnal mammal.

In conclusion: (1) The ability of hedgehogs to master body temperature control over a wide range of ambient temperature is of similar nature to, and not less effective than, that in other eutherians. (2) The adaptive strategies that hedgehog species from different geographical areas use to cope with the particular conditions of the environment follow lines similar to those of other eutherians of corresponding distribution. This point is true also in the case of osmoregulatory capacities (Yaacobi and Shkolnik, 1974).

Table 3. *Evaporative water loss (E) in relation to metabolic rate (M) in hedgehogs at various ambient temperatures (T_a)*

T_a (°C)	20	T1.c.[a]	35	40
Erinaceus				
T_B (°C)	34	34	37.5	38–39
E (mg H_2O g^{-1} h^{-1})	0.8	1.0	2.6	6.7–10
E/M (%)	9.0	22.5	55.0	101–120
Hemiechinus				
T_B (°C)	33.8	33.8	36	37.5–39
E (mg H_2O g^{-1} h^{-1})	1.0	1.0	2.2	4.0–8.2
E/M (%)	9.3	30	48	106–160
Paraechinus				
T_B (°C)	34.2	34.2	35.5	36.2–38.5
E (mg H_2O g^{-1} h^{-1})	0.45	0.55	0.65	3.3– 4.5
E/M (%)	7.6	30	57	108–137

[a]Lower critical temperature: for *Erinaceus*, 29 °C; for *Hemiechinus*, *30 °C*; for *Paraechinus*, 31 °C.
Data from Shkolnik and Schmidt-Nielsen, 1976.

Table 4. *Changes in thermal conductance in relation to difference between ambient and body temperature ($T_B - T_a$)*

$T_B - T_a$ (°C)	Conductance (cal cm^{-2} $°C^{-1}$ h^{-1})				Data source
	30–15	10	5	2–0	
Animal:					
Lepus alleni	0.17	0.2	0.3	0.5–0.7	Dawson and Schmidt-Nielsen (1966)
Erinaceus europaeus	0.28	0.28	0.3	0.6–0.8	Shkolnik and Schmidt-Nielsen (1976)
Hemiechinus auritus	0.24	0.24	0.28	0.6–0.8	Shkolnik and Schmidt-Nielsen (1976)
Paraechinus aethiopicus	0.19	0.19	0.19	0.19	Shkolnik and Schmidt-Nielsen (1976)
Setifer setosus	0.14	0.13	0.17	0.17	Shkolnik (unpub.)
Tenrec ecaudatus			0.25–0.32	0.23	Hildwein (1970) (for summer)
Tachyglossus aculeatus	0.2	0.2	0.2	0.6–0.8	Schmidt-Nielsen et al. (1966)

References

Banhölzer, U. (1976). Water balance, metabolism and heart rate in fennec. *Naturwissenschaften 4*:202–3.

Dawson, T. J. (1973). "Primitive" mammals. In *Comparative Physiology of Thermoregulation*, vol. 3, ed. G. C. Whittow, New York and London: Academic Press.

Dawson, T. J., and Hulbert, A. J. (1970). Standard metabolism, body temperature, and surface areas of Australian marsupials. *Am. J. Physiol. 218*:1233–8.

Dawson, T. J., and Schmidt-Nielsen, K. (1966). Effect of thermal conductance on water economy in the antelope jack rabbit *Lepus alleni. J. Cell Physiol. 67*:463–71.

Harrison, D. L. (1964). *The Mammals of Arabia*, vol. 1. London: Ernest Benn.

Hart, J. S. (1971). Rodents. In *Comparative Physiology of Thermoregulation*, vol. 2, ed. G. C. Whittow, pp. 1–149 New York and London: Academic Press.

Hildwein, G. (1970). Capacités thermorégulatrices d' un mammifère insectivore primitif, le tenrec, leurs variations saisonnière. *Arch. Sci. Physiol. 24*:55–71.

Hildwein, G., and Malan, A. (1970). Capacités thermorégulatrices du hérisson en été et hiver en l'absence d'hibernation. *Arch. Sci. Physiol. 24*:133–43.

Kleiber, M. (1961). *The Fire of Life: An Introduction to Animal Energetics*. New York: Wiley.

Schmidt-Nielsen, K. (1964). *Desert Animals. Physiological Problems of Heat and Water*. New York: Oxford University Press, 277 pp. (Reprinted by Dover Publications, 1979.)

Schmidt-Nielsen, K., Dawson T. J., and Crawford, E. C., Jr. (1966). Temperature regulation in the echidna (*Tachyglossus aculeatus*). *J. Cell Physiol. 67*:63–72.

Shkolnik, A., and Borut, A. (1969). Temperature and water metabolism in two species of spiny mice. *J. Mammal. 50*:245–55.

Shkolnik, A., and Schmidt-Nielsen, K. (1976). Temperature regulation in hedgehogs from temperate and desert environments. *Physiol. Zool. 49*:56–64.

Yaacobi, D., and Shkolnik, A. (1974). Structure and concentrating capacity in kidneys of hedgehogs. *Am. J. Physiol. 226*:948–52.

15

Relations of metabolic rate and body temperature

PIOTR POCZOPKO

The question raised in this paper is: Are the different metabolic rates observed in various groups of mammals directly linked to their different body temperatures? It is well known that many "primitive" mammals have lower body temperatures than so-called higher mammals, and that they commonly also have lower metabolic rates. Could the lower metabolic rates be a direct consequence of the lower body temperatures? Would the difference disappear if their metabolic rates were recalculated, or normalized, to a higher body temperature?

Metabolic rates in poikilotherms and isolated tissues

It is well known that the metabolic rate of poikilotherms varies with body temperature. The body temperature of the frog *Rana esculenta*, for example, fluctuates from about 0 °C in winter when it hibernates under the ice on the bottom of small ponds, to about 25 °C in summer when it swims and basks in warmer water (Jusiak and Poczopko, 1972). The increase in metabolic rate caused by a temperature change can be expressed as the Q_{10} (the change in metabolic rate caused by a 10 °C increase in temperature).

The data in Table 1 show that the metabolic rate increased up to 30 °C, and decreased again at higher temperatures (33 and 36 °C). The Q_{10} did not remain the same throughout the range; it varied from 1.74 in the lowest temperature range to 3.36 between 24 and 30 °C. In the higher range, in which the frogs became torpid and sometimes died, the Q_{10} decreased to below unity.

The rate of oxygen consumption by tissue slices of the mouse and the frog is clearly temperature-dependent, as is the metabolic rate of the intact frog (Jusiak and Poczopko, 1972). However, at a given incubation temperature (e.g., 36 °C), the oxygen consumption of the slices derived from frog liver was almost five times lower than that of mouse liver slices. This shows that temperature alone is not the factor that determines the metabolic rate. This conclusion is further supported by the fact that even in the same species the effect of temperature on metabolic rate may depend on the previous history of the animal. For example, Poczopko and Jusiak (1972) showed that the Q_{10} of the metabolic rate of *Rana esculenta* between 12 and 24 °C was 2.73 in spring but 1.46 in autumn. For liver slices of this frog the spring and autumn values of Q_{10} were 2.68 and 1.16, respectively. The dependence of the metabolic rates of isolated tissues on previous acclimation has been discussed by Prosser (1967) in terms of changes in enzyme concentration.

Metabolic rate and body temperature in adult homeotherms

In this discussion the term standard metabolic rate will mean the rate determined when the animals are at rest, in a postabsorptive state, and in a thermoneutral environment. When we express the standard metabolic rates (in terms of per kg body weight) of several mammalian species that differ considerably in body weight, we find differences that are size-dependent rather than temperature-dependent. For instance, the metabolic rate per unit mass of the cow weighing 600 kg is only 7.6% of that of the mouse weighing 21 g, whereas body temperatures of these two species are almost the same, 38.6 and 37.0 °C, respectively.

When, however, the metabolic rates are expressed in terms of Kleiber's (1932, 1961) "metabolic body size," i.e., $kg^{0.75}$, the dependence on body size disappears. Indeed the metabolic rates of the mouse and the cow expressed per $kg^{0.75}$ are almost identical. On the other hand, the metabolic rate expressed per $kg^{0.75}$ in a dog is approximately 3.78 W $kg^{-0.75}$, whereas in a rabbit it is 2.96 W $kg^{-0.75}$, despite the higher body temperature of the rabbit (by 0.6 °C). Similarly, the data of Harris and Benedict (1919) revealed that a man weighing 75 kg produces 3.39 W $kg^{-0.75}$, whereas a woman of the same body weight produces only about 2.90 W $kg^{-0.75}$, although the normal body temperature does not depend on sex.

Table 1. *Effect of temperature on the metabolic rate in frogs* (Rana esculenta L.)

Ambient temp. (°C)	No. of animals	Metabolic rate (W/kg^{-1})[a]	Q_{10}
5	8	0.26	1.74
12	8	0.39	1.46
24	7	0.62	3.36
30	6	1.28	0.25
33	4	0.84	0.13
36	4	0.46	

[a]Mean body weight 30 g.
Recalculated from Jusiak and Poczopko (1972).

Within the homeotherms one can distinguish several groups of animals with characteristic levels of standard metabolic rate (Lasiewski and Dawson, 1967; Dawson and Hulbert, 1970; Poczopko, 1971; Blaxter, 1972). These groups usually differ also in regard to body temperature. Some workers have expressed the opinion that the differences in metabolic levels can be attributed to differences in the mean body temperature. The data compiled in Table 2 can be used to evaluate the validity of this interpretation. A Q_{10} calculated for the difference in the metabolic rate for marsupials and eutherian mammals, based on the difference in their body temperatures, amounts to 3.9, and between eutherians and nonpasserine birds, 2.7. These are reasonable values. However, the difference in metabolic rates between reptiles and marsupials, if it were based on the difference in body temperature would yield a Q_{10} of 35.4, and between nonpasserine and passerine birds, 409.8. Such Q_{10} values are meaningless and without biological significance, and the "reasonable" Q_{10} values can be considered as no more than incidental.

The above conclusion is supported by metabolic rates of shrews, which are between two and three times as high as in other mammals (Poczopko, 1971), whereas the body temperature of this group is the same as in other eutherians (Gębczyński, 1977). Similarly, the smallest birds, the hummingbirds, have very high metabolic rates although their body temperature is similar to that in other nonpasserine birds (Lasiewski, 1963,1964).

To compare different groups some authors have normalized the observed metabolic rates of animals with different body temperatures to a common body temperature. For this purpose the empirically established Q_{10} for each particular group has been used. A comparison of lizards and five groups of

Table 2. *Mean standard metabolic rates and body temperatures in some groups of Vertebrata*

Groups of animals	Metabolic rate ($kg^{-0.75}$)	Body temperature (°C)
Reptiles	0.33	30.0[a]
Mammals		
Marsupial	2.37	35.5[b]
Eutherian	3.34	38.0[c]
Birds		
Nonpasserine	3.87	39.5[d]
Passerine	7.07	40.5[d]
Shrews		
(5 species)	8.96	38.2[e]
Hummingbirds		
(9 species)	7.07	34.4–41.2[d]

[a]Templeton (1970); [b]Dawson and Hulbert (1970); [c]Poczopko (1971); [d]Lasiewski (1964); [e]Gębczyński (1977).

homothermic animals normalized to a body temperature of 38 °C, has been presented by Dawson and Hulbert (1970). In spite of the normalization, considerable between-group differences in metabolic rates remain. On the one hand this finding supports the hypothesis that factors other than body temperature have considerable effect on the metabolic rate. On the other, however, it must be acknowledged that such a comparison has only theoretical value, for the actual Q_{10} values over the range of the body temperature correction would be different from those established for the same groups at other intervals of body temperature.

Summit metabolism

Table 2 considers only the standard metabolic rates. Metabolic capacity can also be expressed in terms of summit metabolism or maximal working metabolism (Janský, 1965). Summit metabolism achieved during severe cold exposure, caused by shivering and/or nonshivering thermogenesis, may be four to six times standard metabolism (Giaja, 1938). Maximal working metabolism caused by prolonged physical exercise may be as much as ten times the standard rate (Janský, 1965). The four to six times increase in metabolic rate during summit metabolism is, as a rule, unaccompanied by any change in body temperature. During severe physical exercise the body temperature does increase (Kozłowski and Domaiewski, 1972) but this increase is very small in comparison with a change in the metabolic rate.

Metabolism and body temperature during postembryonic development

Changes in both body temperature and metabolic rate occur in many homeotherms during their postembryonic development (Poczopko, 1973; Freeman, 1965; Misson, 1977; Šimkova, 1960). The body temperature of many newborn or newly hatched animals is relatively low and increases to reach the adult level (depending on species) within a few days or a few weeks. The metabolic rate during postembryonic development changes greatly and is not linear, even on a log/log plot (Poczopko, 1973). When the metabolic rates are expressed per $kg^{0.75}$, it appears that in newborn animals the level is similar to that in adults. The metabolism then gradually increases for a while, then decreases again, and stabilizes at the level characteristic for the group of animals to which it belongs.

Several authors who have studied only very young animals have pointed out that the rises in body temperature and in metabolic rate during postembryonic development are parallel (e.g., Freeman, 1965). However, with further development, adult levels of body temperature and of metabolic rate are reached independently and are then fixed. Figure 1 shows this for chickens. After an initial concurrent rise of body temperature and metabolic rate, the body temperature stabilizes at a high level, but the metabolic rate drops and stabilizes later at a lower level. This pattern can also be seen if the data of Poczopko (1969) on the body temperature and of Piekarzewska (1977) on the metabolic rate of developing rabbits are plotted together.

These results indicate that the changes in body temperature and the metabolic rate during postnatal development are not directly interdependent.

Primitive mammals, hibernation, and torpor

The meaning of the term "primitive mammal" is not self-evident. Sometimes a relatively low body temperature is considered to indicate a primitive kind of homeothermy. For this reason marsupials are sometimes described as primitive. I consider that an inability to maintain thermal stability may be a more acceptable index. Therefore, marsupials, which thermoregulate as well as many eutherian mammals (Schmidt-Nielsen, 1964; Dawson and Hulbert, 1970), cannot be considered inferior in this respect. The same may be said of shrews, which have stable body temperatures within the usual eutherian range and very high metabolic rates, yet belong to the Insectivora, which in regard to morphology retain certain conservative traits and therefore have been considered primitive (Dawson, 1973).

For similar reasons hibernators have been classified as primitive because their body temperatures can undergo considerable seasonal fluctuations. However, this condition is well regulated, and the ability to hibernate may be regarded as a specialization rather than a primitive or conservative function.

The ability to enter into the state of hypothermia called hibernation, estivation, or torpidity, occurs widely among mammalian groups in insectivores, rodents, marsupials, and monotremes (Hayward, 1971; Hudson, 1973). The phenomenon also occurs, although less frequently, in birds (Dawson and Hudson, 1970).

When animals enter hibernation, estivation, or torpor, their thermoregu-

Figure 1. Changes in body temperature and metabolic rate of chickens during postembryonic development. T_B = body temperature (from Šimkova, 1960). SMR = standard metabolic rate (from Barott and Pringle, 1946).

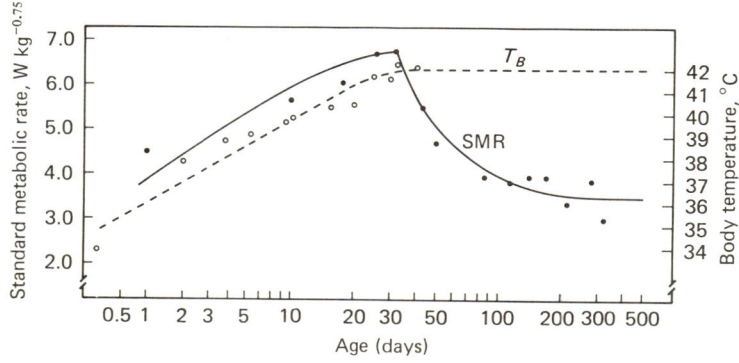

latory response to cold is abolished, and the body temperature drops. As a consequence, both body temperature and metabolic rate are low in the hibernating animal.

In the context of this paper the process of arousal from hibernation is the most interesting. The parallel increase in body temperature and in metabolic rate in the arousing hibernator is similar to that in poikilothermic animals. There is, however, an essential difference. In poikilotherms, which can better be described as ectotherms, an increase in body temperature is the cause of the rise in metabolic rate. In contrast, during the arousal of endothermic hibernators, metabolic rate increases first, causing the rise in body temperature.

Discussion and conclusions

The examples cited here show that even in the relatively simple situations of ectothermic organisms and of isolated tissues the rate of metabolism may vary between species or with seasonal acclimation, variations that may reflect quantitative or even qualitative changes in enzymes (Prosser, 1967).

This reasoning seems equally applicable to homeothermic animals. If the concentration of oxidative enzymes is proportional to the 0.75 power of the body weight, then it follows that the metabolic rate expressed per unit body weight in a mouse is much higher than in the cow, in spite of similar body temperatures. Indeed cytochrome oxidase activity in the body of animals has been shown to vary in proportion to the 0.75 power of the body weight (Janský, 1965). Cytochrome oxidase is the last link in the chain of oxidative reactions and determines the maximal metabolic capability of an animal (Janský, 1965) or of its isolated organs (Janský and Mejsner, 1971). This maximal capability is approximately 10 times higher than that used by an animal at standard conditions (Janský, 1965). Within the range of standard to maximal metabolism, changes in the rate of heat production may be effected with little or no change in body temperature. Thus, we can agree with Mount (1968) that in homeotherms there is no reason to expect any relation between metabolic rate and body temperature.

It seems that even torpid animals, with low body temperatures, have reserve capability to increase the metabolic rate. When the regulatory systems call for shivering and/or nonshivering thermogenesis, the metabolic rate increases, and only then the body temperature begins to rise.

From these examples we may conclude that body temperatures and metabolic rates in particular groups of homeotherms, and in particular in species within these groups, have, in the course of evolution, been fixed at a characteristic level. There is no reason to suspect that differences in metabolic levels are due to differences in body temperature. The thermoregulatory mechanisms have become adjusted accordingly and play their roles efficiently, and "correction" or normalization of the metabolic levels to a common body temperature has only speculative significance. If one asks what the metabolic rate of poikilotherms would be if their body temperatures were equal to that of mammals, the right answer would in most instances be zero; for they would be dead.

It is a pleasure to express my thanks to Dr. John Bligh for his help in the preparation of this paper.

References

Barott, H. G. and Pringle, E. M. (1946). Energy and gaseous metabolism of the chicken from hatch to maturity as affected by temperature. *J. Nutr. 31*:35–50.

Blaxter, K. L. (1972). Fasting metabolism and the energy required by animals for maintenance. In *Festkrift til Professor Dr. Agr. Dr. H. C. Knut Breirem 19–31. Mariendals Boktrykkeri A. s. Gjøvik.*

Dawson, T. J. (1973). Primitive mammals. In *Comparative Physiology of Thermoregulation*, vol. 3, ed. G. C. Whittow, pp. 1–46. New York: Academic Press.

Dawson, T. J. and Hulbert, A. J. (1970). Standard metabolism, body temperature and surface area in Australian marsupials. *Am. J. Physiol. 218*:1233–8.

Dawson, W. R. and Hudson, J. W. (1970). Birds. In *Comparative Physiology of Thermoregulation*, vol. 1, ed. G. C. Whittow, pp. 223–310. New York: Academic Press.

Freeman, B. M. (1965). The relationship between oxygen consumption, body temperature and surface area in the hatching and young chick. *Br. Poult. Sci. 6*:67–72.

Gębczyński, M. (1977). Body temperature in five species of shrews. *Acta Theriol. 22*:521–30.

Giaja, J. (1938). *Homeothermie et thermorégulation.* Paris: Hardman.

Harris, J. A., and Benedict, F. G. (1919). A biometric study of basal metabolism in man. *Carnegie Inst. Washington Publ. 279.*

Hayward, J. S. (1971). Nonshivering thermogenesis in hibernating mammals. In *Nonshivering Thermogenesis*, ed. L. Janský, pp. 119–32, Prague: Academia.

Hudson, J. W. (1973). Torpidity in mammals. In *Comparative Physiology of Thermoregulation*, vol. 3, ed. G. C. Whittow, pp. 97–165. New York: Academic Press.

Janský, L. (1965), Adaptability of heat production mechanisms in homeotherms. *Acta Univ. Carol. Biol. 1*:1–91.

Janský, L., and Mejsner, J. (1971). Nonshivering thermogenesis during arousal from hibernation. In *Nonshivering Thermogenesis*, ed. L. Janský, pp. 139–45, Prague: Academia.

Jusiak. R., and Poczopko, P. (1972). A comparison of the effect of temperature on metabolic rate of tissue slices of the mouse and the frog. *Bull. Acad. Pol. Sci. Ser. Biol. 20*:523–9.

Kleiber. M. (1932). Body size and metabolism. *Hilgardia 6*:315–53.

Kleiber. M. (1961). *The Fire of Life.* New York: Wiley.

Kozłowski. S., and Domaniewski, J. (1972). Thermoregulation during physical exercise in man of diverse physical performance capacity. *Acta Physiol. Pol. 23*:761–72.

Lasiewski. R. C. (1963). Oxygen consumption of torpid, resting, active and flying hummingbirds. *Physiol. Zool. 36*:122–40.

Lasiewski, R. C. (1964). Body temperature, heart and breathing rate, and evaporative water loss in hummingbirds. *Physiol. Zool. 37*:212–23.

Lasiewski, R. C., and Dawson, W. R. (1967). A re-examination of the relation between standard metabolic rate and body weight in birds. *Condor 69*:13–23.

Misson, B. H. (1977). The relationship between age, body temperature and metabolic rate in the domestic fowl (*Gallus domesticus*). *J. Thermal Biol. 2*:107–10.

Mount, L. E. (1968). *The Climatic Physiology of the Pig*. London: Edward Arnold.

Piekarzewska, A. B. (1977). Changes in thermogenesis and its hormonal regulators during the postnatal development of rabbits and guinea pigs. *Acta Theriol.* 22:159–80.

Poczopko, P. (1969). The development of resistance to cooling in baby rabbits. *Acta Theriol.* 14:449–62.

Poczopko, P. (1971). Metabolic levels in adult homeotherms. *Acta Theriol.* 16:1–21.

Poczopko, P. (1973). Niektóre aspekty termoregulacji we wczesnych okresach życia postembrionalnego ptaków i ssaków. *Acta Physiol. Pol.* 24 suppl. 6:101–15.

Poczopko, P., and Jusiak, R. (1972). The effect of ambient temperature and season on total and tissue metabolism in the frog (*Rana esculenta* L.). *Bull. Acad. Pol. Sci. Ser. Biol.* 20:437–41.

Prosser, C. L. (1967). Molecular mechanisms of temperature adaptation. *Am. Assoc. Adv. Sci.* Publ. No. 84, Washington, D.C.

Schmidt-Nielsen, K. (1964). *Desert Animals. Physiological Problems of Heat and Water.* London: Oxford University Press, 277 pp.

Šimkova, A. (1960). Vývoj tělesne teploty kuřat ve vztahu k různemu tepelnému režimu při odchovu. *Živočišna Vyroba* 33:449–60.

Templeton, J. R. (1970). Reptiles. In *Comparative Physiology of Thermoregulation*, vol. 1. ed. G. C. Whittow, pp. 167–221. New York: Academic Press.

16

The smallest insectivores: coping with scarcities of energy and water

STAN LEE LINDSTEDT

Shrews are the smallest of all mammals, the adult size of some species being as little as 1.9 g. They have been considered "primitive" because they possess a number of conservative morphological traits, especially in regard to dentition. They cope successfully, in part by selection of suitable microhabitats (as all animals must do to survive), and in part by employing characteristic mammalian solutions to the demands imposed by environmental conditions. Their physiological characteristics are related primarily to their extremely small body size and seem to be a result of the constraints that body size places on the available solutions.

A small endotherm encounters greater challenges to its energy and water balance than do larger animals, for heat exchange with the surroundings takes place at a rate proportional to surface area, yet heat storage or thermal inertia is directly proportional to body mass. The smallest animals have the highest area-to-mass ratios and therefore encounter the greatest physiological challenges in remaining homeothermic. Furthermore, the smallest endotherms have the least thermal insulation (i.e., the thinnest layer of fur or feathers). Hence, their heat transfer coefficient or thermal conductance (the inverse of insulation) is disproportionately higher than that of larger animals. Consequently, the smallest animals are first to experience the effects of perturbations in their microclimate as their body temperatures are more directly affected by the physical environment (wind, radiation, temperature, etc.) than are those of larger animals. The argument periodically resurfaces that the smallest homeotherms therefore should be confined to the most favorable (i.e., thermoneutral) environment (see Tracy, 1977).

There is another apparent energetic liability of small body size. Fuel and water stores are both scaled in direct proportion to body mass, yet the rates of use or consumption of both fuel and water are scaled more closely in proportion to surface area. For example, with respect to fuel stores, gut mass and body fat content are both roughly constant percentages of mammalian body weight. If we regard them as representing an animal's primary energy stores, then the amount of fuel stored also varies directly with body mass. The resting, thermoneutral rate of fuel consumption in mammals (standard metabolic rate, SMR) is proportional to the 0.75 power of body mass, whereas the daily existence energy (EE) of free-living mammals scales as the 0.67 power of body mass (King, 1974). The period between refuelings is equal to the amount of fuel stored divided by the rate of its consumption. Hence, for mammals, the period between refuelings is roughly proportional to $M^{0.25}$ or $M^{0.33}$. Regardless of the exact exponent considered to represent metabolic use of energy, the smallest mammals must be refueled more frequently; hence, they are the most vulnerable to fluctuating or inconsistent food sources.

One might, therefore, expect to find that the smallest homeotherms reach their physiological limits first when stressed either thermally or by deprivation of food or water. In fact, hummingbirds nesting in the cold environment of the Rocky Mountains of the United States are so near energetic insufficiency that should an incubating female miss a few feeding trips during the daylight hours (e.g., because of a rain storm) she will be forced into a period of hypothermia that same night (Calder, 1975).

What about the smallest mammals, the shrews? Although shrews are often found in cool habitats, they live in protected microclimates (beneath leaf litter, etc.), and they avoid extremely cold temperatures (Pruitt, 1953). Further, it has been suggested that an abundant food supply is most important in determining local distribution of shrews (Getz, 1961); also water must be present in their habitat (Chew, 1951).

Finally, although it is difficult to argue the adaptive value of this trait, the metabolic rate in shrews exceeds that predicted for eutherian mammals of their body size. Morrison and Pearson (1946) were the first to report high resting levels of oxygen consumption in shrews. In the past three decades, metabolism has been measured in many shrews and has consistently been found to exceed weight-predicted values.

It has been suggested that this elevation may be an artifact reflecting the "nervous" nature of shrews. Hawkins et al. (1960) found that, if the shrews were accustomed to the laboratory conditions and measuring apparatus, their levels of metabolism were about equal to those of small mice. However, if one compares their data to the "standard" Kleiber line (Figure 1), not only is the slope slightly more shallow (putting the smaller shrews further away from predicted values than the larger mice), but the reported metabolic levels are more than double the values predicted from the Kleiber equation. However, the SMR of one of the smallest mice, the 7-g pygmy mouse (*Baiomys*), is equal to its predicted value (Hudson, 1974), supporting the validity of the Kleiber line in this size range (Figure 1). More recently, oxygen consumption in the shrewlike marsupials (5- to 10-g animals of the genus

Planigale) has also been found to equal estimates for standard marsupial metabolism (Dawson and Wolfers, 1978). It therefore appears that the elevated metabolism of shrews represents a special characteristic of the shrews, rather than a failure of the equation to apply to the smallest mammals. Further, although the observed SMR of all shrew species consistently exceed weight-predicted values (Figure 1), Vogel (this volume) has found that SMR and body temperature are consistently higher in shrews of the subfamily Soricinae (the red-toothed shrews) than in shrews of the subfamily Crocidurinae (the white-toothed shrews).

Seldom has metabolism been measured in fasted shrews, and, even then, the animals were not likely postabsorptive. Consequently, the levels of metabolism for all shrews would probably be somewhat lower if postabsorptive measurement could be made.

The shrew with the lowest relative resting metabolic rate is the desert shrew (*Notiosorex crawfordi*). Comparing the desert shrew with other shrews, there is a parallel to many other desert mammals which show a depression of standard metabolism when compared to mammals in general (see Hulbert, this volume; Shkolnik, this volume). The rates of metabolism of different species of shrews can be compared directly, and their thermoneutral levels of metabolism can be compared to the Kleiber-predicted value when metabolic power is calculated per $kg^{0.75}$ ("metabolic body weight") (Figure 2). By comparison, the desert shrew can tolerate much higher ambient tem-

Figure 1. Standard metabolic rate (SMR) of small mammals. The solid regression line represents the SMR of "all" mammals (Kleiber, 1961). The broken line represents the SMR calculated for four species of shrews and laboratory mice (Hawkins et al., 1960). The open triangle represents the pygmy mouse *Baiomys taylori* (Bt), and the solid triangles represent seven species of shrews: *Sorex cinereus* (Sc); *Notiosorex crawfordi* (Nc); *Sorex minutus* (Sm); *S. araneus* (Sa); *Neomys fodiens* (Nf); *N. anomalus* (Na); *Blarina brevicauda* (Bb). (See Lindstedt, 1977, for references.)

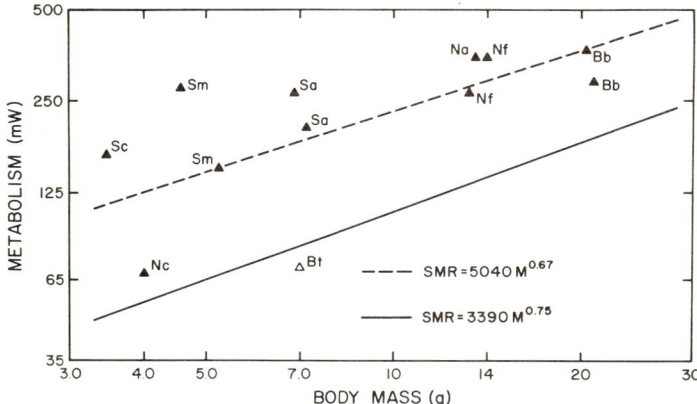

peratures, perhaps because of its lower rate of metabolic heat production in high air temperatures.

The short-tailed and masked shrews rarely experience high ambient temperatures, whereas the desert shrew must cope daily with problems of heat and aridity. Consequently, water acquisition and conservation may be as critical as temperature to its survival. The only water available to these nondrinking shrews is that contained in their insect diet. Insects are between 60 and 75% water and provide more water than an energetically equivalent diet of seeds. However, an insect diet involves the liability of considerable nitrogenous wastes (urea) from protein catabolism, which must be excreted in the urine. Does the additional water of an all-insect diet compensate for the increased urinary water loss caused by this nitrogen load?

If the desert shrew could concentrate urine no more than man, it would require about 1.5 times its daily water input just to excrete the nitrogen contained in its diet. However, the morphology of its kidney is more typical of desert rodent kidneys than that of other shrews, suggesting a capability of producing highly concentrated urine. By minimizing water losses, the desert shrew can remain in a positive water balance without the need for drinking (Figure 3), although at high air temperatures (and low humidity) it may need to feed on more succulent insect prey (Lindstedt, 1977). In other

Figure 2. Relation between metabolic rate and ambient temperature in three species of shrews, the masked shrew (*Sorex cinereus*, 3.5 g), the desert shrew (*Notiosorex crawfordi*, 4 g), and the short-tailed shrew (*Blarina brevicauda*, 20 g). Dividing metabolism by "metabolic body weight" ($kg^{0.75}$) allows all three species to be compared directly in spite of specific metabolic differences. It also permits the inclusion of the Kleiber equation as a single value, predicting the SMR of all mammalian species, though not the temperature at which it occurs. (From Lindstedt, *Physiol. Zool.*, in press.)

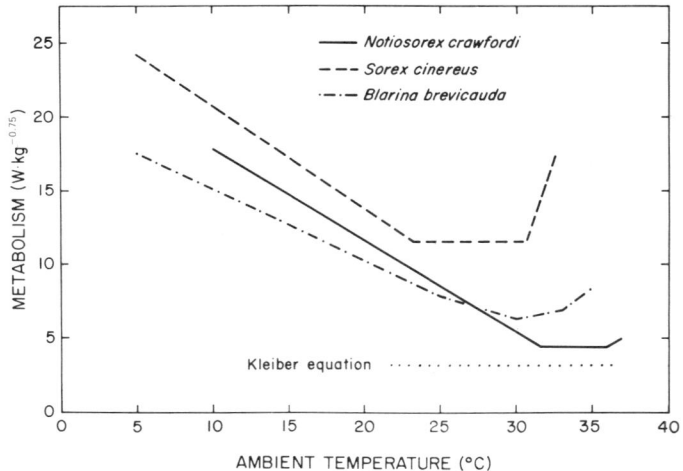

words, this animal demonstrates physiological water conservation not commonly found in shrews (see Chew, 1951).

Because of their small size and consequent high thermal conductance, the oxygen consumption of shrews increases steeply with decreasing air temperature. Additionally, with their need for frequent refueling it is little wonder that shrews are limited in their distribution to areas of abundant food supply. However, what happens in those habitats where food supply fluctuates or is limited? Several species of shrews are capable of undergoing shallow, daily hypothermia when food is restricted.

When fed ad libitum, desert shrews appear to be active both day and night and maintain a constant body temperature of about 37.5 °C. However, if mildly food-stressed, they spend the entire day within their nests with a reduced body temperature. During these periods of hypothermia, their respiratory rate decreases from 200 to about 35 breaths per minute, and their body temperature is reduced by about 9 or 10 °C.

All other shrews that have been found capable of hypothermia are in the subfamily Crocidurinae. Vogel (1974) first reported hypothermia in *Suncus etruscus* and since has suggested that all the Crocidurinae may become hypothermic (Vogel, this volume). Nagel (1977) has measured oxygen consumption of hypothermic musk shrews (*Crocidura russula*). He found that decreasing the ambient temperature resulted in an *increase* in oxygen consumption not only in normothermic but in hypothermic animals as well, indicating a resistance to a further decrease in body temperature.

He never measured a body temperature less than 18.5 °C, nor would any

Figure 3. Estimated water balance in the desert shrew at two air temperatures (25 and 35 °C) and two relative humidities (10 and 50%). Total water balance is expressed as the ratio of water gains to water losses. The water content of food refers to an all-insect diet. (From Lindstedt, *Physiol. Zool.*, in press.)

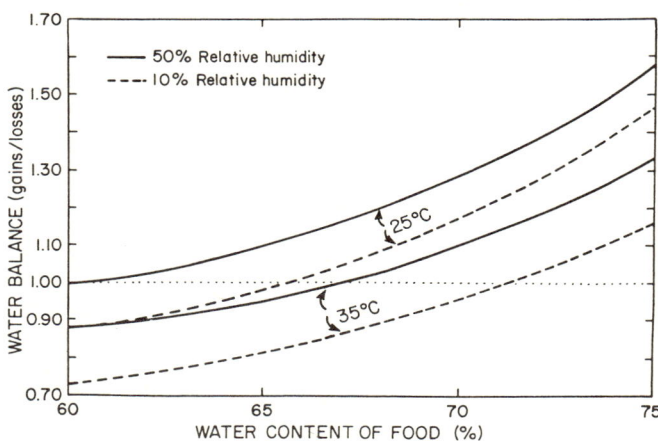

of the animals enter torpor if the air temperature was less than 10 °C. Although an increase in oxygen consumption of torpid mammals is not unusual (see Heller, this volume), it is unusual at such high air temperatures. Hainsworth and Wolf (1970) recorded a very similar response in an 8-g West Indian hummingbird (*Eulampis jugularis*). These animals regulate their body temperatures when hypothermic as well as when normothermic.

The oxygen consumption of hypothermic desert shrews fits the same pattern, but at a much higher body temperature. These animals resist entry into torpor at air temperatures below 20 °C, though they regulate their oxygen consumption as tightly when hypothermic as when normothermic, as is indicated by identical values of heat transfer coefficients (thermal conductance). By regulating at a body temperature of 28 °C, desert shrews are apparently sacrificing very little. Although they are slightly uncoordinated, they respond instantly to sound or touch and probably could easily avoid any nonavian predators. Yet, if the air temperature is between 20 and 25 °C, the energy savings are substantial.

At 25 °C hypothermic shrews operate at a cost of about 20% of normothermic shrews, but at 20 °C the cost is about 50% of remaining normothermic. Perhaps, at air temperatures below 20 °C the costs outweigh the benefits of energy saved, and hypothermia is no longer useful; hence the animals remain normothermic.

However, animals must be concerned with absolute, not relative, energy savings, and the absolute energy savings are constant below 25 °C. Perhaps there is another explanation for what appears to be a rather restricted range of hypothermia. The deserts of the southwestern United States have two dry seasons, late spring and late fall. Though air temperatures during those seasons are mild, insect abundance is low. Hence, hypothermia may occur naturally only during those seasons when nest temperatures are within the observed range of hypothermia.

Summary

Shrews are able to cope with the increased demands for energy and water caused by their small size primarily by living in habitats where food and water are abundant. However, their small size also permits them to utilize thermally protected microhabitats. At least one species has mastered the physiological problems of water frugality to survive in a hot-arid "unshrewlike" desert environment. Several species have the capability of becoming hypothermic, apparently in response to food deprivation. These physiological and behavioral solutions are similar to those found in many "higher" eutherian mammals, and any observed differences seem more a function of the shrews' extremely small body size than a "primitive" taxonomic position.

This work was supported by NSF DEB75–18576, NIH AM05738, and NIH HL 07249.

References

Calder, W. A., III (1975). Factors in the energy budget of mountain hummingbirds. In *Perspectives in Biophysical Ecology*, ed. D. Gates, pp. 431–41. New York: Springer-Verlag.

Chew, R. M. (1951). The water exchanges of some small mammals. *Ecol. Mongr.* 21:215–25.

Dawson, T. J., and Wolfers, J. M. (1978). Metabolism, thermoregulation and torpor in shrew sized marsupials of the genus *Planigale*. *Comp. Biochem. Physiol.* 59A:305–9.

Getz, L. L. (1961). Factors influencing the local distribution of shrews. *Am. Midl. Nat.* 65:67–89.

Hainsworth, F. R., and Wolf, L. L. (1970). Regulation of oxygen consumption and body temperature during torpor in a hummingbird, *Eulampis jugularis*. *Science* 168:368–9.

Hawkins, A. E., Jewell, P. A., and Tomlinson, G. (1960). The metabolism of some British shrews. *Proc. Zool. Soc. London* 135:99–103.

Hudson, J. W. (1974). The estrous cycle, reproduction, and development of temperature regulation in the pygmy mouse, *Baiomys taylori*. *J. Mammal.* 55:572–88.

King, J. R. (1974). Seasonal allocation of time and energy resources in birds. In *Avian Energetics*, pp. 4–85. Cambridge: Nuttall Ornithological Club.

Kleiber, M. (1961). *The Fire of Life*. New York: Wiley.

Lindstedt, S. L. (1977). Physiological ecology of the smallest desert mammal, *Notiosorex crawfordi*. Ph.D. dissertation, University of Arizona. Ann Arbor, Michigan: University Microfilms.

Morrison, P. R., and Pearson, O. P. (1946). The metabolism of a very small mammal. *Science* 104:287–9.

Nagel, A. (1977). Torpor in European white-toothed shrews. *Experientia* 33:1455–6.

Pruitt, W. O., Jr. (1953). An analysis of some physical factors affecting the local distribution of the short-tailed shrew (*Blarina brevicauda*) in the northern part of the lower peninsula of Michigan. *Misc. Publ. Univ. Mich.* 79:1–39.

Tracy, C. R. (1977). Minimum size of mammalian homeotherms: Role of the thermal environment. *Science* 198:1034–5.

Vogel, P. (1974). Kälteresistenz und reversible Hypothermie der Etruskerspitzmaus (*Suncus etruscus*, Soricidae, Insectivora). *Sonderdruck aus A. F. Säugetierkunde Bd.* 39:78–88.

17

Metabolic levels and biological strategies in shrews

PETER VOGEL

Measurements of the energy consumption of shrews have, over the past 30 years, regularly shown far higher values than expected according to their body size. The question that Poczopko was asking in 1971, of whether there is a deviation from the typical eutherian basal metabolism, could finally be answered in the affirmative in 1976 (Vogel, 1976). Several authors are of the opinion that the special physiological condition of shrews is determined by their small size (Gebczynski, 1965, 1971a; Pucek, 1970; P. Gehr, personal communication), but this hypothesis is only partly borne out by the facts. Because among the Soricidae, independently of body size, the subfamily Crocidurinae show markedly lower metabolic rates than the subfamily Soricinae (Vogel, 1976), other factors must be involved. I consider that the different metabolic levels may be explained as a physiological adaptation to differing climatic conditions because, according to zoogeographical and palae-ontological data (Repenning, 1967), the Crocidurinae evolved in the palae-otropical region in contrast to the Soricinae, which evolved in the holarctic region (cf. Table 1).

The difference in energy requirements between the two subfamilies obviously has far-reaching consequences for the entire biology of shrews.

The following paragraphs will classify available information on the energy metabolism of the Soricidae and point out possible correlations with other physiological, ecological, and ethological peculiarities. In doing so, I base my remarks essentially on published material, quoting only particularly relevant work.

Results

Metabolic rates

The first clues to the differing energy requirements among the Soricidae were given by Dryden et al. (1969), Hunkeler and Hunkeler (1970), and Hildwein (1972) in studies on tropical shrews, attributing the relatively low metabolic level to the tropical climate. The comparison of 13 species of different origin showed that these differences were characteristic of the two subfamilies (Vogel, 1976). Taking into account methodologically comparable results from 22 species, the Soricinae yielded a mean energy consumption $W^{-0.75}$ of 69.1 cal h^{-1} and the Crocidurinae a mean energy consumption $W^{-0.75}$ of 44.2 cal h^{-1}, wherein $W^{0.75}$ is the metabolic body size. As the shrews were given the same food and tested at similar ambient temperature, a different basal metabolism may be inferred, although the measurements were not made at standard conditions (among others, below the thermoneutral zone). This last doubtlessly explains our unexpectedly high values for the two smallest Crocidurinae, *Suncus etruscus* and *Crocidura bottegi*, which, with their short-haired fur, even at 25°C must considerably increase their metabolic heat production. According to Fons and Sicart (1976), the zone of thermal neutrality for shrews lies between 30 and 35 °C. Within this range, the metabolism $W^{-0.75}$ of *C. russula* and *S. etruscus* is, as expected, 30 to 35 cal h^{-1}, whereas for *Sorex araneus* and *Sorex minutus* it is 55 to 75 cal h^{-1} (Gebczynski, 1965, 1971a). This information is summarized in Figure 1.

The mean metabolic values in Figure 1 are fairly divergent. This divergency is due not only to different conditions of measurement, but also to obviously different metabolic levels within the Soricinae. Thus the tribe Soricini (*Sorex* and *Microsorex*) is characterized by extremely high values, whereas the tribes Blarinini (*Blarina* and *Cryptotis*) and Neomyini (*Neomys* and *Notiosorex*) show rather more moderate values.

Body temperature

Today, information on body temperature is also available. Measurements by Morrison et al. (1959) on *Sorex cinereus* gave an average of 38.8 °C (within a range of 36.1 to 40.9 °C). Gebczynski (1977) found for five species of Soricinae (*Sorex minutus, Sorex caecutiens, Sorex araneus, Neomys anomalus, Neomys fodiens*) average body temperatures between 37.3 and 39.1 °C, and in the majority of cases (14 out of 19 age–season groups studied) a range of 38.0

Table 1. *Classification of the Soricidae*

Family	Subfamily	Tribes	Genera
Soricidae	Soricinae	Soricini	*Sorex*
			Microsorex
		Blarinini	*Blarina*
			Cryptotis
		Neomyini	*Neomys*
			Notiosorex
	Crocidurinae	Crocidurini	*Crocidura*
			Suncus

After Repenning (1967).

to 38.7 °C. *Sorex palustris,* according to Calder (1969), has a body temperature of 39.7 ± 0.4 °C.

Nagel (1977) mentions a range of 34 to 38 °C for European Crocidurinae. According to H. Frey (personal communication), the rectal temperature of animals out of torpor amounts to 34.7 °C for *Suncus etruscus* and 36.0 °C for *Crocidura russula.*

Data for *Suncus murinus* are conflicting: 33.7 °C at narcosis (Dryden and McAllister, 1970) as well as 38 to 39 °C for animals with a fixed thermocouple in the rectum (Hasler and Nalbandov, 1974), whereby stress cannot be excluded. A verification (Frey, personal communication) gave a mean of 35.3 °C.

To summarize, we agree with Frey on a difference of about 3 °C between the two subfamilies, the body temperature being clearly higher in the Soricinae, which could partially account for the higher metabolic level.

Torpor

The capacity for torpor as a means of energy economy is an important concept. With the exception of *Notiosorex* (Lindstedt, this volume), this ability seems to be lacking in all the Soricinae examined (Morrison et al., 1959; Gebczynski, 1971b).

Figure 1. Energy consumption of Soricinae and Crocidurinae. Basal metabolism, according to Kleiber, $M_b = 16.2 \ W^{0.75}$ cal h^{-1}. Graph from Vogel (1976) with added values from the literature for *Suncus etruscus* (1, 2), *Crocidura russula* (3), both measured at 29 °C (from Fons and Sicart, 1976), *Crocidura f. occidentalis* (4) at 29 °C (from Hildwein, 1972), and *Suncus murinus* (5, 6) at 25 °C (from Dryden et al., 1974).

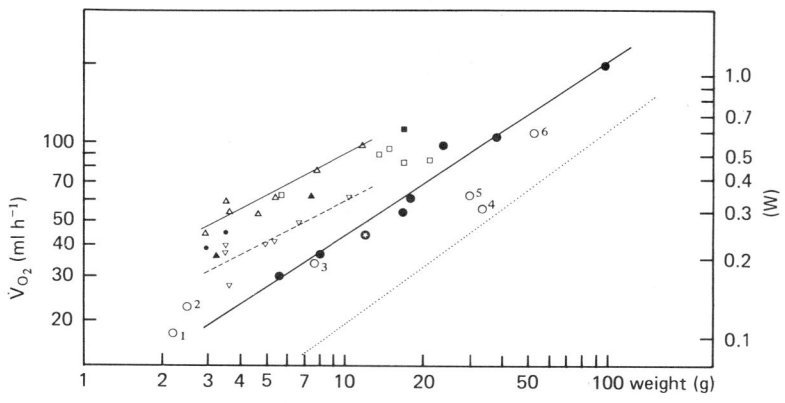

Soricinae

▲ ■ *Sorex* and *Neomys*; from author
□ Blarinini, Neomyini, mean
— ▲ Soricini, mean
--- ▽ Soricini, minimum

Crocidurinae

— ● tropical Crocidurinae; from author
◉ European *Crocidura russula*
● *Crocidura bottegi* and *Suncus etruscus*
○ Crocidurinae; from literature
······ basal metabolism after Kleiber

All Crocidurinae examined, on the other hand, can more or less easily be induced to reduce their metabolism and become torpid (Vogel, 1974: *Suncus etruscus, Crocidura russula;* Hutterer, 1977: *C. suaveolens;* Nagel, 1977: *C. leucodon*). Torpor also occurs in tropical species, such as the African *C. jouvenetae,* and therefore should not be considered as adaptation to a temperate climate (Vogel et al., unpublished). Moreover, it can also occur spontaneously in a warm ambient temperature to animals fed ad libitum (Frey and Vogel, 1978).

Blood characteristics
The particularly high metabolic level of the Soricinae leads one to expect unusual blood parameters. Thus maximum values for *Sorex minutus* (W = 3 g) for winter months are (Wolk, 1974): hematocrit index = 57%, red blood cells (RBC mm^{-3}) = $25.57 \times 10^6 \pm 3.9$ and RBC diameter of only 4.3 μm. The interesting point in our analysis is the comparison between representatives of both subfamilies, which is made possible thanks to the findings of Bartels et al. (1978).

This comparison (Figure 2) shows on the one hand how the values vary with body size, and on the other, that there is a significant difference between the two subfamilies. This also explains why the maximum values are not found in the smallest mammal (*Suncus etruscus*), but in the smallest tested member of the Soricinae (*Sorex minutus*).

Activity
Crowcroft (1957) observed a connection between the activity rhythm of the Soricinae (characterized by their short-term cycles) and their extraordinary energy need. Thus the animals are also active in the daytime. Although the European Crocidurinae present much the same activity pattern in cold weather, they can considerably diminish their daytime activity during warm weather. Consequently they correspond in their behavior to the tropical members which are active exclusively at night (Vogel et al., unpublished). The activity patterns of the shrews thus seem unmistakably related to the differing metabolic levels.

Population density and social behavior
Territorial behavior and home range size depend, among other factors, on food need and availability. Consequently population density and social behavior may be influenced by energy considerations.

Therefore we shall compare the relatively stable winter population densities of Soricinae and Crocidurinae, where the extreme values during the period of minimum energy availability are particularly interesting. The maximum population density of *Sorex araneus* amounts to 1.8 to 16.3 animals per hectare (Croin Michielson, 1966; Nosek et al., 1972; Pernetta, 1977), whereas the values for *Sorex minutus* (Croin Michielson, 1966; Nosek et al.,

1972), *Sorex cinereus, Sorex arcticus,* and *Blarina brevicauda* (Buckner, 1966) do not reach this maximum. In the case of the Crocidurinae, unusual population densities of 77 to 100 animals per hectare were established for *Crocidura russula* (Genoud, 1978), and have been confirmed in further test fields. The same author also points out that the studied Crocidurinae, in contrast to the Soricinae, only show a low territoriality.

Whereas the Soricinae, as a rule, have a strictly solitary existence, group behavior has been observed for the Crocidurinae under laboratory conditions (Vogel, 1969), and latest unpublished results of experimental field studies show that members of *Crocidura russula* often live in pairs.

Longevity

Admittedly Dehnel (1949) and Borowski and Dehnel (1953) have shown that *Sorex araneus, Sorex minutus,* and *Neomys fodiens* die, in free wild life, in the autumn of their second calendar year at the latest (i.e. their maximum life span does not exceed 18 months). However, comparative data for the Croci-

Figure 2. Hematological parameters of shrews in relation to body weight (all values from autumn measurements). (●) Soricinae: Sm = *Sorex minutus;* Sa = *Sorex araneus;* Nf = *Neomys fodiens* (from Wolk, 1974). (o) Crocidurinae: Se = *Suncus etruscus;* Cr = *Crocidura russula* (from Bartels et al., 1978). Hb = hemoglobin content; Hct = hematocrit index, RBC = number of red blood cells, RBC diam. = diameter of red blood cells.

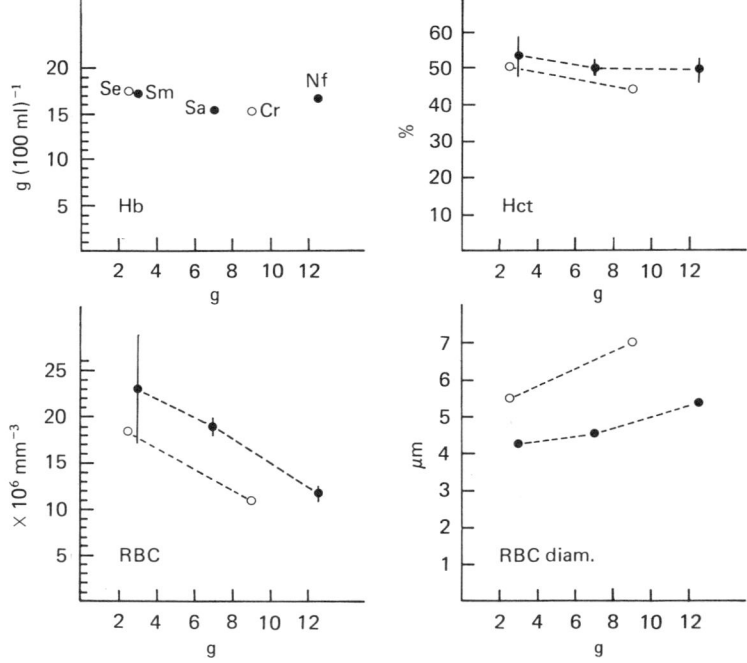

durinae are lacking. We are therefore obliged to base our comparison on laboratory results. Here we find striking differences between the subfamilies, correlated with the metabolic levels (Vogel, 1972a). In the meantime, the latest data on ten species have been collated by Hutterer (1977): for every two species of equal body size compared, the maximum longevity of the Crocidurinae is about one year longer than that of the Soricinae.

From the population standpoint, only the expected life span of animals reaching maturity can be taken into account. It should be shorter in the Soricinae, as their extreme metabolic level may have fatal consequences in the event of momentary lack of food.

Reproduction biology and early development

In case of differing longevities, differing rates of reproduction should logically be expected. In fact, the average litter size for Soricinae is higher than 5, whereas for the Crocidurinae it is below 5 (Vogel, 1972a). Thus Brambell (1935) in *Sorex araneus* found a litter size of 6.5 (N = 51); whereas Price (1935) in *Crocidura russula* found it to be 3 (N = 670), Vogel (1972a) in the same species, 3.5 (N = 86), and Dryden (1968) in *Suncus murinus*, 1.8 (N = 45).

Striking differences concerning the gestation period were observed by Dryden (1969) and by Vogel (1972a). It lasts 20 to 23 days in the Soricinae (*Sorex, Neomys, Cryptotis, Blarina*), and 26 to 31 days in the Crocidurinae (*Crocidura, Suncus*). We find differing degrees of maturity among newborn animals that correspond to the differences in the gestation period. Thus the state of newborn Crocidurinae corresponds to that of typical murid young. Newborn Soricinae, on the other hand, are far less developed. Their skull ossification (number of centers) is closer to that of newborn marsupials than to that of typical eutherian nidicolous species (Vogel, 1972b) and represents a unique case among placentals.

The longer period of gestation and the higher degree of development at birth in the Crocidurinae could well be connected with the regression in litter size; the small number of young permits longer intrauterine development. Hellwing (1971) found a corresponding relation on an intraspecific level. Small litters are correlated with a gestation period lasting 1 or 2 days longer.

Postjuvenile development

In the examined Crocidurinae, juvenile development after leaving the nest continues until sexual maturity is attained. As a rule this occurs at 2 to 3 months; in *Suncus murinus*, at 30 days at the earliest (Dryden, 1969); for *Crocidura russula*, after 45 to 50 days (Hellwing, 1971). In temperate zones, animals born in late summer or in autumn show a slackening of development during winter; that is, the increase in weight continues, but very slowly.

With few exceptions (*Blarina brevicauda;* Pearson, 1944), the Soricinae only attain sexual maturity in the second calendar year of life. Prior to this, the animals in wintertime go through a so-called winter depression (Dehnel, 1949), known as "Dehnel's phenomenon." This is not only an arrested development, but really a strong regression. With a decrease of up to 40% in body weight, there is a diminution in the size of the organs, which, for instance, in the case of the weight of the brain, amounts to 20%. This winter depression, which has also been demonstrated in the genera *Neomys* and *Blarina* (for details see Pucek, 1970), is more marked as the winter climate becomes colder. This temporary decline in the size of the body leads, in consequence, to an increase in metabolic rate but also to a decrease in absolute energy consumption together with a decrease in the amount of daily food per head, which, according to Mezhzherin (1964) and Pucek (1970), represents an adaptation to circumpolar winter conditions. In fact, Buckner's findings (1964) suggest that the small species of shrew are more efficient in finding the small soil arthropods. With about a 100% difference in the body size of young-adult animals in January as compared to old-adult animals in summer, the absolute saving of energy is evident, and, in the case of minimal food availability, its ecological significance is plausible.

Discussion

Ecophysiological significance

Within the last few years the study of soricids has shown many physiological, ecological, and ethological characteristics that, when considered individually, are as difficult to understand as isolated stones in a mosaic. Comparison of the two subfamilies shows divergences among the Soricidae that permit us to see a correlation of the data. The hypothesis that the different metabolic levels are the consequence of the evolution of the Soricinae in a temperate climate and of the Crocidurinae in a tropical climate (Vogel, 1976) seems to find confirmation because criteria such as longevity and reproduction rate behave in accordance with the dependence on climate postulated by F. Bourliere (personal communication) and illustrated by Hunkeler and Hunkeler (1970), Hildwein (1972), and Thiollay (1976).

Considered in this way, the different characteristics represent different biological strategies in Eisenberg's sense (this volume), where metabolism and thermoregulation obviously play a leading part. The fact that in cold climatic regions there exist only Soricinae, whereas, on the other hand, in tropical regions there are almost exclusively Crocidurinae, suggests the decisive survival value of both strategies.

Exceptions among the Soricinae, such as *Notiosorex*, which possesses torpor capacity and lives in desert regions, or *Chimarrogale*, which as a semiaquatic species can be found down to the tropics (Borneo and Sumatra), are striking examples of secondary adaptation made possible by lack of specialized competition.

Evolutionary interpretation

One final important question is that of whether the one strategy can be described as primitive and the other as evolved. In order to examine this question we base our judgment on criteria whose validity has been confirmed: tooth pigmentation, tooth and mandibular condyle morphology (Repenning, 1967), hair structure (Vogel and Köpchen, 1978), litter size, and degree of development at birth (Portmann, 1938; Vogel, 1972b).

In the Soricinae, conservative characters remain in red tooth pigmentation, tooth morphology, litter size, and degree of development at birth. On the other hand the separate faces of mandibular condyle and the complex hair structure are evolved peculiarities.

In the Crocidurinae, the mandibular condyle conformation and hair structure have remained conservative, whereas the lack of tooth pigmentation, tooth morphology, the decrease in litter size, and the degree of development at birth represent evolved peculiarities.

This coexistence of conservative (plesiomorphic) and evolved (apomorphic) characteristics can only be explained if derived from a less specialized, common ancestor that combined the primeval peculiarities and that, according to palaeontological data, was already characterized by red-pigmented teeth.

Alongside these morphological characteristics, any statement on the metabolic levels must of necessity be speculative. We could begin by accepting Dawson's hypothesis (1972) that increased body temperature is a late phylogenetic attainment. The higher metabolism of the Soricinae would thus be a secondary feature. In spite of this logical deduction, it seems to me that the opposite interpretation is more likely in the case of shrews.

In our speculations we now proceed to the role of dental pigmentation, which certainly requires careful investigation. According to Dötsch and Königswald (1978), the red tooth color is due to an iron compound deposited in the outer layer of enamel. According to Halse (1974), such inclusions in the front surfaces of the incisors of rodents increase the resistance of the enamel. The conclusion is obvious: In the soricids the pigment could protect against abrasion caused by excessive use and thus might be a sign of high energetic need. Because the common ancestors were already provided with red teeth, according to this assumption they should have had a high metabolic level, most probably corresponding to that of the Neomyini and Blarinini of the present. Both the extreme metabolic level of the Soricini among the Soricinae and the relatively low level of the Crocidurinae should then be regarded as the most specialized forms of two divergent biological strategies.

The interpretation that the level of the Crocidurinae is a secondary decrease may appear improbable, yet this possibility is supported by the analogous special case of the American desert shrew *Notiosorex*.

Conclusion

In many respects shrews can be regarded as models of primitive eutherians. Soricids, like marsupials, are poorly developed at birth, and the anatomy of adults corresponds to that of early mammals, having scarcely changed since the Oligocene period. Other insectivores are morphologically more specialized, for example, the burrowing Talpidae and Chrysochloridae and the quilled Tenrecidae and Erinaceidae. However, alongside these conservative characteristics, there are also physiological specializations. Thus, the term "primitive" must be used with caution.

The energy metabolism of shrews can give us insights into their evolution. Soricinae ("hot shrews") have high metabolic rates compared to Crocidurinae ("cold shrews"). This difference may be related to their different origins (holarctic region for the former, palaeotropical region for the latter) and thus expresses an ecophysiological adaptation to differing climatic conditions. The specific metabolic requirement obviously has far-reaching consequences, so that in all probability blood parameters, feeding, population density, and social behavior, as well as longevity, reproduction biology, and juvenile development, are affected. In this regard the two subfamilies show two divergent biological strategies that remain evident even in the common area of distribution and under the same climatic conditions as in western and central Europe.

Therefore the Soricidae are very interesting not just from a metabolic point of view; because of the existence of extreme and moderate forms, they provide outstanding material for the study of ecophysiological interrelationships.

This study was supported by grants 3.515.71, 3.821.72, 3.413-074 from the Swiss National Science Foundation.

References

Bartels, H., Bartels, R., Baumann, R.-M., Fons, R., Jürgens, K.-D., and Wright, P. (1978). La fonction respiratoire du sang et poids relatif de certains organes chez deux espèces de Musaraignes: *Crocidura russula* et *Suncus etruscus* (Mammifère. Soricidae). *C. R. Acad. Sci. (Paris) 286D*:1195–8.

Borowski, S., and Dehnel, A. (1953). [Angaben zur Biologie der Soricidae.] *Ann. Univ. Mariae Curie-Sklodowska Sect. C* 7:305–448. (In Polish; German and Russian summaries.)

Brambell, F. W. R. (1935). Reproduction in the common shrew (*Sorex araneus* L.). *Phil. Trans. R. Soc. London B 225*:1–62.

Buckner, C. H. (1964). Metabolism, food capacity and feeding behavior in four species of shrews. *Can. J. Zool. 42*:259–79.

Buckner, C. H. (1966). Populations and ecological relationships of shrews in tamarack bogs of southeastern Manitoba. *J. Mammal.47*:181–94.

Calder, W. A. (1969). Temperature relations and underwater endurance of the smallest homeothermic diver, the water shrew. *Comp. Biochem. Physiol. 30*:1075–82.

Croin Michielson, N. (1966). Intraspecific and interspecific competition in the shrews *Sorex araneus* L. and *S. minutus* L. *Arch. Néerl. Zool. 17*:73–174.

Crowcroft, W. P. (1957). *The Life of the Shrew.* London: Reinhardt, 166 pp.

Dawson, T. J. (1972). Primitive mammals and patterns in the evolution of thermoregulation. In *Essays on Temperature Regulation,* eds. J. Bligh and R. E. Moore, pp. 1–18. Amsterdam-London: North Holland.

Dehnel, A. (1949). Studies on the genus *Sorex* L. *Ann. Univ. Mariae Curie-Sklodowska Sect. C 4*:17–102.

Dötsch, C., and Königswald, W. v. (1978). Zur Rotfärbung von Soricidenzähnen. *Z. Säugetierkd. 43*:65–70.

Dryden, G. L. (1968). Growth and development of *Suncus murinus* in captivity on Guam. *J. Mammal. 49*:51–62.

Dryden, G. L. (1969). Reproduction in *Suncus murinus. J. Reprod. Fert. Suppl. 6*:377–96.

Dryden, G. L. and McAllister, H. Y. (1970). Sustained fertility after CdCl injection by a non-scrotal mammal, *Suncus murinus* (Insectivora, Soricidae). *Biol. Reprod. 3*:23–30.

Dryden, G. L., Baumann, T. R., Conaway, C. H., and Anderson, R. R. (1969). Thyroid hormone secretion rate and biological half-life ($t_{1/2}$) of L-thyroxine-[131] I in the musk shrew (*Suncus murinus*). *Gen. Comp. Endocrinol. 12*:536–40.

Dryden, G. L., Gebczynski, M., and Douglas, E. L. (1974). Oxygen consumption by nursling and adult musk shrew. *Acta Theriol. 19:* 453–61.

Fons, R., and Sicart, R. (1976). Contribution à la connaissance du métabolisme énergétique chez deux Crocidurinae: *Suncus etruscus* (Savi, 1822) et *Crocidura russula* (Hermann, 1870) (Insectivora, Soricidae). *Mammalia 40*:299–311.

Frey, H. and Vogel, P. (1978). Etude de la torpeur chez *Suncus etruscus* (Savi, 1822) (Soricidae, Insectivora) en captivité. *Rev. Suisse Zool. 81*:23–36.

Gebczynski, M. (1965). Seasonal and age changes in metabolism and activity of *Sorex araneus* Linnaeus 1758. *Acta Theriol. 10*:303–31.

Gebczynski, M. (1971a). The rate of metabolism of the lesser shrew. *Acta Theriol. 16*:329–39.

Gebczynski, M. (1971b). Oxygen consumption in starving shrews. *Acta Theriol.16*:288–92.

Gebczynski, M. (1977). Body temperature in five species of shrews. *Acta Theriol. 22*:521–30.

Genoud, M. (1978). Etude d'une population urbaine de musaraignes musettes (*Crocidura russula* Hermann, 1870). *Bull. Soc. Vaud. Sci. Nat. 74*:25–34.

Halse, A. (1974). The mineral phase of rodent incisor enamel with special reference to the iron-containing layer. *Norske Tannlaegeforen. Tid. 84*: 138–43.

Hasler, M. J., and Nalbandov, A. V. (1974). Body and peritesticular temperatures of musk shrews (*Suncus murinus*). *J. Reprod. Fert. 36*:397–9.

Hellwing, S. (1971). Maintenance and reproduction in the white toothed shrew, *Crocidura russula monacha* Thomas, in captivity. *Z. Säugetierkd.36*:103–113.

Hildwein, G. (1972). Métabolisme énergétique de quelques mammifères et oiseaux de la forêt équatoriale. II. Résultats expérimentaux et discussion. *Arch. Sci. Physiol. 26*:387–400.

Hunkeler, C., and Hunkeler, P. (1970). Besoins énergétiques de quelques Crocidures (Insectivores) de Côte d'Ivoire. *Terre Vie 24*:449–56.

Hutterer, R. (1977). Haltung und Lebensdauer von Spitzmäusen der Gattung *Sorex* (Mammalia, Insectivora). *Z. angewandte Zool. 64*:353–67.

Mezhzherin, B. A. (1964). Dehnel's phenomenon and its possible explanation. *Acta Theriol. 8*:95–114.

Morrison, P., Ryser, F. A., and Dawe, A. R. (1959). Studies on the physiology of the masked shrew *Sorex cinereus*. *Physiol. Zool. 32*:256–71.

Nagel, A. (1977). Torpor in the European white-toothed shrews. *Experientia 33*:1455–6.

Nosek, J., Kozuch, O., and Chmela, J. (1972). Contribution to the knowledge of home range in common shrew *Sorex araneus* L. *Oecologia 9*:59–63.

Pearson, O. P. (1944). Reproduction in the shrew (*Blarina brevicauda* Say). *Am. J. Anat. 75*: 33–93.

Pernetta, J. C. (1977). Population ecology of British shrews in Grassland. *Acta Theriol. 22*:279–96.

Poczopko, P. (1971). Metabolic levels in adult homeotherms. *Acta Theriol.16*:1–21.

Portmann, A. (1938). Die Ontogenese der Säugetiere als Evolutionsproblem. I. Die Ausbildung der Keimblase. II. Zahl der Jungen, Tragzeit und Ausbildungsgrad der Jungen bei der Geburt. *Biomorphosis 1*:49–66, 109–26.

Price, M. (1935). The reproductive cycle of the water shrew, *Neomys fodiens bicolor* Shaw. *Proc. Zool. Soc. London 123*:599–621.

Pucek, Z. (1970). Seasonal and age change in shrews as an adaptative process. *Symp. Zool. Soc. London 26*:189–207.

Repenning, C. A. (1967). Subfamilies and genera of the Soricidae. *Geol. Sur. Prof. Pap. 565*:1–74.

Thiollay, J.-M. (1976). Besoins alimentaires quantitatifs de quelques oiseaux tropicaux. *Terre Vie 30*:229–45.

Vogel, P. (1969). Beobachtungen zum intraspezifischen Verhalten der Hausspitzmaus (*Crocidura russula* Hermann, 1870). *Rev. Suisse Zool. 76*:1079–86.

Vogel, P. (1972a). Beitrag zur Fortpflanzungsbiologie der Gattungen *Sorex, Neomys* und *Crocidura* (Soricidae). *Verh. Naturforsch. Ges. Basel 82*:165–92.

Vogel, P. (1972b). Vergleichende Untersuchung zum Ontogenesemodus einheimischer Soriciden (*Crocidura russula, Sorex araneus* und *Neomys fodiens*). *Rev. Suisse Zool. 79*:1201–332.

Vogel, P. (1974). Kälteresistenz und reversible Hypothermie der Etruskerspitzmaus (*Suncus etruscus*, Soricidae, Insectivora). *Z. Säugetierkd. 39*:78–88.

Vogel, P. (1976). Energy consumption of European and African shrews. *Acta Theriol. 21*:195–206.

Vogel, P. and Köpchen, B. (1978). Besondere Haarstrukturen der Soricidae (Mammalia, Insectivora) und ihre taxonomische Deutung. *Zoomorphologie 89*:47–56.

Wolk, E. (1974). Variations in hematological parameters of shrews. *Acta Theriol. 19*:315–46.

18

The respiratory system of the smallest mammal

EWALD R. WEIBEL, HELGARD CLAASSEN,
PETER GEHR, SENADA SEHOVIC, and
PETER H. BURRI

The first mammals appear to have been small and insectivorous. Among the living mammals the smallest insectivores belong to the family of shrews (Soricidae), and it could be imagined that shrews might show some conservative features possessed by the first mammals. One of the functions most fundamental for survival is energy metabolism, which is closely linked with respiration and food ingestion. In the present study we shall concentrate on the respiratory system of shrews, examining the gas exchange apparatus in the lung where O_2 is being loaded onto erythrocytes, as well as the mitochondria of muscle cells where O_2 is being consumed in the process of oxidative phosphorylation. The question is that of how the size of these structures is related to O_2 demand, and whether – in comparison to other mammals – any conservative features can be detected.

Of particular interest is the Etruscan shrew (*Suncus etruscus*), the smallest living mammal, whose adult body mass averages no more than 2 g. This very small size evidently creates a number of problems, particularly with respect to energy demand. It is therefore no surprise that the metabolic requirements of this very small insectivore are very high on a per unit weight basis when compared to other mammals. An Etruscan shrew needs to ingest food at least every hour, and consumes about six times its own body weight in insects every day (Morrison et al., 1959). Accordingly, O_2 consumption is extraordinarily high, averaging 0.4 ml O_2 min^{-1} g^{-1} body mass (Vogel, 1976), a value that is four times higher than in the mouse, or six times higher than even maximal weight-specific O_2 consumption in man.

In searching for conservative features in the respiratory system of the Etruscan shrew, we need to remember that at least in higher mammals the lung and the mitochondrial mass of muscle cells are capable of quantitative adaptation as the energy demand is changed (Weibel, 1979). On the one hand, maximal O_2 consumption is proportional to the mass of mitochondria in skeletal muscle (Hoppeler et al., 1973; Howald and Poortmans, 1975); and on the other, the size of the pulmonary gas exchange apparatus is adapted to the flow of O_2 required to satisfy the O_2 needs of the organism (Weibel, 1973). Bearing in mind the very high mass-specific metabolic requirements of the Etruscan shrew, we would therefore expect both the relative size of the pulmonary gas exchange apparatus and the size of the mitochondrial complement of muscle cells to be particularly high in this small insectivore if it were as well adapted to functional requirements as higher mammals.

Mitochondria in muscle cells

In muscle cells O_2 is consumed in the mitochondria in the process of ATP generation by oxidative phosphorylation, specifically in the respiratory chain of the inner mitochondrial membrane. The capacity of muscle cells to consume O_2 should thus be limited by the number of respiratory chain units available, and this should be proportional to the size of the inner membrane area or to the mitochondrial volume contained in the muscle cell. The latter parameter will be studied in three different muscles of the Etruscan shrew and compared to corresponding estimates for the rat.

Skeletal muscle

Figure 1 shows a section of skeletal muscle from the hind leg of an Etruscan shrew. It shows that two types of muscle fibers are present which differ in terms of the arrangement and relative number of mitochondria: in one of them (A) relatively few small mitochondria occur in an intermyofibrillar position; in the other (B), the mitochondria are larger and more numerous, and there are, in addition to the intermyofibrillar mitochondria, large packets of mitochondria densely arranged beneath the sarcolemma. A comparison with skeletal muscle from other species suggests that the fiber designated B corresponds to the class of aerobic (or "red") muscle fibers which produce their ATP primarily by oxidative phosphorylation; in this fiber type mitochondria occupy about 25% of the volume of the muscle cell as compared to approximately 8% in aerobic fibers of the rat or man. In muscle fibers of type A, mitochondria occupy 9% of the muscle cell volume; this is considerably higher than the relative mitochondrial volume in anaerobic (or "white") fibers (~2%) in other mammals. Unfortunately nothing is known yet about the physiology of the various fiber types in shrew muscles.

On the average it is found that in locomotory muscles of *Suncus etruscus* mitochondria make up 22% of the muscle cell, which is four times more than in similar muscle cells of the rat (Table 1).

Heart and diaphragm

As a correlate to high metabolic needs, heart rate and breathing rate are extraordinarily high in *Suncus etruscus*, namely 1000 to 1200 min^{-1} and 300 min^{-1}, respectively (Morrison et al., 1959; Weibel, 1972). Sustaining these contraction rates requires a high flow of ATP and accordingly necessitates a large complement of mitochondria because heart and diaphragm operate predominantly on aerobic metabolism. Indeed, the shrew's myocardial cells (Figure 2) are shown to contain a larger volume of mitochondria than any other known vertebrate muscle cell; they make up 45% of the cell volume (Table 1), whereas myofibrils occupy only 40% of the cell volume. The

Figure 1. Electron micrograph of two types of muscle cells (A and B) and capillaries (C) from the hind leg of *Suncus etruscus*. Mitochondria occur dispersed between the myofibrils (f) as "intermyofibrillar" mitochondria (mif) in both fiber types, and as subsarcolemmal packets (mis) beneath the sarcolemma (S) in fibers of type B. Mitochondria are less frequent and smaller in fiber A.

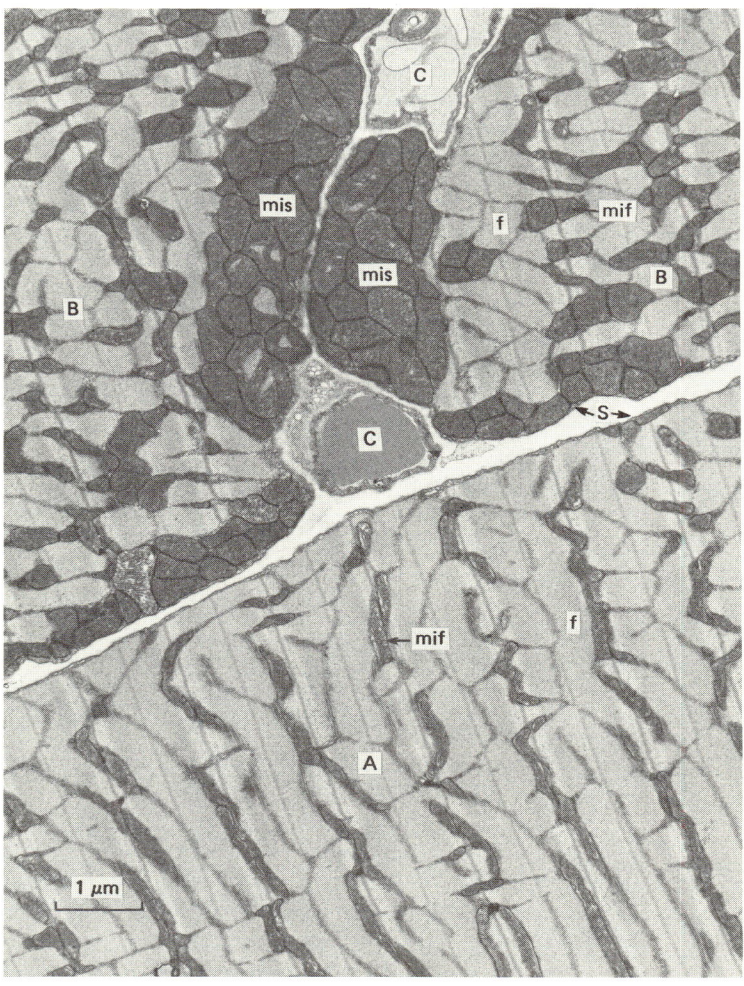

relative mitochondrial volume in diaphragmatic muscle fibers is 35% (Figure 3, Table 1); it is interesting that this corresponds to the mitochondrial volume density in myocardial cells of the rat (Table 1), whose heart rate is likewise about 300 min^{-1}. Comparing the mitochondrial complements in heart and diaphragm for shrew and rats, one finds that in both instances the shrew muscle contains 1.5 times more mitochondria, whereas its heart and breathing rates are 3.5 times greater.

Muscle capillaries

Preliminary studies of capillarization in various muscles have shown that the capillary density is directly proportional to the mitochondrial mass present in the muscle cell (Weibel, 1979). Figures 1–3 show that capillaries progressively increase in density in the shrew from skeletal muscle to diaphragm

Table 1. *Comparison of morphometric properties of respiratory system in* Suncus etruscus *and the white laboratory rat*

Property[a]	Suncus etruscus	Rat
Body weight (W) (g)	2.56	203.0
O_2 consumption (ml min^{-1})	0.95	9.05
$\dot{V}_{O_2}/W$ (ml min^{-1} g^{-1})	0.37	0.045
Heart		
Heart rate (min^{-1})	1050	300
V(mi)/V(cell) (%)	45	31
N(cap)/N(fibers)	3.2	1.8
V (mi)/S(cap) (μ^3/μ^2)	3.7	4.2
Diaphragm		
Breathing rate (min^{-1})	300	85
V(mi)/V(cell) (%)	34	22
N(cap)/N(fibers)	2.2	1.9
V(mi)/S(cap) (μ^3/μ^2)	4.0	6.1
Leg muscle		
V(mi)/V(cell) (%)	22.5	5.8
N(cap)/N(fibers)	1.7	1.9
V(mi)/S(cap) (μ^3/μ^2)	4.8	5.0
Lung		
Volume (ml)	0.102	6.11
Barrier thickness (μm)	0.26	0.33
S(alv)/W(cm^2 g^{-1})	63	19
V(cap)/W(cm^3 g^{-1})	0.0045	0.0028
$D_L(O_2)/W$(ml O_2 min^{-1} mm Hg^{-1} g^{-1})	0.0102	0.0054

[a] V = volume; S = surface; N = number; D_L = diffusing capacity; mi = mitochondria; cap = capillaries; alv = alveoli.

and myocardium. For the rat and shrew muscles reported in Table 1, there are about two capillaries per muscle fiber except for the shrew heart, where there are more. From the data shown in Table 1, one can further conclude that the capillaries offer about $1\mu m^2$ of surface to 5 μm^3 of muscle mitochondria. It is particularly interesting to note that this ratio appears to be a very general rule and applies to all muscles so far investigated in widely different

Figure 2. Electron micrograph of myocardial cells and capillaries (C) from *Suncus etruscus*. Mitochondria (mi) are dispersed as small groups between the myofibrils (f). Note sarcolemma (S), part of intercalated disc (ID) and thin capillary endothelium (EN).

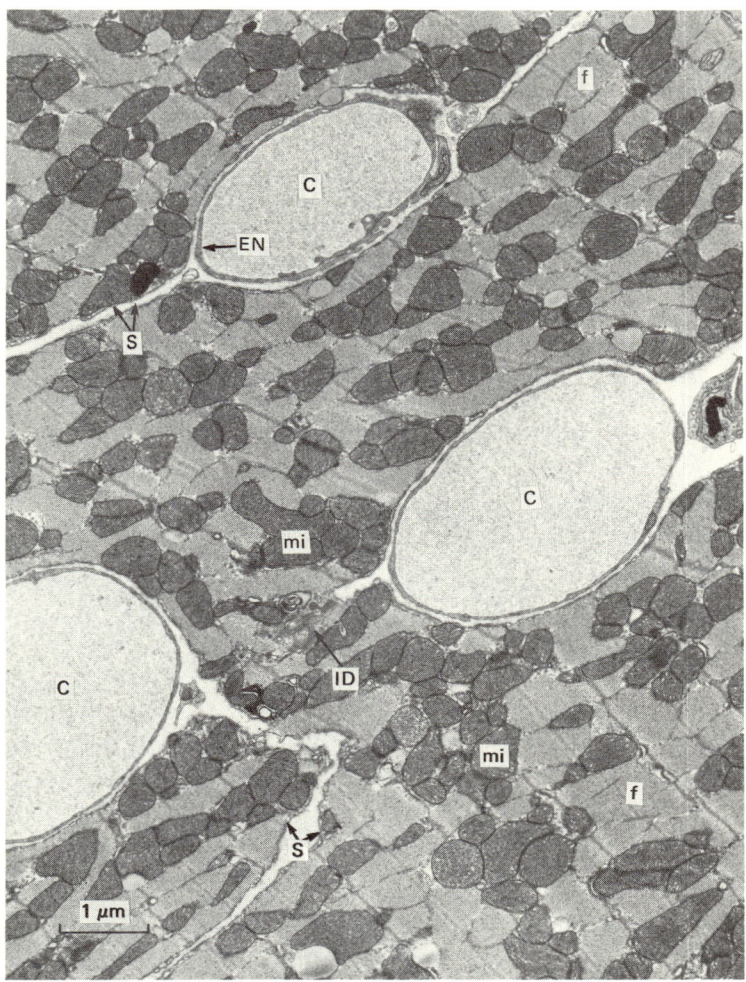

species with mitochondrial volume densities in the muscle cells ranging from less than 2% to 45% (Weibel, 1979).

The lung

In principle, the fine structure of the lung of *Suncus etruscus* does not differ from that of other mammalian lungs (Figure 4): the septa between air spaces contain a single meshwork of capillaries, and the air–blood barrier is made

Figure 3. Muscle cells from diaphragm of *Suncus etruscus* contain both inter-myofibrillar (mif) and subsarcolemmal (mis) mitochondria. Note nucleus (N) beneath sarcolemma (S), capillaries (C) with erythrocytes (EC), and lipid droplets (L) often closely associated with mitochondria.

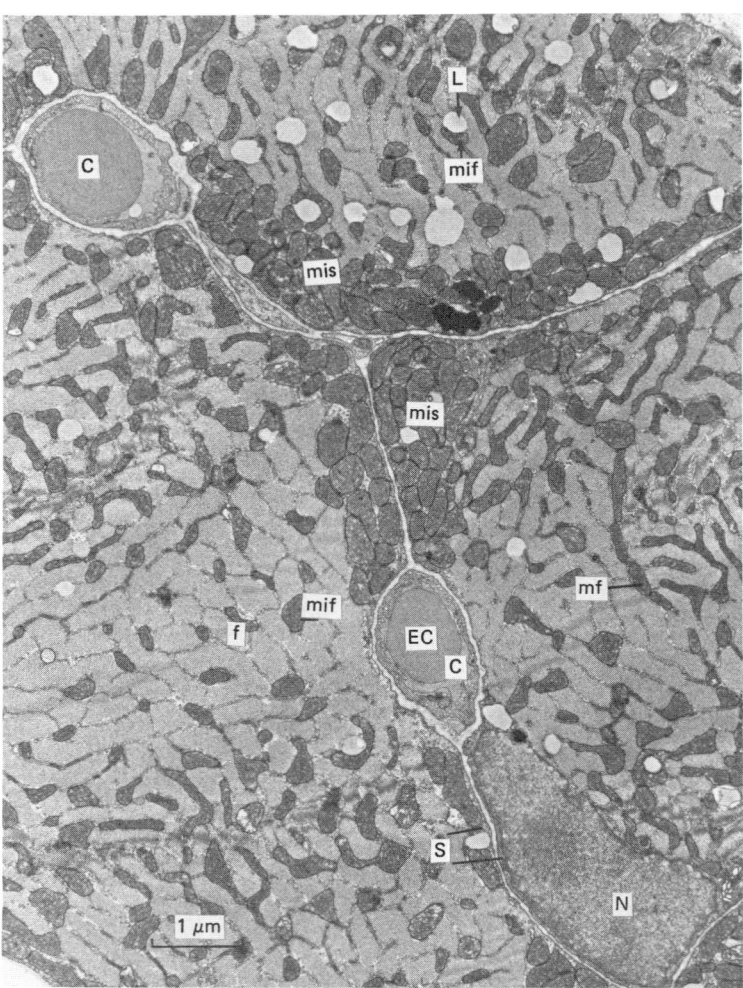

of the three layers – endothelium, interstitium, and epithelium; the epithelium is a mosaic of the squamous type I and cuboidal type II cells, and is overlaid by a thin surface-lining layer which can be demonstrated by vascular perfusion fixation.

There are, however, important quantitative differences, as evidenced by comparing low-power scanning electron micrographs of shrews and larger mammals (Figure 5). The "alveoli" are very small; indeed, the architecture of lung parenchyma resembles a densely interlaced system of blood capillar-

Figure 4. Alveolar capillaries (C) of Etruscan shrew lung showing erythrocyte (EC) and thin barrier made of type I epithelial cells (EP), endothelium (EN), and interstitium (IN), separating blood from alveolar air (A). Type II epithelial cells are not shown on this micrograph.

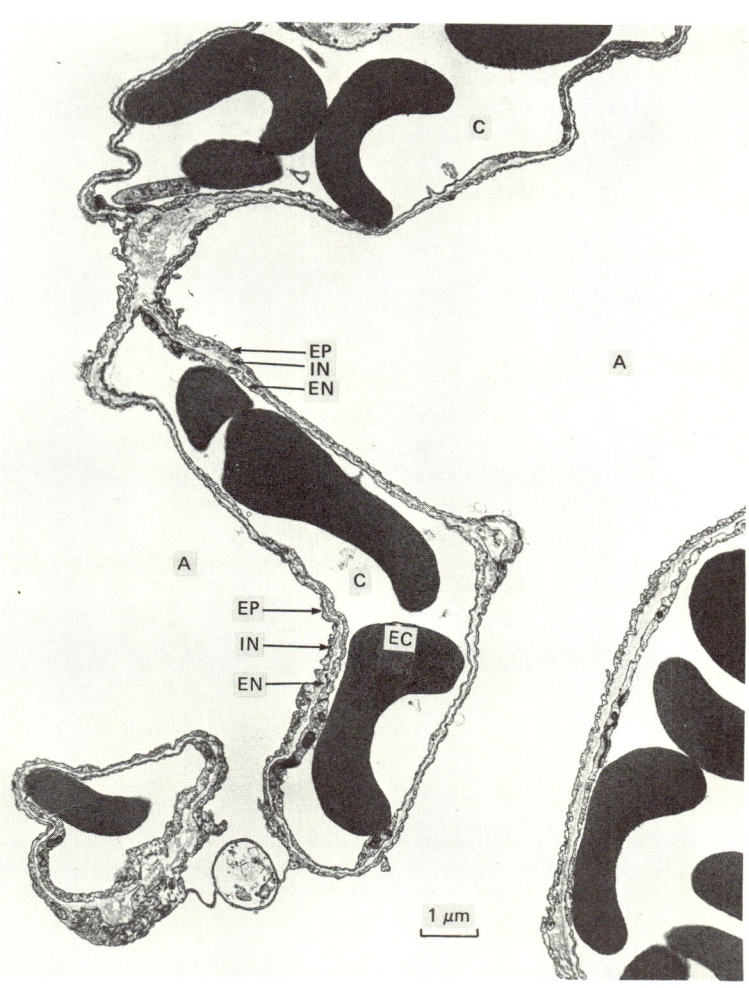

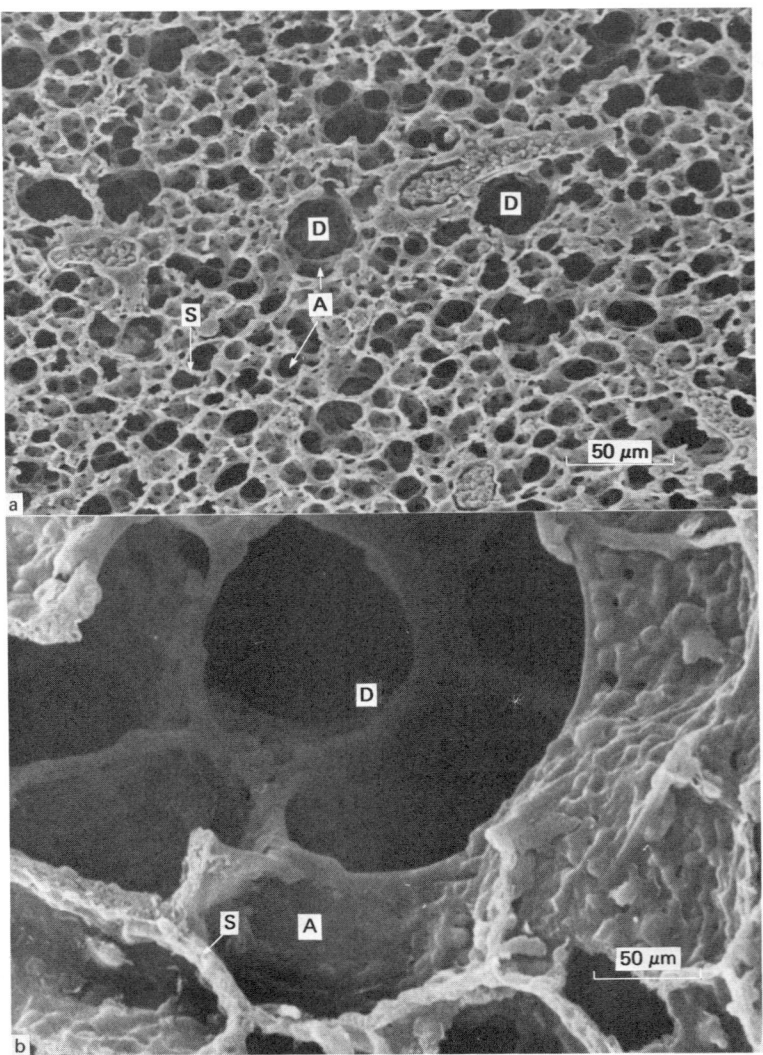

Figure 5. Comparison of lung parenchyma from Etruscan shrew (a) with that of human lung (b) at same magnification. Very tiny alveoli (A) surround small alveolar ducts (D) in the shrew lung; these structures are very much larger in the human lung. The alveolar septa (S) are also much finer in the shrew.

ies and a maze of narrow scalloped air channels. It is thus not surprising that the density of gas-exchanging surfaces in the unit lung volume is particularly high in the shrew: 1830 cm^2 cm^{-3} as compared to 550 cm^2 cm^{-3} in the dog, or 775 cm^2 cm^{-3} in the rat, for example. The capillaries are small, mainly because of the small size of erythrocytes, whose diameter is about 4 μm in shrews (Wolk, 1974), and the air–blood barrier is very thin, its harmonic mean being of the order of 0.26 μm or about 55% of that of the dog.

Because the major part of the left thoracic cavity is occupied by the very large heart, the size of the lung of the Etruscan shrew is relatively small, measuring no more than 0.1 ml at near-full inflation. This is slightly less than one would expect from an extrapolation of the allometric relation of lung volume and body weight. Because of the extraordinarily dense arrangement of alveolar capillaries, the alveolar gas exchange surface is very large; it exceeds the extrapolated allometric correlation by 83% (Figure 6).

Figure 6. Allometric plot of alveolar surface area, S_a for mammalian lungs with Etruscan shrew data stressed, as full dots. They lie clearly above the regression line. JWM = juvenile white mouse.

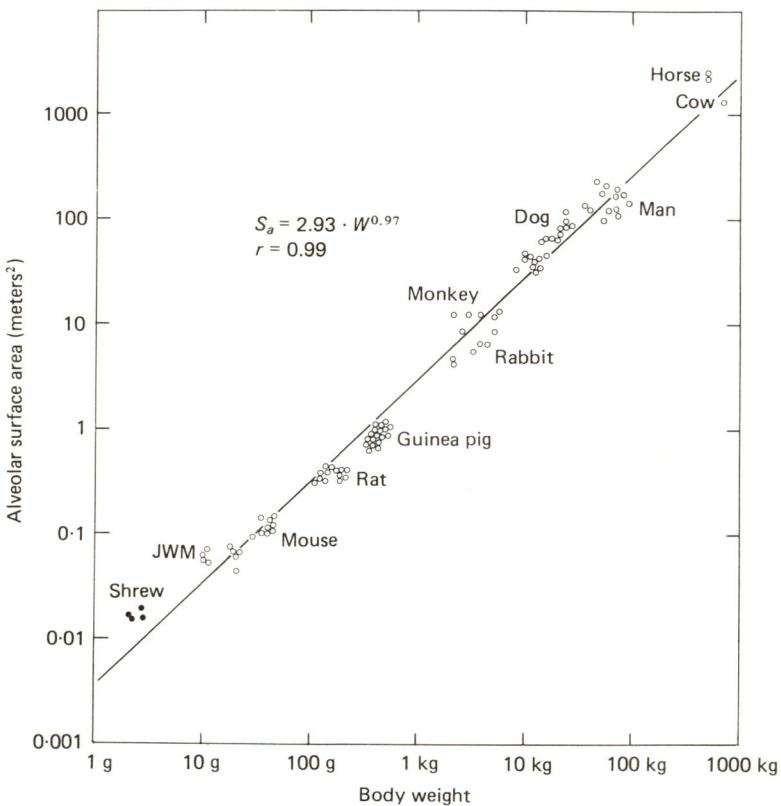

Accordingly, the pulmonary diffusing capacity for O_2 is found to be very high, namely about twice that of the rat if the differences in body weight are considered (Table 1). However, if D_L is calculated by considering the effect of the small erythrocytes in terms of facilitating O_2 binding, then D_L is 3.3 times greater in the shrew than in the rat.

Discussion and conclusions

Has this study revealed any "conservative" traits of the respiratory system of this smallest insectivore? We think not. In all respects the size of the metabolic machinery appears adequately adapted to the extraordinary metabolic requirements of these animals, making use of structural patterns identical to those in any other mammal; the differences are merely quantitative. Furthermore, some of the critical quantitative characteristics – such as the proportionality between mitochondrial mass and capillary surface in muscles –are identical to those found in other mammals. One might interpret the small erythrocyte size, as it seems to prevail in all shrews, as a conservative trait; however, in terms of gas exchange it seems to offer distinct advantages, and, furthermore, in lower vertebrates erythrocytes tend to be larger, rather than smaller than in mammals.

We can conclude that the strategies for matching the size of the respiratory system to the metabolic needs are the same in shrews as in more advanced mammals. Perhaps respiration is too basic a function and too important for survival for conservative traits to be perpetuated in the structural makeup of the respiratory system.

This work was supported by the Swiss National Science Foundation grant number 3.394.74.

References

Hoppeler, H., Lüthi, P., Claassen, H., Weibel, E. R., and Howald, H. (1973). The ultrastructure of the normal human skeletal muscle. A morphometric analysis on untrained men, women and well-trained orienteers. *Pflügers Arch.* 344:217–32.

Howald, H., and Poortmans, J. R. (1975). Metabolic adaptation to prolonged physical exercise. In: *Proc. Second Int. Symp. Biochem. Exercise,* Magglingen, 1973. Basel: Birkhäuser.

Morrison, P. R., Pierce, M., and Ryser, F. A. (1959). Food consumption and body weight in the masked and short-tail shrews. *Am. Midl. Nat.* 57:493–500.

Vogel, P. (1976). Energy consumption of European and African shrews. *Acta Theriol.* 21:195–206.

Weibel, E. R. (1972). Morphometric estimation of pulmonary diffusion capacity. V. Comparative morphometry of alveolar lungs. *Respir. Physiol.* 14:26–43.

Weibel, E. R. (1973). Morphological basis of alveolar-capillary gas exchange. *Physiol. Rev.* 53:419–95.

Weibel, E. R. (1979). Oxygen demand and the size of respiratory structures in mammals. In *Evolution of Respiratory Processes*, eds. C. Lenfant and S. C. Wood, pp. 289–346. Vol. 13 of *Lung Biology in Health and Disease*, ed. C. Lenfant. New York: M. Dekker.

Wolk, E. (1974). Variations in the hematological parameters of shrews. *Acta Theriol.* *19*:315–46.

19

Energetics of locomotion: primitive and advanced mammals

C. RICHARD TAYLOR

Many biologists believe that the locomotory system improved during the evolution of mammals (Hildebrand, 1974; Romer, 1970). Energetically more efficient modes of locomotion with lower mileage costs are usually assumed to be among the most important results of this improvement. In this paper I have attempted to evaluate the validity of this belief by reviewing data from living vertebrates. I have taken two approaches: (1) I have considered insectivores and lizards as "living models" of the first mammals and compared the energetic cost of their locomotion to that of more "advanced" and "specialized" cursorial mammals, and (2) I have compared the energetic cost of highly specialized modes of locomotion with the more primitive modes from which they evolved.

Insectivores probably provide the best living model of the locomotory systems of the earliest mammals. The fossil record indicates that the skeletal system of the first mammals was very similar to that of living small insectivores (Jenkins and Parrington, 1976), and it seems reasonable to assume that their muscular systems were also very similar. The papers of both Armstrong and Weibel (this volume) show that the muscular system appears to be essentially the same in insectivores and other groups of mammals. They conclude that it is a conservative system that has changed very little during the evolution of vertebrates. Therefore, energetic data on the locomotion of insectivores would appear to provide a good approximation of the energetics of the earliest mammals. Lizards possess a locomotory system similar to that possessed by the immediate ancestors of the earliest mammals. Thus a comparison of the energetic cost of their locomotion with that of the insectivores might help us to understand what changes, if any, occurred as mammals evolved from reptiles.

Another way to evaluate whether "a lower energetic cost" has been selected for during the evolution of mammalian locomotion is to compare the energetic cost of two highly specialized modes of locomotion – brachiation and hopping – with that of the more primitive modes from which they evolved – branch walking and quadrupedal running.

Enough data on the energetics of locomotion in various vertebrates have been collected during the last 10 years to make these comparisons possible. Measurements have been made on: lizards (Bakker, 1972; Moberly, 1968), monotremes (Edmeades and Baudinette, 1975; Crompton et al., 1978), marsupials (Baudinette et al., 1976; Dawson and Taylor, 1973; Crompton et al., 1978), insectivores (Crompton et al., 1978), and a wide variety of advanced mammals (see Taylor, 1977).

Components of cost of locomotion

There are two components to the cost of locomotion in most terrestrial vertebrates: an incremental cost and a zero speed cost (Taylor et al., 1970; Taylor, 1977). For reasons that are not well understood, the rate of oxygen consumption increases linearly with speed in most forms of terrestrial locomotion. This is extremely convenient because the slope of the relationship between metabolic power input (usually measured as the rate of oxygen consumption) and speed is a constant for each animal. This constant is the incremental cost of locomotion. It is usually expressed as a mass specific cost (T_{run}) having the units of energy used per g body mass per distance traveled. This cost is independent of the speed at which the distance is traveled. Taylor et al. (1970) found that incremental cost changes in a regular manner with body size and can be predicted from the body mass (m). Their empirical relationship is:

$$T_{run} = 8.46 \, m^{-0.4} \qquad (1)$$

where T_{run} has the units ml O_2 g^{-1} km^{-1}, and m is body mass in grams (1 ml O_2 g^{-1} km^{-1} ≈ 20.1 J kg^{-1} m^{-1}). This original equation was derived using data from an odd assortment of mammals including mice, rats, and dogs. The data base has expanded considerably during the last nine years, yet the relationship still appears to apply reasonably well both for quadrupeds, such as rodents with generalized locomotory systems, and for those with highly adapted cursorial systems, such as horses, gazelles, and cheetahs (Taylor, 1977). This relationship provides a good baseline against which incremental cost of locomotion in insectivores, lizards, brachiators, and hoppers can be compared.

One obtains the second component of energetic cost of locomotion, zero cost, by extrapolating the relationship between cost and speed to zero speed. This component of cost also varies in a regular manner with body size. It approximates a simple multiple of resting metabolism measured under standard conditions (≈ 1.5 times). Therefore, mass specific zero cost (W_{zero}) can be predicted by multiplying Kleiber's equation for mass specific resting metabolism by 1.5 or:

$$W_{zero} = 6m^{-0.25} \qquad (2)$$

where W_{zero} has the units ml O_2 g^{-1} h^{-1}, and m is body mass in grams.

A priori arguments

It has been argued that both components of cost of locomotion changed as mammals evolved from reptiles and during the evolution of mammals. The reptilian ancestors of mammals, like many modern-day lizards, possessed a sprawling wide-track gait. Presumably they "lay on their bellies" most of the time, lifting their bodies off the ground only when they wanted to move. Locomotion required a continuous "push-up" to keep the body off the ground. This has been assumed to be an energetically costly and slow way to move about. Romer (1970, p. 192) sums up this view: "Primitive land verte-brates . . . had a sprawling posture, with the limbs extending far out from the side of the body. . . . This pose . . . is wasteful of energy, for much muscular effort is used up merely in keeping the body off the ground." A major change from a sprawling to an upright gait occurred with the appear-ance of the first mammals, and there has been a tendency toward a more upright gait as mammals have evolved more specialized cursorial modes of locomotion. It has been argued that less muscular effort would have been required for support during locomotion as mammals became more upright. Also the incremental cost of locomotion should have decreased as the ani-mals changed from a more sprawling to a more upright gait. Swinging the limbs beneath the animal, rather than rowing them to the side, should enable the animal to move along a straighter line (and thereby to travel further for a given muscular effort).

Heath (1968) arrived at a different conclusion in his a priori arguments. He proposed that the resting metabolism (and therefore zero cost) increased as posture became upright. He argued that continuous muscular force gen-eration would have been required for an upright posture, instead of just when the animal got up off its belly to move (as was the case with the mammal's reptilian ancestors). These a priori arguments are testable using energetic measurements from living lizards with a sprawling gait and com-paring these data with data from mammals possessing various degrees of upright posture.

Quadrupedal locomotion along the ground

How large are the differences in incremental cost between lizards, mammals with a sprawling gait, and highly adapted cursorial mammals whose upright posture provides a pillarlike support? Bakker (1972) was able to train lizards of a variety of sizes to walk and run on a treadmill while he measured their steady state oxygen consumption as a function of speed. His data, as well as those from iguanas (Moberly, 1968), show that if differences exist between incremental cost of locomotion in lizards and mammals, they are too small to resolve out of the biological noise (Figure 1). Our models for primitive mammals, small living insectivores, also have incremental costs that are in-distinguishable from those predicted by Equation 1, as do monotremes and marsupials (Figure 1). These mammals all possess the more primitive sprawling type of mammal gait. The solid dots in Figure 1 include data from cheetahs, gazelles, and horses, some of the most highly specialized cursorial mammals. These values were not available when Equation 1 was calculated. They are certainly not lower than that predicted by Equation 1.

In summary, there do not appear to be distinguishable differences in the incremental cost of running between lizards, insectivores, monotremes, marsupials, and highly specialized cursorial mammals (Figure 1).

Are there any differences in zero speed cost? Bakker (1972) noted that the "zero speed cost" was less in a lizard than in a mammal. This reflects a basic difference between reptilian and mammalian energetics which has been discussed in another paper in this volume (see Taylor, in this volume). Insectivores, marsupials, and monotremes also have a lower "zero speed cost" (see Figure 2 of Taylor, "Evolution of mammalian homeothermy," in this volume). They used less energy than "typical" mammals over the entire range of speeds for which data are available. It has been noted, however, that body temperatures of these species varied from 28 °C (setifer) to 37.5

Figure 1. Incremental cost of quadrupedal locomotion in lizards (open squares), monotremes (open diamonds), marsupials (closed triangles), insectivores (closed squares), and other "typical" mammals (solid circles) plotted as a function of body size on logarithmic coordinates. The solid line is the relationship for quadrupeds reported by Taylor et al. (1970). The equation for this line is given as Equation 1 in the text. Sources of data are as follows: lizards – Bakker (1972), Moberly (1968); monotremes – Edmeades and Baudinette (1975), Crompton et al. (1978); marsupials – Baudinette et al. (1976), Crompton et al. (1978); insectivores – Crompton et al. (1978); "advanced" mammals – Taylor (1977).

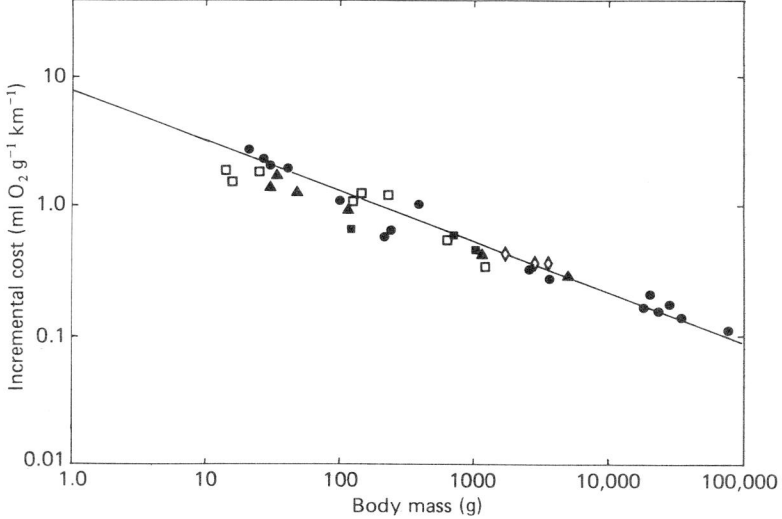

°C (hedgehog). If we normalize the predicted zero speed cost of each animal to a temperature of 38 °C (using a Q_{10} of 2.5), two distinct relationships emerge. Zero speed costs of reptiles and insectivores are about one-third to one-fifth that of other mammals. It is interesting that zero speed cost of marsupials and monotremes is similar to that of other mammals when one uses this normalization.

We are therefore able to distinguish two differences in the energetics of quadrupedal locomotion among reptiles, our living "primitive" mammals, and "advanced" mammals. First, animals with a body temperature lower than the mammalian norm of 37 to 39 °C have a lower zero cost. Second, if one uses a Q_{10} of 2.5 to normalize for differences in body temperature, the variability of zero cost is greatly reduced, and two separate functions relating zero cost and size emerge – one for reptiles and insectivores and another for the remaining mammals. Zero cost of reptiles and insectivores, like their resting metabolism, is one-third to one-fifth that of other mammals (for a discussion of this difference, see Taylor in this volume).

Arboreal locomotion

What about arboreal locomotion? A primitive mode of arboreal locomotion is "branch walking" as observed in lorises, and the most specialized mode is brachiation, as observed in spider monkeys and gibbons. It has frequently been argued that brachiation is an energetically cheaper way to move. There is even a good mechanistic reason for expecting a lower cost – a swinging pendulum (Fleagle, 1974). Parsons and I (Parsons and Taylor, 1977) decided to find out how much energy was saved by brachiating. We built a motor-driven "rope-mill" (similar to a treadmill) and trained lorises to walk along the rope and spider monkeys to brachiate beneath it.

We found lorises consumed oxygen at about the same rate while walking along the top of the rope as would be predicted for a quadrupedal mammal moving along the ground (Table 1). It was interesting that it didn't make

Table 1. *Energetic cost of two forms of arboreal locomotion ("primitive" branch walking; "advanced" brachiation)*

Type of locomotion	Animal	Zero velocity cost (ml O_2 g^{-1} h^{-1})[a]	Incremental cost (ml O_2 g^{-1} km^{-1})[a]
Branch walking	Loris	−24%	+14%
Brachiation	Spider monkey	+23%	+65%
Walking on treadmill	Spider monkey	+20%	+19%

[a]Calculated as percent deviations from values predicted for ordinary quadrupedal locomotion along the ground (equations 1 and 2 in text); values for spider monkey walking along a treadmill included for comparison.
Data for arboreal locomotion from Parsons and Taylor (1977).

any difference to the cost whether the loris walked along the top of the rope or walked suspended beneath it.

The highly specialized spider monkey, however, was a surprise. It used significantly more energy to brachiate at any speed then it did to walk at the same speed on a treadmill. The data from the loris indicate that this higher cost is not due to a difference between a suspended and an upright mode of locomotion per se. The higher cost is probably the result of the much greater vertical excursions of the center of mass of the brachiator compared to the walker. It would seem that even though large amounts of energy are saved during brachiation via the swinging pendulum mechanism (Fleagle, 1974), the mechanical work involved is so much greater than that involved in walking that more metabolic energy is required for movement. A mechanical analysis of work involved in brachiating, similar to that discussed by Heglund in this volume, is necessary to resolve these matters. In any case, we are forced to conclude that as a locomotory mechanism, "specialized" brachiation appears energetically more costly than "primitive" branch walking.

Saltatory locomotion

Perhaps the most specialized mode of terrestrial locomotion is bipedal hopping. Dawson and Taylor (1973) obtained a very surprising result when they measured the energetics of this type of locomotion. Once the kangaroo started to hop, its oxygen consumption decreased with increasing speed. Neither the incremental cost nor the zero speed cost has much meaning for these large kangaroos. The predicted energetic cost for a quadruped of the same size was less than the measured cost for a kangaroo for speeds up to 18 km h^{-1}. At faster speeds, however, hopping was a cheaper way to move.

Large amounts of energy could be saved by this hopping mechanism if kangaroos moved long distances at high speed. But why would a kangaroo (or for that matter any animal) move long distances at high speed? Dr. Mervyn Griffiths, an Australian zoologist familiar with the natural history of marsupials, suggested two good reasons (in the discussion that followed the oral presentation of this paper). During the dry season, rain on the Australian plains is very localized. Great distances may separate these localized "rain showers." The fresh green grass that sprouts immediately after the rain is limited, and it is available only to the first kangaroos that arrive. Cheap, high speed locomotion might have a strong selection value as kangaroos compete among themselves for this patchy resource. Also, bush fires are common. Cheap high speed locomotion might help the kangaroos to keep ahead of these fires, offering considerable survival value. Importantly, both Griffiths and T. J. Dawson reported that kangaroos do normally move long distances in nature at high speeds (25 to 30 km h^{-1}). Thus, there do seem to be some reasons for kangaroos acquiring a cheap high-speed mechanism of locomotion, and their hopping mechanism does appear to save them large amounts of energy.

This unusual energy-saving mechanism, however, appears to be limited to large hoppers. Dawson (1976) found no savings from hopping in the 30-g Australian hopping mouse. Recently Burke, Parker, and I (unpublished observations from our laboratory) have compared energetics of hopping in 1- to 3-kg marsupials (*Bettongia* sp., rat kangaroos) and rodents (*Pedetes* sp., springhares). We found their energetics to be similar to a quadruped rather than the big red kangaroo. Small hoppers don't seem to save any energy by hopping.

We must conclude that hopping is an energetically cheap way to move at high speeds for large hoppers (>10 kg), but not for smaller ones. Also, there does not seem to be much difference in cost of hopping between a rodent and a marsupial of about the same size (1 to 3 kg).

Summary and conclusions

My initial question was whether mammals as they evolved developed energetically cheaper modes of locomotion. The answer appears to be no. If anything, there may be a trend for the cost of locomotion to increase during mammalian evolution. Insectivores (our living model for the earliest mammals) and lizards (our model for the reptilian ancestors of mammals) both seem to move more cheaply than the advanced cursorial mammals with their highly adapted locomotory systems. "Advanced" brachiation appears to require more energy to move a given distance than "primitive" branch walking. Only in large marsupial hoppers have we found an example where cost of locomotion has decreased. From this analysis we can learn not to place undue emphasis on the energetic cost of locomotion when we try to interpret the relative advantages of the different modes of locomotion that mammals have evolved.

This paper was supported by NIH grants 2 RO1 AM 18140 and 2 RO1 AM 18123 and NSF grant PCM75-22684.

References

Bakker, R. T. (1972). Locomotor energetics of lizards and mammals compared. *Physiologist 15:*278.

Baudinette, R. V., Nagle, K. A., and Scott, R. A. D. (1976). Locomotory energetics in a marsupial (*Antechinomys spenceri*) and a rodent (*Notomys alexis*). *Experientia 32:*583.

Crompton, A. W., Taylor, C. R., and Jagger, J. A. (1978). Evolution of homeothermy in mammals. *Nature 272:*333–6.

Dawson, T. J. (1976). Energetic cost of locomotion in Australian hopping mice. *Nature 259:*305–7.

Dawson, T. J., and Taylor, C. R. (1973). Energetic cost of locomotion in kangaroos. *Nature 246:*313–4.

Edmeades, R., and Baudinette, R. V. (1975). Energetics of locomotion in a monotreme, the echidna *Tachyglossus aculeatus*. *Experientia 31:*935.

Fleagle, J. (1974). Dynamics of a brachiating siamang. *Nature 248*:259–60.

Heath, J. E. (1968). The origins of thermoregulation. In *Evolution and Environment* (Yale Peabody Museum Centennial Symposium), ed. E. Drake, pp. 259–78. New Haven: Yale University Press.

Hildebrand, M. (1974). *Analysis of Vertebrate Structure.* Philadelphia: W. B. Saunders.

Jenkins, F. A., Jr. and Parrington, F. R. (1976). Post cranial skeletons of the Triassic mammals Eozostrodon, Megazostrodon and Erythrotherium. *Phil. Trans. R. Soc. London B273*:387–431.

Moberly, W. R. (1968). The metabolic responses of the common iguana, *Iguana iguana*, to walking and diving. *Comp. Biochem. Physiol. 27*:21–32.

Parsons, P. E., and Taylor, C. R. (1977). Energetics of brachiation versus walking: A comparison of a suspended and an inverted pendulum mechanism. *Physiol. Zool. 50*:182–8.

Romer, A. S. (1970). *The Vertebrate Body*, 4th ed. Philadelphia: W. B. Saunders.

Taylor, C. R. (1977). The energetics of terrestrial locomotion and body size in vertebrates. In *Scale Effects in Animal Locomotion*, ed. T. J. Pedley, pp. 127–41. London: Academic Press.

Taylor, C. R., Schmidt-Nielsen, K., and Raab, J. L. (1970). Scaling of energetic cost of running to body size in mammals. *Am J. Physiol. 219*:1104–7.

20

Physiological responses to locomotion in marsupials

R. V. BAUDINETTE

Marsupials have a separate evolutionary history from eutherian mammals of about 130 million years (Lillegraven, 1969). It is misleading however, to regard all marsupials as "primitive" and to treat them as a homogeneous group. Some (e.g., *Didelphis*, the Virginia opossum) have a long fossil record, and appear to have retained many structural features that represent an early stage of evolution. However, other members of the Marsupialia (e.g., the kangaroos, family Macropodidae) represent comparatively recent radiations and show marked speciation only in the last 10 million years.

Despite such heterogeneity, marsupials exhibit three physiological parameters that are considered to be primitive or conservative characters. These are levels of resting or maintenance metabolism, body temperature, and heart rate. In this paper I will consider the correlation between low metabolic levels at rest and the physiological responses to exercise in marsupials.

Power input during locomotion

The energy demands of locomotion in eutherian mammals were measured by Taylor et al. (1970) by examining rates of oxygen consumption ($\dot{V}_{O2}$) as animals ran on a treadmill. The relation between $\dot{V}_{O2}$ and speed was found to be linear, but the gradients (slopes) of the lines increased with decreasing body mass. Because each liter of oxygen used in oxidative metabolism can be equated with about 20.1 kJ of energy, this implies that the metabolic power required for running increases linearly with speed, but the rate of increase is greater in smaller than in larger animals. From such experiments a simple empirical equation has been derived to predict the relation between $\dot{V}_{O2}$ and speed from the body mass (m) of the animal. It is of the form $E_{\text{run}} = am^b$ where E_{run} represents the "incremental cost of transport" (Taylor, 1977) and a and b are empirically derived constants. E_{run} has units of J kg^{-1} meter^{-1} and represents the amount of energy to move a unit mass over a unit distance.

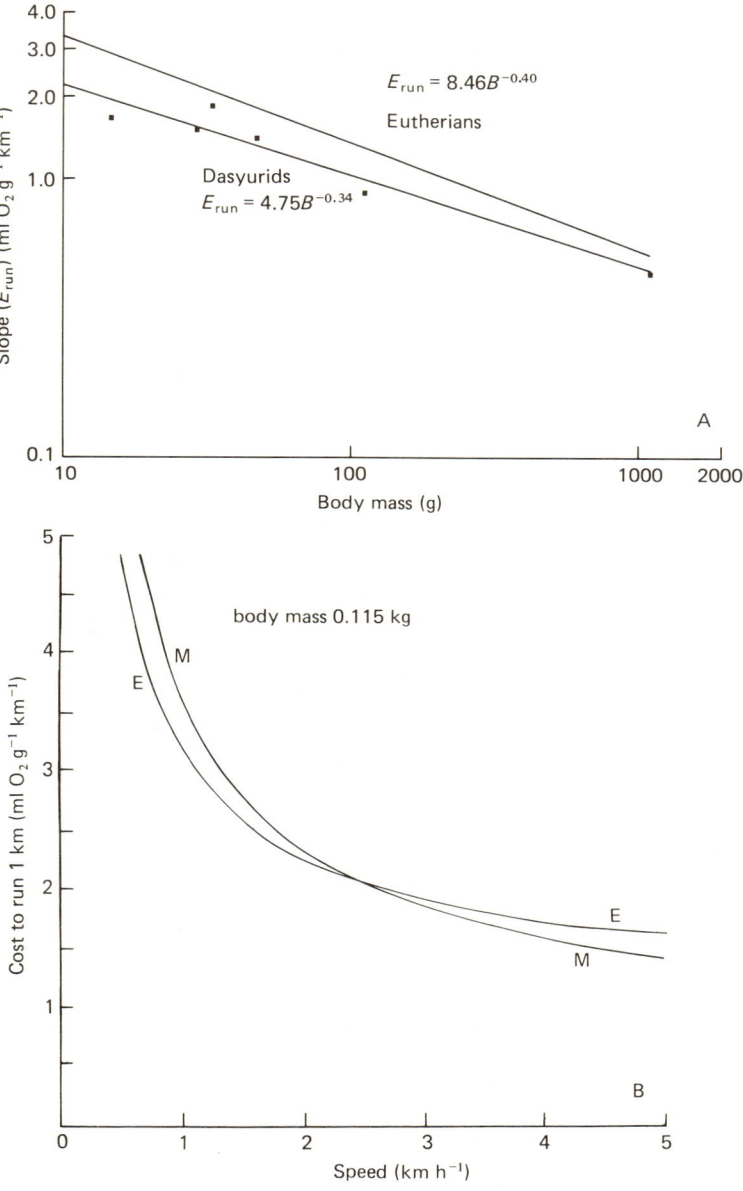

Figure 1. Locomotory energetics in dasyurid marsupials and eutherians. (A) The slopes of the relationships between oxygen consumption and speed (E_{run}) plotted as a function of body mass. The eutherian data are from Taylor et al. (1970); the marsupial data from Baudinette et al. (1976a). (B) The energy cost to run 1 km as a function of speed. M denotes data for *Dasyuroides byrnei* (Baudinette et al., 1976a); E denotes a hypothetical eutherian of equal mass, calculated from the equation of Taylor et al. (1970).

*Given the different rates of maintenance metabolism
in eutherians and marsupials, are there differences
in metabolism during exercise?*

To answer this, individuals from five species of the marsupial family Dasyuridae were used to measure rates of oxygen consumption ($\dot{V}_{O2}$) while running on a treadmill (Baudinette, et al., 1976a). The dasyurids are carnivorous or insectivorous quadrupedal marsupials that are native to Australia. In animals ranging from 0.015 to 1.12 kg in body weight, $\dot{V}_{O2}$ increased linearly with running speed, and as in eutherians, the rate of increase was generally greater in the smaller animals. Table 1 compares the allometric relation between E_{run} and speed in dasyurids with those for other vertebrates. The exponent in these equations, a reflection of the effect of size within the group, is statistically similar in all quadrupeds, but is significantly different from that found in birds. The constant term (a) does vary among the groups, and a comparison between the equations for eutherians and dasyurid marsupials (Figure 1A) suggests that the latter have lower power increments with speed. Marsupials that move bipedally cannot be compared in this way, as their rate of oxygen consumption remains constant as speed increases (Dawson and Taylor, 1973). (Hopping locomotion is considered in detail in this volume by Alexander.)

A second comparison between marsupials and eutherians involves the energy required to run a given distance. In Figure 1B the energy required to run 1 km for a dasyurid, *Dasyuroides byrnei* (Baudinette et al., 1976a), is compared with that for eutherians of equal body mass. At low speeds, it is energetically more costly for the marsupial, but above speeds of 2.5 km h^{-1} this position reverses. A possible explanation could be a greater

Table 1. *The incremental cost of transport (E_{run}) as a function of body mass (m) in vertebrates*

Phyletic group	No. of species	Est. equation $E_{run} = am^b$	Source
Reptiles	5	$5.52m^{-0.34}$	Bakker (1972) [a]
Reptiles (*Lacerta*)	4	$5.7m^{-0.48}$	Cragg (1975) [b]
Birds	7	$2.46m^{-0.20}$	Fedak et al. (1974)
Dasyurid marsupials	5	$4.75m^{-0.34}$	Baudinette et al. (1976a)
Eutherians	6	$8.46m^{-0.40}$	Taylor et al. (1970)

[a] The equation for reptiles has been recalculated from Bakker (1972).
[b] The data of Cragg (1975) is from four members of the lizard genus *Lacerta* and is reported in Hughes (1977).

rate of energy usage in contracting marsupial muscle at low speeds, or, alternatively, it could represent a behavioral difference in an experimental situation. At present we have no information to support either of these hypotheses.

Are maximum metabolic levels similar in marsupials and eutherians?

In Table 2, equations describing maximal values of oxygen consumption during locomotion ($\dot{V}_{O2}$ max) are compared to levels at rest ($\dot{V}_{O2}$ std). A dimensionless quotient may be produced by dividing such equations, and the resulting residual exponent describes the manner in which this quotient varies with body mass. This procedure is termed "allometric cancellation," and its use is discussed in detail by Stahl (1962). The residual exponents are small, indicating that the scope to increase $\dot{V}_{O2}$ above standard levels is probably independent of body mass; that is, both maximum and standard levels of $\dot{V}_{O2}$ scale to approximately the same power function of body weight. The quotient, a measure of this capacity for increase, is similar in eutherians and marsupials, and after normalization of the marsupial data (using a Q_{10} of 2.5) from a resting body temperature of 35.5 to 38 °C, the difference is insignificant in animals of a similar size range. Therefore marsupials and eutherians are similar in their ability to increase power input above resting or maintenance levels.

In summary, the differences in locomotory energetics between eutherians and quadrupedal marsupials appear small. The relation between $\dot{V}_{O2}$ and speed shows a qualitatively similar variation with body mass. The maximum levels of $\dot{V}_{O2}$ and the ratios of $\dot{V}_{O2}$ max/$\dot{V}_{O2}$ std are the same in both groups. However, the power increments with speed in dasyurids are slightly less than in eutherians, and locomotion is energetically more costly at low running speeds.

Body temperature during locomotion

Is there also a difference in body temperature between eutherians and marsupials during exercise?

During locomotion in the laboratory, rectal temperatures up to 41 °C have been observed in the red Kangaroo (Dawson et al., 1974). Similar unpublished results from our laboratory show temperatures between 39 and 40 °C for the brush-tailed possum and the Tasmanian devil. There is to date no evidence that the 2 to 3 °C difference in body temperature between eutherians and marsupials at rest is reflected in a similar difference during exercise. In some placentals (e.g., Thompson's gazelle) rectal temperatures up to 43 °C have been reported during exercise (Taylor and Lyman, 1972). However, in these animals brain temperatures are maintained at lower levels by counter-current heat exchange between the carotid rete and cavernous sinus. This arrangement has not been described from any marsupial, but the

Table 2. *Dimensionless ratios describing the ability of mammals to increase rates of oxygen consumption and heart rates over standard levels*

Ratio [a]	Group	Numerical value	Residual exponent	Source of maximum value	Source of standard levels
Max $\dot{V}_{O_2}$/Std $\dot{V}_{O_2}$	Eutherian mammals (0.029–0.906 kg)	7.6	0.02	Pasquis et al. (1970)	Kleiber (1947)
Max $\dot{V}_{O_2}$/Std $\dot{V}_{O_2}$	Eutherian mammals (6.3–677 kg)	11.5	0.04	Pasquis et al. (1970)	Kleiber (1947)
Max $\dot{V}_{O_2}$/Std $\dot{V}_{O_2}$	Eutherian mammals (0.007–260 kg)	8.3	0.05	Taylor et al. (1978)	Kleiber (1947)
Max $\dot{V}_{O_2}$/Std $\dot{V}_{O_2}$	Dasyurid marsupials (0.15–1.12 kg)	8.7	0.08	Baudinette et al. (1976a)	MacMillen and Nelson (1969)
Max f_H/Std f_H	Eutherian mammals (0.032–67.1 kg)	1.6	0.06	Baudinette (1978)	Stahl (1967)
Max f_H/Std f_H	Marsupials (0.112–26.7 kg)	3.5	0.08	Baudinette (1978)	Kinnear and Brown (1967)
Max $\dot{V}_{O_2}$/Std $\dot{V}_{O_2}$ [b]	Dasyurid marsupials (0.15–1.12 kg)	6.9	0.08	Baudinette et al. (1976a)	MacMillen and Nelson (1969)
Max f_H/Std f_H [b]	Marsupials	2.8	0.08	Baudinette (1978)	Kinnear and Brown (1967)

[a] Max $\dot{V}_{O_2}$ and Std $\dot{V}_{O_2}$ denote maximum and standard levels of oxygen consumption for the specified group of animals. Max f_H and Std f_H likewise denote heart rates.

[b] Using corrected (38 °C) values for Std $\dot{V}_{O_2}$ and Std f_H.

vascular architecture in marsupials shows a concurrent branching of arteries and veins that contrasts with the typical anastomosing capillary network of eutherians (see Johnson, 1977). The functional significance of this specialization, especially its possible role in temperature stability of the brain during locomotion, would seem worthy of investigation.

If peak body temperatures are similar during exercise in both groups, the lower resting body temperatures should permit heat storage to play a more important role in the heat balance of marsupials. Figure 2 shows a comparison of the heat produced that is stored during runs at 22 °C for several animals. Clearly, even small marsupials rely on this strategy of heat balance rather more than eutherians, suggesting that these animals may be adapted for running short sprints rather than longer distances.

Nonevaporative heat loss during locomotion

The fraction of heat lost from an animal by nonevaporative heat dissipation is dependent on the relative surface area involved and the temperature gradient between this surface and the environment. Other factors being

Figure 2. The percentage of the heat produced that is stored during 15–30 min. runs at 22 °C as a function of running velocity. Eutherian data are from Taylor (1974); sources for the marsupial data are listed in Table 3. (a) *Sminthopsis crassicaudata*, 0.015 kg; (b) *Dasyuroides byrnei*, 0.115 kg; (c) *Dasyurus viverrinus*, 1.12 kg; (d) *Trichosurus vulpecula*, 1.85 kg; (e) *Setonix brachyurus*, 2.8–3.1 kg; (f) *Megaleia rufa*, 18.26 kg.

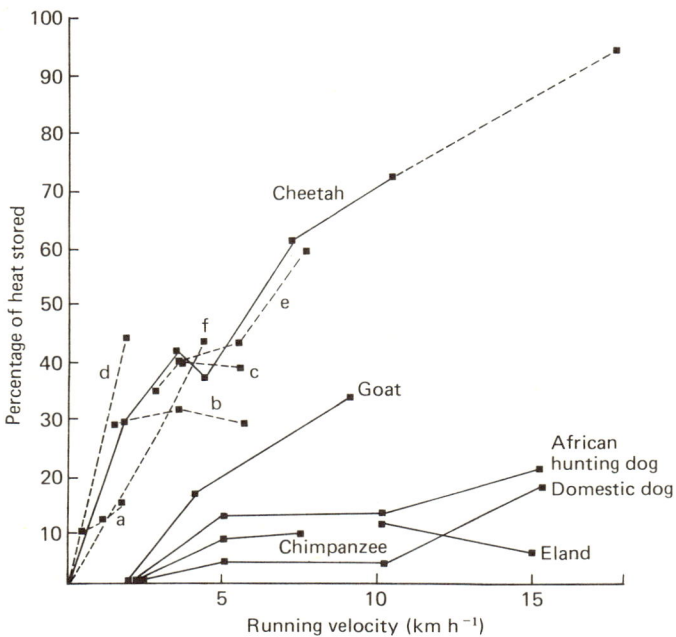

equal, we would expect heat loss via this route to be greatest in smaller animals with high surface temperatures. This expectation is realized in small eutherians where Taylor (1974) reported that kangaroo rats and white rats lose about 90% of the heat produced during exercise by nonevaporative means during 15- to 30-min runs at 22 °C. The contribution of this avenue of heat loss to the overall balance became proportionally less in larger animals and with increases in running speed.

The available marsupial data (Table 3) show that the smaller marsupials rely less on heat loss by nonevaporative means than eutherians. For example, the domestic dog loses about 70% of its heat production by non-evaporative means while running at low speeds (Taylor, 1974), but the marsupials *Dasyurus viverrinus* and *Trichosurus vulpecula* lose a lower fraction even though their body mass is obviously much less. The conclusion from these comparisons is that marsupials rely less on nonevaporative cooling during locomotion.

Evaporative heat loss and sweating

In marsupials at rest under high external heat levels, panting appears to be the dominant form of evaporative cooling (see, e.g., Dawson et al., 1969). Licking the fur is also known to occur, and Needham et al. (1974) have described a vascular arrangement in the forelimbs of the red kangaroo that would appear to enhance cooling by this means. Sweating while at rest has only been reported in the rat kangaroo *Potorous tridactylus* (Hudson and Dawson, 1975), where active glands were observed in rings around the sparsely haired tail, and sweating rates of 600 to 650 g H_2O $(meter^2 h)^{-1}$ were observed. Sweating at rest does not occur in the Tasmanian devil (Hulbert and Rose, 1972), the bandicoot (Hulbert and Dawson, 1974), or the brush-tailed possum (Dawson, 1969).

Sweating has been observed only during exercise in the red kangaroo and resulted in an increase in cutaneous evaporation to 12 times the levels measured at rest (Dawson et al., 1974). The sweating rate in this species is 170 g H_2O $(meter^2 h)^{-1}$ a little less than half the reported maximum rates in cattle and horses (Allen and Bligh, 1969). In the quokka, Baudinette (1977) reported sweat gland activity during exercise on the proximal-ventral aspect of the tail, on the lower legs, and on the upper surface of the hind feet. Recent unpublished observations from our laboratory showed rates of cutaneous evaporation in the brush-tailed possum that indicate sweating in response to exercise, but in the Tasmanian devil sweating cannot be demonstrated.

At present, few generalizations can be made on the role of sweating in marsupials. In the resting potoroo, one of the smallest members of the family Macropodidae, sweating occurs from the tail at rates about double those seen in large eutherians. In the red kangaroo, the quokka, and the brush-tailed possum, it is seen during exercise-induced heat loads, but does

Table 3. *Heat dissipation during locomotion in marsupials*

Species	Body wt. (kg)	Speed (km h^{-1})	Percentage of total heat production			Source
			Stored	Lost by nonevaporative routes	Lost by evaporation	
Family Dasyuridae						
Sminthopsis crassicaudata	0.015	0.6	10	28	62	Baudinette et al. (1976a)
		1.2	12	26	62	
		1.8	16	24	60	
Antechinomys spenceri	0.030	2.0	28	–[a]	–	Baudinette et al. (1976b)
Dasyuroides byrnei	0.115	1.5	29	28	43	Baudinette et al. (1976a)
		3.4	31	32	37	
		5.5	29	38	33	
Dasyurus viverrinus	1.120	2.8	35	6	59	Baudinette et al. (1976a)
		3.4	41	3	56	
		5.5	38	6	56	
Family Phalangeridae						
Trichosurus vulpecula	1.85	rest	0	73	27(21 respiratory 6 cutaneous)	Dawson (1969)
		2.0	44	21	35(8 respiratory 27 cutaneous)	Bell, Baudinette, and Nicol (unpublished)
Family Macropodidae						
Setonix brachyurus	2.8–3.1	2.8	39	–	–	Baudinette (1977)
		5.6	43	–	–	
		7.6	60	–	–	
Megaleia rufa	18.26	rest	0	67	33 (17 respiratory 16 cutaneous)	Dawson et al. (1974)
		4.0	43	14	43(17 respiratory 26 cutaneous)	

[a]Dash indicates no data available.

not occur in response to external heat loads. Tasmanian devils show a third pattern in that sweating cannot be elicited during locomotion or in response to high external heat loads.

Cardiovascular performance during locomotion

Heart rates during exercise

There is a marked difference in the resting heart rates (f_H std) of marsupials and eutherians. An equation describing the minimum heart rates of 14 species of marsupials was given as $f_H$std $= 106m^{-0.27}$, where m is the body mass in kg (Kinnear and Brown, 1967). A comparison with the eutherian equation, $f_H$std $= 241m^{-0.25}$ (Stahl, 1967), suggests that marsupials have average minimum heart rates of a little less than half those predicted for eutherians of similar body mass. Are these reduced rates at rest also apparent during exercise?

Baudinette (1978) measured heart rates (f_H) during exercise in individuals from six marsupial species and from four eutherian species; the body weights ranged from 0.032 to 67.1 kg.

Values of f_H showed a linear increase with running speed in both groups. The mean values of maximum steady state heart rates ($f_H$max) during locomotion are represented by the equation $f_H$max $= 375m^{-0.19}$, a relation that statistically represents both marsupials and eutherians (Figure 3A). A comparison of this equation with those for the resting heart rates of eutherians (Stahl, 1967) and marsupials (Kinnear and Brown, 1967) leads to two conclusions (Table 2). First, the ability of both eutherians and marsupials to increase heart rates above standard levels is independent of body mass. Second the ratio of $f_H$std was found to be greater in marsupials, a reflection of lower values at rest. This difference remains even if heart rates are normalized to a common body temperature of 38 °C.

Other cardiovascular parameters

The oxygen affinity of hemoglobin in seven species of marsupials has been reported by Bland and Holland (1977). The relationship between this parameter and body mass is qualitatively similar to that reported for eutherians by Schmidt-Nielsen and Larimer (1958) and by Bartels (1964). Similarly, a study of cardiovascular properties and hematological parameters in members of the family Macropodidae failed to demonstrate any real differences between this group and other mammals (Maxwell et al., 1964).

There has only been one study concerned with circulatory adjustment during exercise in marsupials. Baudinette et al. (1978) measured oxygen consumption, heart rate, and arterio-venous oxygen differences $(a-v)C_{O2}$, during locomotion in the brush-tailed possum. The increased delivery of oxygen during locomotion was met by increases in cardiac output ($\dot{V}_b$) and $(a-v)C_{O2}$. At low running speeds the latter parameter seems to increase at a proportionately greater rate, but at higher speeds increased oxygen de-

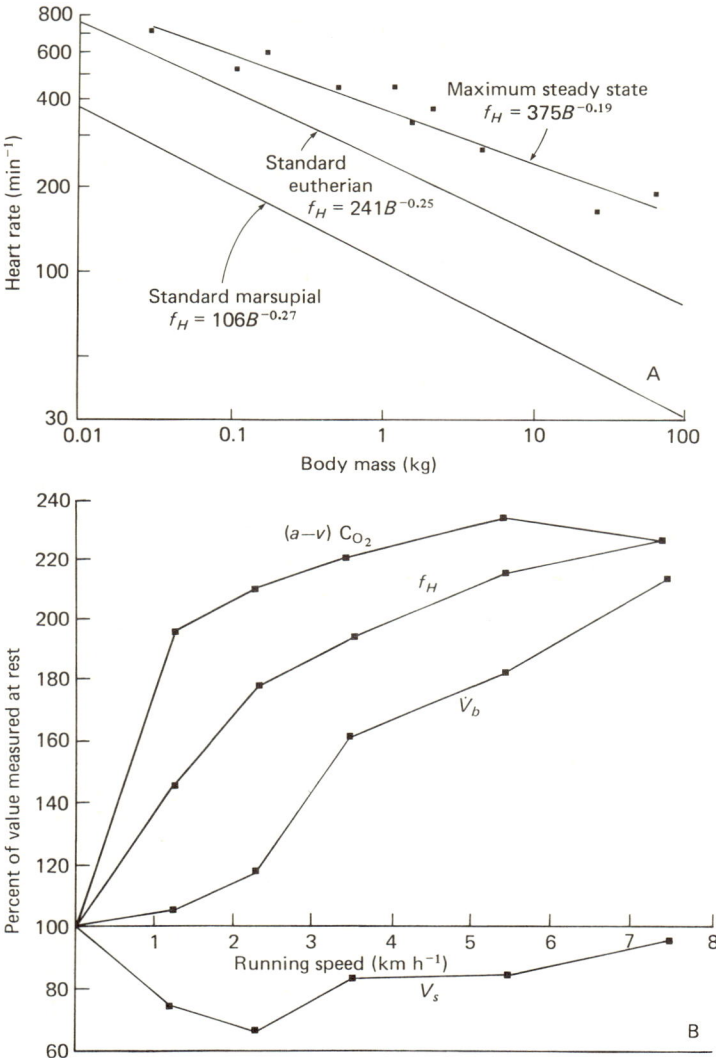

Figure 3. Cardiovascular performance during locomotion in marsupials. (A) The relation between maximum steady-state heart rates and body mass. Predicted standard values are from Stahl (1967) for eutherians and Kinnear and Brown (1967) for marsupials. (B) The relative changes in stroke volume (V_s), cardiac output (V_b), heart rate (f_H), and arterial–venous oxygen difference $(a-v)$ C_{O2} with running speed in the brush-tailed possum. Data are from Baudinette et al. (1978).

mands are met primarily by increases in cardiac output. At the maximum speed at which the animals were running, cardiac output and arterio-venous oxygen difference were both approximately double the resting level. Throughout a range of running speeds, stroke volume remained relatively constant. A plot of the relative changes in cardiovascular parameters is shown in Figure 3B. There are again few comparative data to support generalizations; however, in the possum $\dot{V}_b$ rises about 7 liters for every liter increase in oxygen consumption, an observation also reported for several eutherians (see Baudinette et al., 1978). The relatively constant values for stroke volume have also been observed in eutherians (e.g., Rushmer, 1959). At present, therefore, we have no evidence to suggest that marsupials and eutherians differ in their cardiovascular adjustment to exercise.

Conclusion

Do marsupials and eutherians show differences in their locomotory energetics, and do these differences reflect the lower resting metabolic rates of the former? There do appear to be slight differences; quadrupedal marsupials use metabolic power at a slightly greater rate when running at low speeds, but their minimum energy requirement to transport a unit of mass over a distance appears less than for a eutherian. Marsupials appear to rely rather more on heat storage during exercise, and less on nonevaporative means of heat dissipation. Sweating occurs during locomotion in at least some small members of the group, a response to exercise usually only associated with larger ungulates and primates.

Only the difference in heat storage can be clearly related to any of the conservative traits considered above. Body temperatures in both groups of animals appear similar during exercise, but the lower resting level provides marsupials with a greater storage capacity. Clearly the explanation for the low metabolic levels of marsupials must be related to the energy budgets of the animals at rest rather than during exercise.

References

Allen, T. E., and Bligh, J. (1969). A comparative study of the temporal patterns of cutaneous water vapour loss from some domesticated mammals with epitrichial sweat glands. *Comp. Biochem. Physiol.* 31:347–63.

Bakker, R. T. (1972). The locomotor energetics of lizards and mammals compared. *Physiologist* 15:76.

Bartels, H. (1964). Comparative physiology of gas transport in mammals. *Lancet (London)* 11:599–604.

Baudinette, R. V. (1977). Locomotory energetics in a marsupial, *Setonix brachyurus*. *Aust. J. Zool.* 25:423–8.

Baudinette, R. V. (1978). Scaling of heart rate during locomotion in mammals. *J. Comp. Physiol.* 127:337–342.

Baudinette, R. V., Nagle, K. A., and Scott, R. A. D. (1976a). Locomotory energetics in dasyurid marsupials. *J. Comp. Physiol. 109:*159–68.

Baudinette, R. V., Nagle, K. A., and Scott, R. A. D. (1976b). Locomotory energetics in a marsupial (*Antechinomys spenceri*) and a rodent (*Notomys alexis*). *Experientia 32:*583–4.

Baudinette, R. V., Seymour, R. A., and Orbach, J. (1978). Cardiovascular responses to exercise in the brush-tailed possum. *J. Comp. Physiol. B, 124:*143–7.

Bland, D. K. and Holland, R. A. (1977). Oxygen affinity and 2,3-diphosphoglycerate in blood of Australian marsupials of differing body size. *Respir. Physiol. 31:*279–90.

Cragg, P. A. (1975). Respiration and body weight in the reptilian genus *Lacerta:* A physiological, anatomical and morphometric study. Ph.D. thesis, Bristol University.

Dawson, T. J. (1969). Temperature regulation and evaporative water loss in the brush-tailed possum, *Trichosurus vulpecula*. *Comp. Biochem. Physiol. 28:*401–7.

Dawson, T. J., and Taylor, C. R. (1973). Energetic cost of locomotion in kangaroos. *Nature 246:*313–4.

Dawson, T. J., Denny, M. J. S., and Hulbert, A. J. (1969). Thermal balance of the macropodid marsupial *Macropus eugenii* Demarest. *Comp. Biochem. Physiol. 31:*645–53.

Dawson, T. J., Robertshaw, D., and Taylor, C. R. (1974). Sweating in the kangaroo: A cooling mechanism during exercise, but not in the heat. *Am. J. Physiol. 227:*494–8.

Fedak, M. A., Pinshow, B., and Schmidt-Nielsen, K. (1974). Energy cost of bipedal running. *Am. J. Physiol. 227:*1038–44.

Hudson, J. W., and Dawson, T. J. (1975). Role of sweating from the tail in the thermal balance of the rat-kangaroo *Potorous tridactylus*. *Aust. J. Zool. 23:*453–61.

Hughes, G. M. (1977). Dimensions and the respiration of lower vertebrates. In *Scale Effects in Animal Locomotion*, ed. T. J. Pedley, New York: Academic Press.

Hulbert, A. J., and Dawson, T. J. (1974). Thermoregulation in perameloid marsupials from different environments. *Comp. Biochem. Physiol. A, 47:*591–616.

Hulbert, A. J., and Rose, R. W. (1972). Does the devil sweat? *Comp. Biochem. Physiol. 43A:*219–22.

Johnson, J. I., Jr. (1977). Central nervous system of marsupials. In *The Biology of Marsupials*, ed. D. Hunsaker, New York: Academic Press.

Kinnear, J. E., and Brown, G. D. (1967). Minimum heart rates of marsupials. *Nature 215:*1501.

Kleiber, M. (1947). Body size and metabolic rate. *Physiol. Rev. 27:*511–41.

Lillegraven, J. A. (1969). The latest Cretaceous mammals of the upper part of the Edmonton Formation of Alberta, Canada, and a review of the marsupial–placental dichotomy in mammalian evolution. *Univ. Kans. Palaeontol. Contrib. Artic. 50,*122pp.

MacMillen, R. E. and Nelson, J. E. (1969). Bioenergetics and body size in dasyurid marsupials. *Am. J. Physiol. 217:*1246–51.

Maxwell, G. M., Elliott, R. B., and Kneebone, G. M. (1964). Hemodynamics of kangaroos and wallabies. *Am. J. Physiol. 206:*967–70.

Needham, A. D., Dawson, T. J., and Hales, J. R. S. (1974). Forelimb blood flow and saliva spreading in the thermoregulation of the red kangaroo, *Megaleia rufa*. *Comp. Biochem. Physiol. 49A:*555–65.

Pasquis, P., Lacaisse, A., and Dejours, P. (1970). Maximal oxygen uptake in four species of small mammals. *Respir. Physiol. 9:*298–309.

Rushmer, R. F. (1959). Constancy of stroke volume in ventricular responses to exercise. *Am. J. Physiol. 196:*745–50.

Schmidt-Nielsen, K., and Larimer, J. L. (1958). Oxygen dissociation curves of mammalian blood in relation to body size. *Am. J. Physiol. 195:*424–8.

Stahl, W. R. (1962). Similarity and dimensional methods in biology. *Science 137:*205–12.

Stahl, W. R. (1967). Scaling of respiratory variables in mammals. *J. Appl. Physiol.* 22:453–60.

Taylor, C. R. (1974). Exercise and thermoregulation. in *Environmental Physiology. MTP Int. Rev. Science.* Physiology Series One, vol. 7. ed. D. Robertshaw, pp. 163–84. London: Butterworth.

Taylor, C. R. (1977). The energetics of terrestrial locomotion and body size in vertebrates. In *Scale Effects in Animal Locomotion,* ed. T. J. Pedley, pp. 127–46. New York: Academic Press.

Taylor, C. R., and Lyman, C. P. (1972). Heat storage in running antelopes: Independence of brain and body temperatures. *Am. J. Physiol.* 222:114–7.

Taylor, C. R., Seeherman, H. J., Maloiy, G. M. O., Heglund, N. C., and Kamau, J. M. Z. (1978). Scaling maximum aerobic capacity ($\dot{V}_{O2}$ max.) to body size in mammals. *Fed. Proc.* 37:473.

Taylor, C. R., Schmidt-Nielsen, K., and Raab, J. L. (1970). Scaling of energetic cost of running to body size in mammals. *Am. J. Physiol.* 219:1104–7.

21

Mechanics of locomotion in primitive and advanced mammals

NORMAN C. HEGLUND

Working toward an optimum

It is a common notion that the driving force behind evolutionary change is "improvement"; that each successive animal is somehow better than its predecessors. In this context we might expect that over the eons the locomotory system of mammals has been improved by reducing the energy required for locomotion in much the same manner as an engineer might improve the design of an automobile to increase its fuel economy. An engineer can increase the fuel economy of an automobile by building more efficient engines, and by reducing the mechanical power required for driving; for example, by using stronger and lighter materials, reducing wind resistance, driving at a more constant velocity, and so on. Apparently not all these options were available during the course of mammalian evolution because the engines (the muscles and tendons) and structural materials (the bones) are basically the same in advanced and primitive mammals (see papers by Alexander, Armstrong, and Cavagna in this volume). Therefore we would expect any major advances in mammalian locomotion would have to be achieved through a reduction in the mechanical power required by locomotion, either by a "better" arrangement of the structures or by different mechanisms for using them (i.e., gaits). How can we test this idea?

A simple way would be to compare mechanical power requirements of locomotory systems ranging from generalized to highly specialized forms. I have selected two quadrupeds that have a limb morphology similar to that of the earliest mammals, the hedgehog and the tenrec. For a cursorial mammal I will use data from the dog; and, finally, as an example of a highly specialized form of bipedal locomotion I will use the kangaroo.

Quantifying the mechanics of locomotion

We must know how mechanical power is used by animals during locomotion, and how it can be measured, in order quantitatively to compare these "advanced" and "primitive" mammalian locomotory systems.

Power is the rate at which work is done. For example, the work involved in lifting a body is the product of the mass of the body times the height of the lift times the gravitational constant. This work is independent of the speed at which the body is lifted. However, because power is the rate at which work is done, the average power expended during the same lift would be ten times greater if the lift were accomplished in one-tenth the time. In order for us easily to compare the different animals at a particular speed, and also to simplify our calculations of power output, only steady-speed locomotion will be considered. In other words, we will not deal with the power required for the animal to accelerate its mass from rest to its average speed of locomotion.

For convenience of measurement the mechanical power required during locomotion is often divided into three parts. First is the power needed to overcome friction due to wind resistance, friction against the ground, and friction in the body; second is the power required to move the limbs relative to the body; and third is the power required to move the body as a whole relative to the ground.

The frictional losses have been shown to be small at all but the highest speeds in terrestrial locomotion (Pugh, 1971). For example, in humans, which present a large frontal area to the air during running, wind resistance accounts for less than 2% of the total mechanical power expended at $10 \ \text{kmh}^{-1}$, and less than 8% at $30 \ \text{kmh}^{-1}$ (Cavagna and Kaneko, 1977; Hill, 1927). The work done on friction against the ground is nearly nil unless the animal is slipping (e.g., running on sand), and the work done on friction within the body has been estimated at less than 2% of the total power requirements (Cavagna et al., 1971; Cavagna and Kaneko, 1977) for humans.

The power required for moving the limbs relative to the body has long been thought to be the bulk of the power required during locomotion (Hill, 1950). Functional anatomists have frequently argued that the reason cursorial animals evolved lightened and lengthened distal limb segments was to reduce their mass and thus the energy required to accelerate and decelerate the limbs during each step. The most common example is the evolutionary reduction and lengthening in the horse's forelimb digits: from the five-toed condylarth through the four-toed *Hyracatherium* and the three-toed *Miohippus* to the contemporary one-toed *Equus* (Hildebrand, 1974). The actual mechanical power expended in the movements of the limbs relative to the body has been quantified by cine analysis of man by Cavagna and Kaneko, of kangaroo by Alexander and Vernon, and of a number of large and small animals in unpublished studies done at Harvard University. This power is a relatively small fraction of the total power expended during locomotion in kangaroos at all speeds (Alexander and Vernon, 1975), and in all animals at low speeds (Cavagna and Curadini, 1975; Cavagna and Kaneko, 1977), and will not be dealt with in this paper.

It is frequently assumed that once a constant average speed is achieved, the body (actually the center of mass) moves steadily and on the level, requiring no additional energy input to maintain the average speed. This is not true; even if the average speed is constant, the center of mass moves up and down and accelerates and decelerates during each step, resulting in changes in the total mechanical energy of the center of mass and thus requiring that work be done on the system (Cavagna et al., 1976). In a system such as a galloping dog, when the animal galloping at a constant average speed "takes off," it is actually accelerating itself forward and upward doing positive work. It follows then that if the average speed and height are to be constant, the animal must, upon landing, decelerate itself forward and downward doing negative work. For example, in a 20-kg dog that is galloping at 15 km h^{-1}, the center of mass goes up and down about 5 cm, and changes its speed in the forward direction about 0.05 meter s^{-1} in each stride. These movements of the center of mass can be measured most conveniently by using a force-platform, a device that measures the forces exerted on the ground by an animal over a period of one or more strides (Cavagna, 1975). From the measured forces we can calculate the kinetic energy of the center of mass in the forward and vertical directions, as well as the gravitational potential energy of the center of mass, all as a function of time. The total energy of the center of mass can then be obtained by summing the kinetic and potential energies at each instant. The positive work required to carry out these movements is then calculated as the sum of the increments in the total energy curve of the center of mass over an integral number of strides.

There are two implications of this method of calculating muscular work that may not be apparent. First of all, any possible transfers between the potential and kinetic energy of the center of mass are allowed in the calculation of the total energy. And second, any decrements in the total energy curve are treated as energy dissipated as heat. In this way we are calculating the minimal amount of positive mechanical work that must have been accomplished by the muscles and tendons for movements of the center of mass in each stride. The energy source for this work includes both metabolic energy via the muscles and elastic energy stored in the elastic elements during the previous negative work phase of the stride (i.e., from energy that, in fact, was not dissipated as heat).

Two energy-conserving mechanisms and three gaits

Two basic energy-conserving mechanisms have been found that maximize "fuel economy" during constant-speed locomotion (Cavagna et al., 1977). First, there can be an exchange between the forward kinetic and the gravitational potential energy of the center of mass in a manner similar to a pendulum or an egg rolling end over end. Any energy that is exchanged is energy that does not have to be supplied by the muscles or tendons. And second, there can be an exchange between the kinetic energy and the elastic potential energy, presumably in the muscles and tendons, in a manner similar to a bouncing ball. The site of the storage and recovery of elastic energy is the subject of the papers by Alexander and Cavagna in this volume. These two energy-conserving mechanisms are utilized in three mechanically distinguishable gaits.

The pendulum mechanism is used to varying degrees in the bipedal or quadrupedal walk: The gravitational potential energy and forward kinetic energy are out of phase, allowing the interchange of up to 70% of the kinetic and potential energy that would otherwise have to be provided by the muscles (Cavagna et al., 1976, 1977).

The bouncing ball mechanism is utilized in the quadrupedal trot and the bipedal run (a trotting quadruped can be thought of as equivalent to two bipeds running to the same cadence but out of step, one behind the other). The gravitational potential energy and forward kinetic energy are in phase, so there can be no pendulum mechanism; however, there appears to be considerable storage and recovery of elastic energy (see Alexander and Cavagna in this volume).

Both the pendulum and the bouncing ball mechanisms are utilized in the gallop, a gait that is unique to the quadrupeds. At low galloping speeds the pendulum mechanism is used, interchanging up to 40% of the required

Figure 1. The weight-specific mechanical power required for movements of the center of mass during walking and trotting is very similar for dissimilar animals. The solid circles are the walk points from a tenrec (822 g), assuming the maximum possible exchange between kinetic and potential energy (see text); solid squares are the walk values assuming none of the possible transfers occurred. The crossed circles are measured tenrec trot points; half-solid circles are measured trot points for a hedgehog (269 g). The lower line is from a walking dog (5 kg) (from Cavagna et al., 1977); the upper line is trotting in the same dog. In spite of different locomotory systems and a nearly 20× range in size, the power output is similar for all these animals.

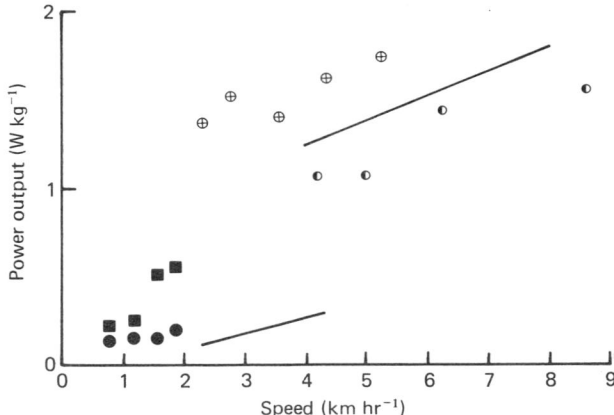

energy. As the galloping speed increases, the mechanics become less pendular and more elastic (Cavagna et al., 1977).

Primitive to advanced: Any "improvements"?

Assuming the locomotion of the tenrec and the hedgehog is similar to that of the earliest mammals, has the last 200 million years of evolution resulted in any "improvements" in mammalian locomotion? The answer is apparently no. Certainly the gaits utilized by the tenrec and the hedgehog are mechanically very similar to some of the gaits of "advanced" mammals. The tenrec walk shows the same out of phase relationship between forward kinetic and gravitational potential energy, and thus utilizes the pendulum mechanism demonstrated in advanced mammaliam bipedal and quadrupedal walking. Trotting in the hedgehog and the tenrec, and hopping in the kangaroo, are all equivalent to running or trotting in an advanced biped or quadruped because the signatures of the mechanical energy versus time tracings are often so similar that they cannot be distinguished unless the magnitude scales are considered.

The definitive test, however, is that the weight-specific mechanical power required for movements of the center of mass during steady-speed locomotion (as measured on the force-platform) is nearly indistinguishable between the generalized locomotory systems (the hedgehog and the tenrec) and the more specialized locomotory system (the dog) in spite of significant differences in body size (Figure 1).

Figure 2. The weight-specific power required for movement of the center of mass is greater for a hopping 20-kg kangaroo (top line) than for a galloping 18-kg dog (bottom line). The data are from Cavagna et al., 1977.

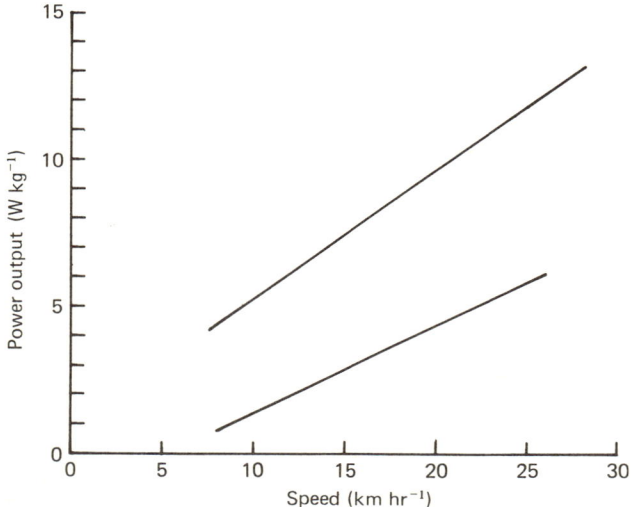

Contrary to what we would expect, the highly specialized saltatory loco-motion of the kangaroo required more power for movements of the center of mass than any other animal studied to date (Figure 2). Although it may seem contradictory that an animal's locomotion can be highly specialized and at the same time require a large mechanical power output (presumably suffering an increased "mileage cost" as a consequence), this problem is resolved by the fact that power input may not be directly linked to power output; that is, the fraction of power input that actually results in power output (efficiency) is not a constant. The kangaroo increases this fraction by taking advantage of large transfers beween kinetic and elastic potential en-ergy, and as a result has a very high efficiency (Cavagna et al., 1977) at high speeds. The ability to move cheaply long distances at high speeds may have meant the difference between extinction (as in the case of all the other large terrestrial marsupials) and continued survival in Australia (Dawson and Tay-lor, 1973).

Conclusion

Any preconceived ideas about "improvements" in the power expended in mammalian locomotion during evolution do not seem to be supported by these studies. Certainly the similarities between the generalized primitive locomotion of the tenrec and the advanced cursorial locomotion of the dog are striking: (1) The walking gait shows an exchange between the vertical potential and forward kinetic energies of the center of mass; (2) the trotting gait also shows the same mechanical energy versus time relationships; and (3) the weight-specific power required to move the center of mass is inde-pendent of either size or animal for the quadrupeds. Apparently the best ideas on how to improve terrestrial locomotion were already used before the mammals evolved.

This paper was supported by NSF grant PCM75-22684.

References

Alexander, R. McN., and Vernon, A. (1975). The mechanics of hopping by kangaroos (Macropodidae). *J. Zool. (London)* 177:265–303.

Cavagna, G. A. (1975). Force platforms as ergometers. *J. Appl. Physiol.* 38:174–8.

Cavagna, G. A. and Curadini, L. (1975). Internal work in locomotion: Effect of walking and running speed on kinetic energy of upper and lower limbs. *IRCS Med. Sci.* 3:294.

Cavagna, G. A., and Kaneko, M. (1977). Mechanical work and efficiency in level walking and running. *J. Physiol. (London)* 268:467–81.

Cavagna, G. A., Komarek, L., and Mazzoleni, S. (1971). The mechanics of sprint running. *J. Physiol. (London)* 217:709–21.

Cavagna, G. A., Thys, H., and Zamboni, A. (1976). The sources of external work in level walking and running. *J. Physiol. (London)* 262:639–57.

Cavagna, G. A., Heglund, N. C., and Taylor, C. R. (1977). Mechanical work in terrestrial locomotion: Two basic mechanisms for minimizing energy expenditure. *Am. J. Physiol. 233(5)*:R243–61.

Dawson, T. J., and Taylor, C. R. (1973). Energetic cost of locomotion in kangaroos. *Nature 246*:313–14.

Hildebrand, M. (1974). *Analysis of Vertebrate Structure*, p. 18. New York: Wiley.

Hill, A. V. (1927). The air resistance to a runner. *Proc. R. Soc. B102*:380–5.

Hill, A. V. (1950). The dimensions of animals and their muscular dynamics. *Sci. Prog. 38(150)*:209–30.

Pugh, L. G. C. E. (1971). The influence of wind resistance in running and walking and the efficiency of work against horizontal or vertical forces. *J. Physiol. (London) 213*:255–76.

22

Elasticity in the locomotion of mammals

R. McN. ALEXANDER

Kangaroos share many conservative features with other marsupials, but their hind legs and their fast hopping gait give the impression of high specialization. This paper considers whether the principles of fast locomotion are similar in kangaroos and more typical mammals, and whether the legs and gait of kangaroos are superior or inferior to those of fast placental mammals.

General principles

Imagine a vehicle on wheels with frictionless axles, traveling over level rigid ground in a perfect vacuum. Once it has accelerated to a desired speed, the engine can be turned off, and it will continue indefinitely at constant velocity without consuming any fuel. Because its velocity is constant, its kinetic energy is constant. Because the height of its center of mass is constant, its gravitational potential energy is constant.

Now imagine a perfectly elastic ball bouncing along, again on level rigid ground in a perfect vacuum. It will continue bouncing indefinitely, and its height and velocity, at corresponding stages of successive bounces, will always be the same. In this case kinetic energy and potential energy both fluctuate, and so does elastic strain energy (energy due to elastic deformation under stress). When the ball hits the ground, kinetic energy is converted to elastic strain energy, which is reconverted to kinetic energy in the elastic recoil. Kinetic energy is converted to potential energy as the ball rises off the ground, and potential energy is converted to kinetic energy as it falls. The total mechanical energy of the ball, the sum of kinetic, potential, and elastic strain energy, remains constant.

A running animal has mechanical energy consisting of the same three components, but the total fluctuates in the course of each stride. Consequently muscles must do work to maintain the animal's motion.

Muscles do positive work when they shorten against a resistance, for instance when I lift a suitcase. They do negative work when they extend while exerting tension, as happens when I lower a suitcase to the ground (as distinct from dropping it). Both positive and negative work performance consume metabolic energy, though positive work is much more costly (Margaria, 1976). When in the course of locomotion the total mechanical energy of the body increases, muscles must do positive work, and when it decreases, muscles must do negative work. This accounts for most of the energy cost of running.

These statements depend on the assumption that the ground is rigid, which is often approximately true. They also depend on the assumption that work done against air resistance is negligible. Air resistance probably accounts for 13% of the energy cost of human running at maximum speed and less at lower speeds (Pugh, 1971). It is probably important for running mammals in general, only near their maximum speeds (see Alexander, 1976).

In some situations one muscle may do work against another. One may do positive work while another does an equivalent quantity of negative work so that metabolic power is consumed but the mechanical energy content of the body is unchanged. Such a wasteful situation is illustrated in Figure 1A, which shows a simple leg with joints at hip and knee. The foot remains on the ground exerting a vertical force while the hip moves horizontally from position (i) through (ii) to (iii). Between positions (i) and (ii) the reaction of the ground on the foot exerts an anticlockwise moment about the hip and a clockwise moment about the knee, so muscles x and y must exert tension. Muscle x shortens doing positive work, as the hip extends, but y extends, doing negative work, as the knee bends. The positive and negative work cancel each other, and the potential and kinetic energy of the body remain

Figure 1. (A) Three successive positions of a schematic leg, with the foot on the ground. The ground exerts a vertical force on the foot throughout. x, y, and z are active muscles. (B) The same, but the force changes direction so as always to be in line with the hip.

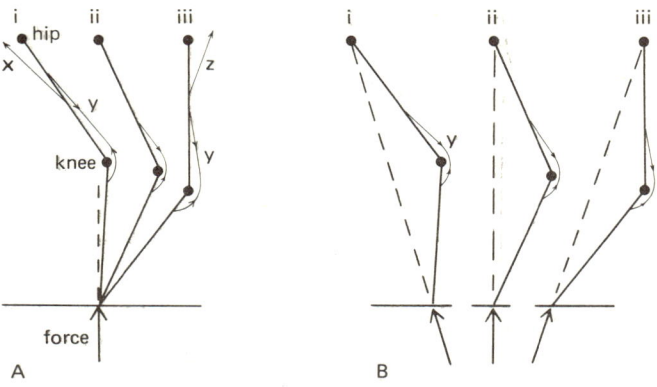

unchanged. Similarly, between positions (ii) and (iii) the active muscles are y, which at this stage does positive work, and z, which does negative work.

Such wastage of metabolic energy is avoided in Figure 1B, in which the force exerted by the foot is kept in line with the hip so that muscles x and z can remain inactive throughout the step. Muscle y is active throughout. It does negative work between positions (i) and (ii) (decelerating the body) and positive work between (ii) and (iii) (accelerating it again). This could in principle be achieved without metabolic energy cost if the muscle were replaced by a perfectly elastic spring.

It will be shown that tendons serve as springs in the legs of running mammals. Experiments have been done on tendons from the feet of sheep to discover how nearly they are perfectly elastic (Cuming et al., 1978). The tendons were subjected to cycles of stretch and recoil, involving stresses similar to those that occur in galloping and at a frequency a little higher than would occur in galloping. It was found that the rebound resilience (the energy recovered in an elastic recoil, divided by the work previously done deforming an elastic body) was only 0.62. For a perfectly elastic spring it would be 1.00. Values of 0.97 and 0.91 have been measured for the elastic proteins resilin (from insects; Jensen and Weis-Fogh, 1962) and abductin (from molluscs; Alexander, 1966). The value of 0.62 for tendon is surprisingly low, and Dr. R. F. Ker has obtained much higher values in experiments with other tendons (personal communication). The tendons used in the initial investigation are not continuous but have (extremely short) muscle fibers interpolated in them.

Hopping of kangaroos

The bipedal hop of kangaroos is a simple gait, conveniently easy to analyze. It has just two phases, a contact phase when both hind feet are on the ground, and a floating phase when they are off the ground. The fluctuations of kinetic and potential energy involved have been assessed in two ways, by kinematic analysis of film (Alexander and Vernon, 1975) and by integration of force-platform records (Alexander and Vernon, 1975; Cavagna et al., 1977). Results from both methods are shown in Figure 2. Potential energy has a minimum in the contact phase and a maximum in the floating phase. Kinetic energy has been divided into three parts:

1. Horizontal external kinetic energy, due to the horizontal component of the velocity of the center of mass

2. Vertical external kinetic energy, due to the vertical component of the velocity of the center of mass

3. Internal kinetic energy, due to movement of parts of the body relative to the center of mass

The most important of these is (1), which is high and constant in the floating phase of each stride and has a minimum in the contact phase. It fluctuates because the force on the ground is not kept vertical but changes direction so as always to be more or less in line with the center of mass. If it were kept vertical, energy would be wasted, as explained in the discussion of Figure 1. Also the body would pitch, but only through a small angle.

Figure 2B refers to slow hopping, and Figure 2A to fast hopping of the same kangaroo. They show that as speed increases, potential energy fluctuations diminish a little, but horizontal external kinetic energy fluctuations increase greatly.

Fluctuations of elastic strain energy must be assessed next (Alexander and Vernon, 1975). Figure 3A shows the principal leg muscles that must be active during the contact phase. The length of each of these muscles can be calculated for successive stages of the stride by measurements on successive frames of a cine film. The forces the muscles exert can be calculated from force-platform records, making plausible assumptions about the sharing of effort between muscles. For example, at the instant illustrated in Figure 3B the force-platform registered a force of 430 N at 89° to the horizontal. Presumably half this force, 215 N, was exerted by the right foot. The reaction exerted by the ground on the right foot is drawn in Figure 3B acting

Figure 2. Graphs of potential and kinetic energy against time during hopping for (A) and (B) a young red kangaroo *(Macropus)* and (C) a wallaby *(Protemnodon)*. Total mechanical energy is shown divided into potential energy, horizontal external kinetic energy, and vertical external kinetic energy ([black, not shown for (C)]. (A) and (B) were obtained by kinematic analysis of film, and (C) was from a force-platform record. Speeds were (A) 6.2 meters s^{-1}, (B) 2.7 meters s^{-1}, and (C) 2.4 meters s^{-1}. From Alexander and Vernon (1975).

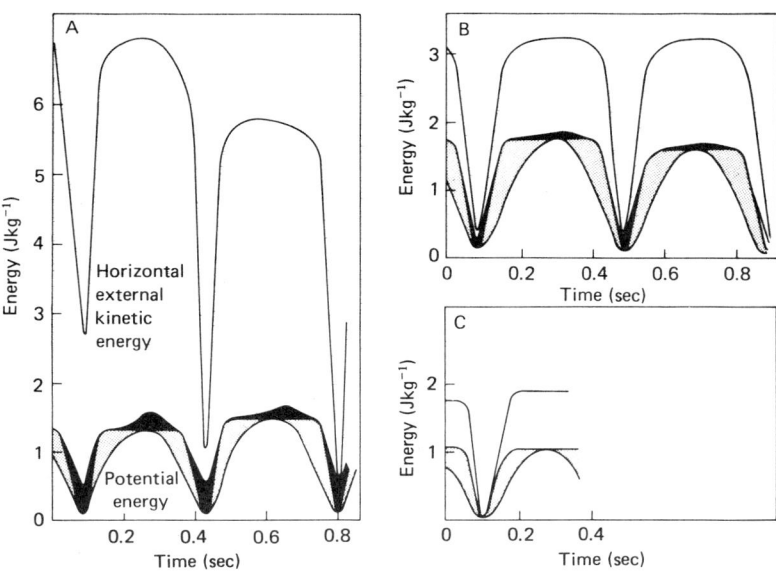

through the centroid of the sole of the foot. Its line of action is 135 mm from the ankle joint, and the tendons of the gastrocnemius and plantaris muscles are 33 mm from the joint. These muscles must balance the moments about the ankle, and to do so must exert between them $215 \times 135/33$ = 880 N (the weight of the foot and the inertia force on the foot are small enough to ignore in this calculation). The gastrocnemius and plantaris muscles differ in size. If equal stresses act in both, the total force of 880 N must be made up of 490 N exerted by the gastrocnemius and 390 N by the plantaris. The possibility of unequal stresses will be considered later. Similar calculations can be made for the other muscles.

Figure 4 has been obtained by such calculations. It shows that the extensor muscles of the hip exert tension and shorten throughout the contact phase; they do positive work. The extensors of the knee exert tension and extend, doing negative work. The gastrocnemius and plantaris extend as the force on them increases and shorten again as the force diminishes; they behave very like springs stretching and recoiling elastically.

Figure 4 shows that at the instant when potential and kinetic energy were least, the gastrocnemius exerted 480 N and the plantaris 370 N. Both muscles are pennate with very long tendons (Figure 3A). The gastrocnemius tendon was 250 mm long, in the animal being discussed, and the plantaris tendon, 420 mm. Calculations using the forces, the dimensions of the tendons, and Young's modulus for tendon, indicate that each tendon would be stretched 11 mm. Thus 11 mm of the 20-mm extension shown in Figure 4A, B is probably elastic extension of the tendons.

Forces of (480 + 370) N stretching tendons by 11 mm in each of two legs store elastic strain energy amounting to 9.4 J, which is 0.9 J (kg body weight)$^{-1}$ in this case. If this quantity of elastic strain energy is stored while potential and kinetic energy have their minima in Figure 2C, the fluctuations of total

Figure 3. (A) A diagram of the hind leg of a kangaroo, showing some of the muscles. Tendons are represented by thick lines. The tendons of the gastrocnemius and plantaris muscles are believed to have an important role as springs. (B) An outline traced from a film of a wallaby hopping across a force-platform showing the force acting on the sole of the right foot at this instant.

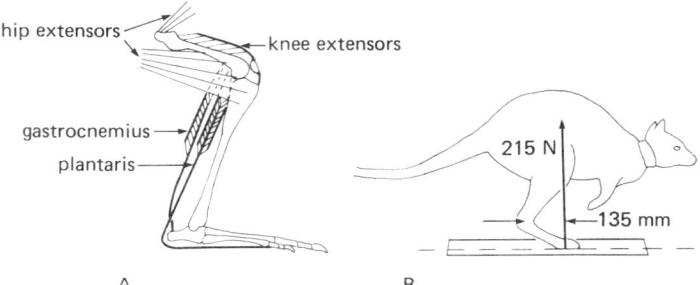

hip extensors — knee extensors
gastrocnemius —
plantaris —

215 N
135 mm

A   B

mechanical energy are greatly reduced (as in Figure 5A). The negative work required of the muscles in each stride is reduced by 0.9 J kg^{-1} and the positive work by 0.9R J kg^{-1}, where R is the rebound resilience of tendon. Even if R is no greater than 0.62 (see above), considerable metabolic energy is saved.

The gastrocnemius and plantaris muscles have muscle fibers 20 to 25 mm long, which should be capable of stretching elastically while exerting isometric stress by 0.4 to 0.8 mm, but any further extension at this stress must involve detaching cross-bridges (Alexander and Bennet-Clark, 1977). This elastic extension is so small compared to the extension of the tendons, which exert the same force, that elastic strain energy stored in the muscle fibers can be neglected (see also Morgan et al., 1978). Even if stresses several times isometric occur (Cavagna, this volume), the strain energy stored in the muscle fibers will be much less than that stored in the tendons.

Figure 4. Graphs of the force exerted by (A) the gastrocnemius, (B) the plantaris, and (C) the extensor muscles of the knee of a wallaby, against changes in length of the muscles. (D). A graph of the moment exerted by the extensor muscles of the hip, about the hip, against the angle of the hip. The graphs refer to the contact phase of the hop of Figure 2(C). Successive points represent measurements at intervals of 10 ms except for a single interval of 40 ms, indicated by a broken line, spanning the period when vibrations following impact of the foot with the platform made it difficult to make reliable measurements. The open symbols show when potential and external kinetic energy were minimal. From Alexander and Vernon (1975).

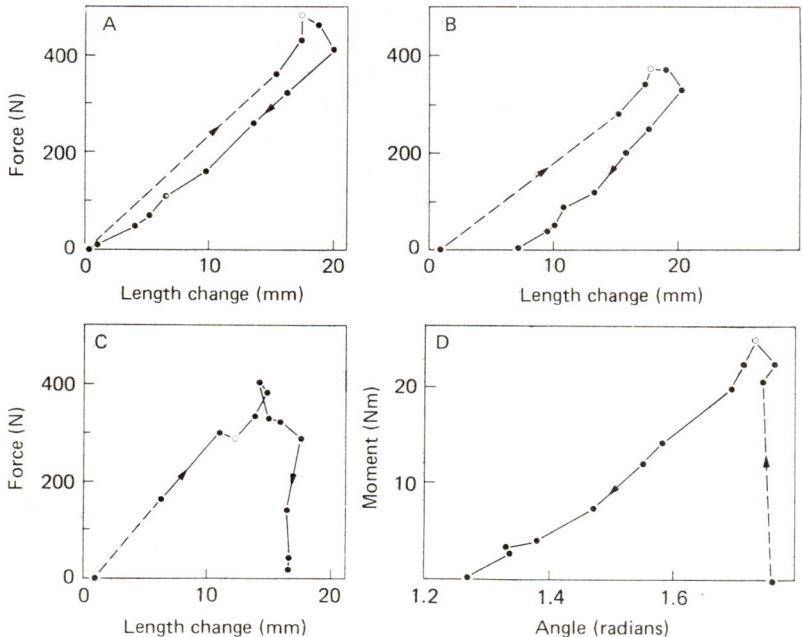

The metabolic energy needed for hopping would have been reduced still more if the tendons had stretched by 20 mm instead of 11 mm. This would have compensated for the fluctuations of potential and external kinetic energy, making the total mechanical energy almost constant (Figure 5B). The tendons would have stretched by 20 mm if they had been 11/20 of their actual thickness and would still have been strong enough for hopping at the rather low speed represented in Figures 2C and 4. However they would have stretched too far (Figure 5C) at higher speeds, at which larger forces act because the feet are on the ground for a smaller fraction of the time. They would probably have broken at the highest speeds.

Kangaroos keep their stride frequency almost constant and increase their speed by increasing their stride length. Hence the feet are on the ground for a smaller fraction of the time at high speeds and must exert larger forces at high speeds. The amplitude of the horizontal external kinetic energy changes is roughly proportional to the (vertical) force exerted in the middle of the contact phase; the elastic strain energy stored in a tendon is proportional to the square of the force on the tendon; and the amplitude of the potential energy changes diminishes a little as speed increases. Hence there can only be one speed at which elastic strain energy compensates exactly for the fluctuations of potential and kinetic energy if the stresses in the gastrocnemius and plantaris muscles are kept equal. It seems that in these circumstances compensation would be closest to perfect at high speeds.

Compensation at lower speeds could be improved by making the stresses in the two muscles unequal, and the work of Armstrong (this volume) suggests that this may be done. If for instance the gastrocnemius exerted the whole of the force of 880 N required at the instant of Figure 3B, its tendon would stretch by about 20 mm, and compensation would be near-perfect. The data of Alexander and Vernon (1975) show that this would involve high but not impossible stresses in the muscle and its tendon.

Alexander and Vernon (1975) calculated from the fluctuations of potential and kinetic energy in Figure 2A, B that if no energy had been saved by elastic storage, the metabolic power requirement for hopping would have

Figure 5. Schematic graphs of mechanical energy against time for a hopping kangaroo. Elastic strain energy e is shown separately from kinetic and gravitational potential energy $k + p$. The fluctuations of e compensate precisely for the fluctuations of $k + p$ only in (B): in (A) they are too small and in (C) too large.

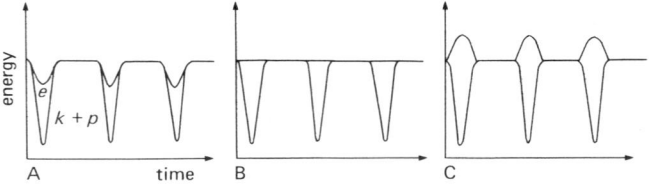

been 24 W kg^{-1} at 2.7 meters s^{-1} and 36 W kg^{-1} at 6.2 meters s^{-1}. Dawson and Taylor (1973) measured the oxygen consumption of hopping kangaroos and obtained values indicating a power requirement of about 20 W kg^{-1} at all speeds. This confirms the conclusion reached on mechanical grounds, that tendon elasticity reduces the metabolic power required for hopping a little at low speeds and very substantially at high speeds.

Comparisons with other mammals

It seems to be a general rule that mammals running at velocity u take strides of length about $2.3\ h\ (u^2/gh)^{0.3}$, where g is the acceleration of free fall and h is the height of the hip joint from the ground (Alexander, 1977a). Stride lengths of a wide variety of species including kangaroos have been found to lie within ±30% of the values given by this expression, irrespective of gait.

Kangaroos hop only at speeds above about $0.8\ (gh)^{0.5}$; at lower speeds they shuffle along on all four legs and the tail. Quadrupedal mammals walk at speeds below about $0.8\ (gh)^{0.5}$, trot or rack at speeds a little above this, and canter and then gallop at still higher speeds (Alexander, 1977a). Hopping must be compared to trotting and faster quadrupedal gaits, and to the human run.

In human running the feet are set down alternately, not simultaneously as in hopping. There are two short floating phases instead of one long one in each stride. A trotting quadruped is like two men running one behind the other; both the fore quarters and the hind quarters have two short floating phases in each stride.

The total horizontal external kinetic energy gained and lost in each stride is the same for hopping as for running or trotting. Much more potential energy is gained and lost in one long floating phase than in two short ones. Hence fluctuations of (kinetic + potential) energy are greater in hopping than in running or trotting at the same speed. Hopping can be expected to require more metabolic power than running or trotting (Alexander, 1977a). The difference should be relatively large at low speeds at which potential energy fluctuations may be larger than kinetic energy fluctuations (Figure 2B,C). It should be small at high speeds at which potential energy fluctuations are relatively unimportant. Measurements of oxygen consumption have confirmed that kangaroos use more metabolic power than quadrupedal mammals of the same mass at low hopping speeds, and about as much as quadrupeds at higher speeds (Figure 6.6 of Goldspink, 1977b).

At high speeds quadrupedal mammals gallop. They more the two fore feet roughly in synchrony, so as to resemble two kangaroos hopping one behind the other. This suggests that hopping and galloping should use about the same metabolic power, at any given speed. However, a galloping quadruped is not at all precisely like two kangaroos, as the muscles of the back are involved in galloping.

The arrangement of muscles shown in Figure 3A is not peculiar to kangaroos. It is found in the hind legs of dogs, antelopes, and indeed all digiti-grade and unguligrade mammals known to me. In every case the extensor muscles of the hip are predominantly parallel-fibered with long muscle fibers and short tendons. The extensors of the knee are predominantly pennate with moderately short muscle fibers and a long tendon of insertion. The gastrocnemius and plantaris are also pennate with short muscle fibers and long tendons. Fore limbs are very different in structure, but they too have pennate distal muscles with short muscle fibers and long tendons which must store elastic strain energy in the contact phase.

Alexander (1977b) estimated the elastic strain energy stored in the tendons of the legs of antelopes galloping at maximum speed. He considered the fore and hind quarters as independent bipeds and showed that enough elastic strain energy must be stored to save a large fraction of the metabolic energy that would otherwise be required. Just how large that fraction was would depend critically on how effort was shared between muscles. It would be larger if the plantaris muscle exerted greater stresses than the gastrocnemius, than if the stresses in the two muscles were equal.

Metabolic energy is needed simply to activate muscles even when neither positive nor negative work is done (Goldspink, 1977a). To exert a given force, a given cross-sectional area of muscle must be activated, irrespective of the length of the muscle fibers. However, the energy required for activation is presumably proportional to the volume activated and so is greater for long fibers than for short ones. Long muscle fibers may be needed to allow a required range of movement at a joint, but if they are not (for instance if tendon elasticity allows the required movement), there is an advantage in having short fibers. Table 1 shows that various placental mammals are more

Table 1. *Lengths of the muscle fibers of the principal extensor muscles of the ankle of a wallaby* (Protemnodon) *and some other mammals*

Animal	Body mass (kg)	Lengths of muscle fibers (mm)		
		Gastrocnemius lateralis	Gastrocnemius medialis	Plantaris
Macropodidae				
Protemnodon	11	21	25	25
Canidae				
Canis	26	16	26	10
Vulpes	8	23	29	17
Bovidae				
Gazella	29	10	11	3
Rhynchotragus	5	11	11	5

Data from Alexander (1974, 1977b), Alexander and Vernon (1975), and unpublished observations by the author and Mr. A. S. Jayes.

highly adapted for economical locomotion than kangaroos, in having shorter fibers in the plantaris muscle. In the extreme case of the camel *(Camelus dromedarius)*, the plantaris muscle has no muscle fibers at all but has become a stout ligament running from the femur to the phalanges.

Kangaroo hopping at high speeds deserves more study. The young kangaroo hopping at 6.2 meters s^{-1} (Figure 2A) seemed to be going as fast as it could on the treadmill, but it might have been able to hop faster in other circumstances. It would be interesting to make field studies of fast hopping, like a recent study of galloping antelopes (Alexander et al., 1977).

Summary

Tendon elasticity plays the same energy-saving role in the hopping of kangaroos as in the fast gaits of quadrupedal mammals such as dogs and antelopes. However, hopping is more expensive of energy than trotting, especially at low speeds, and kangaroos could probably hop with less expenditure of energy if they had shorter fibers in certain muscles. Field studies of fast hopping are desirable.

References

Alexander, R. McN. (1966). Rubber-like properties of the inner hinge ligament of Pectinidae. *J. Exp. Biol. 44:*199–230.
Alexander, R. McN. (1974). Mechanics of jumping by a dog *(Canis familiaris). J. Zool. (London) 173:*549–73.
Alexander, R. McN. (1976). Mechanics of bipedal locomotion. In *Perspectives in Experimental Biology*, vol. 1, ed. P. Spencer Davies, pp. 493–504. Oxford: Pergamon.
Alexander, R. McN. (1977a). Terrestrial locomotion. In *Mechanics and Energetics of Animal Locomotion*, eds. R. McN. Alexander and G. Goldspink, pp. 168–203. London: Chapman and Hall.
Alexander, R. McN. (1977b). Allometry of the limbs of antelopes (Bovidae). *J. Zool. (London) 183:*125–46.
Alexander, R. McN., and Bennet-Clark, H. C. (1977). Storage of elastic strain energy in muscle and other tissues. *Nature (London) 265:*114–17.
Alexander, R. McN., and Vernon, A. (1975). The mechanics of hopping by kangaroos (Macropodidae). *J. Zool. (London) 177:*265–303.
Alexander, R. McN., Langman, V. A., and Jayes, A. S. (1977). Fast locomotion of some African ungulates. *J. Zool. (London) 183:*291–300.
Cavagna, G. A., Heglund, N. C., and Taylor, C. R. (1977). Walking, running and galloping: Mechanical similarities between different animals. In *Scale Effects in Animal Locomotion*, ed. T. J. Pedley, pp. 111–25. New York: Academic Press.
Cuming, W. G., Alexander, R. McN., and Jayes, A. S. (1978). Rebound resilience of tendons in the feet of sheep *(Ovis aries). J. Exp. Biol. 74:*75–81.
Dawson, T. J., and Taylor, C. R. (1973). Energy cost of locomotion in kangaroos. *Nature (London) 246:* 313–4.
Goldspink, G. (1977a). Muscle energetics and animal locomotion. In *Mechanics and Energetics of Animal Locomotion*, eds. R. McN. Alexander and G. Goldspink, pp. 57–81. London: Chapman and Hall.
Goldspink, G. (1977b). Energy cost of locomotion. In *Mechanics and Energetics of Ani-*

mal Locomotion, eds. R. McN. Alexander and G. Goldspink, pp. 153–67. London: Chapman and Hall.

Jensen, M., and Weis-Fogh, T. (1962). Biology and physics of locust flight. V. Strength and elasticity of locust cuticle. *Phil. Trans. R. Soc. London B245:*137–69.

Margaria, R. (1976). *Biomechanics and Energetics of Muscular Exercise.* Oxford: Clarendon Press.

Morgan, D. L., Proske, U., and Warren, D. (1978). Measurements of muscle stiffness and the mechanism of elastic storage of energy in hopping kangaroos. *J. Physiol. (London) 282:*253–61.

Pugh, L. G. C. E. (1971). The influence of wind resistance in running and walking and the efficiency of work against horizontal or vertical forces. *J. Physiol. (London) 213:*255–76.

23

Elastic storage: role of tendons and muscles

GIOVANNI A. CAVAGNA, G. CITTERIO, and
P. JACINI

Elastic storage and recovery of energy: a general mammalian phenomenon

The mechanisms of terrestrial locomotion are remarkably similar among very different animals, both advanced and conservative (primitive) (Cavagna et al., 1977; Heglund, this volume). During running, hopping, and trotting as well as galloping, locomotion takes place through a succession of "bounces" in which mechanical energy is stored and recovered thanks to the elastic properties of tendons and muscles. In most cases, the energy saved by elastic storage increases with the speed of running (Dawson and Taylor, 1973; Cavagna and Kaneko, 1977; Cavagna et al., 1977). This paper discusses the roles of both muscles and tendons in elastic storage and recovery and why this storage increases with speed.

The positive mechanical work done by the muscles per unit distance traveled per unit body mass during locomotion (W_{tot}), their energy expenditure (En.Exp.), and the ratio between positive work and energy expenditure are given in Figure 1 for a hopping kangaroo and a running man. During each step active muscles are first forcibly stretched and perform negative work (during the deceleration of the body downward and forward) and immediately afterward the muscles shorten actively and perform positive work (during the push upward and forward). The similarity of the mechanisms of hopping and running explains the similar relationship between W_{ext} (the work involved in accelerating and lifting the center of mass during each stride) and running speed (Cavagna et al., 1977). W_{int} in Figure 1 indicates the positive work done by the muscles to increase the kinetic energy of the limbs relative to the center of mass of the body. Fluctuations of internal kinetic energy are relatively unimportant in the kangaroo so that $W_{ext} \simeq W_{tot}$ (Alexander and Vernon, 1975); however, W_{int} is large in man (Cavagna and Kaneko, 1977). Thus, the mechanical work done per unit distance ($W_{tot} = W_{ext} + W_{int}$) decreases slightly with speed in the kangaroo, whereas it increases in man. The energy expenditure per unit distance decreases markedly with speed in the kangaroo (Dawson and Taylor, 1973), whereas it remains about constant in man (Margaria, 1938).

Although total mechanical work output and energy expenditure are very different in kangaroo and in man, the ratios between them are similar. In both species the overall efficiency of positive work performed (W_{tot}/En.Exp.) is greater than the maximal efficiency of the transformation of chemical energy into mechanical work by the contractile component of the muscles

Figure 1. *Below:* internal, W_{int}, external, W_{ext}, and total, $W_{tot} = W_{int} + W_{ext}$, mechanical positive work done per unit distance and unit body weight as a function of running speed of man (right; from Cavagna and Kaneko, 1977 and Cavagna et al., 1976) and hopping of kangaroo (left; from Cavagna et al., 1977). The energy expenditure, En.Exp., was based on the oxygen consumption measurement by Margaria (1938) for man and by Dawson and Taylor (1973) for kangaroo. The dotted part of the En.Exp. curve indicates that an oxygen debt may be necessary to meet the mechanical power output. *Above:* efficiency of positive work production measured as W_{tot}/En.Exp. The interrupted line indicates the maximal efficiency of the transformation of chemical energy into mechanical work by the contractile component of muscle (0.25).

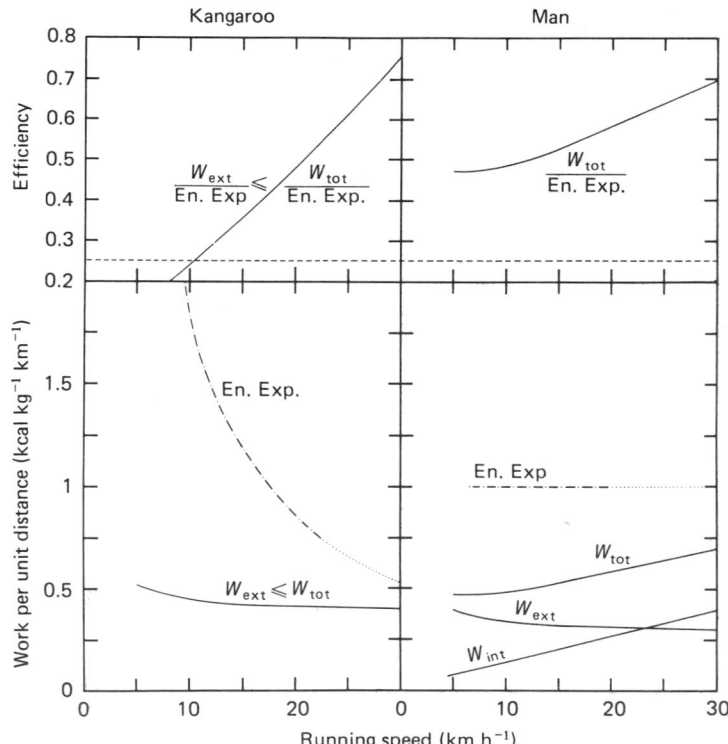

(0.25). This has been taken as evidence that a signficant part of positive work is derived from the elastic energy stored (W_{el}) during the phase of the step in which the muscles perform negative work (Cavagna and Kaneko, 1977; Cavagna et al., 1977). In other words:

$$\frac{W_{tot}}{En.Exp.} = \frac{W_c + W_{el}}{En.Exp.} \tag{1}$$

where W_c is the positive work due to the transformation of chemical energy and (W_c/En.Exp.)$\leqslant$0.25. The values of these ratios (Equation 1) range from 0.2 to 0.7 in the kangaroo and from 0.5 to 0.7 in man.

Greater elastic storage at higher speeds

The increase in efficiency (W_{tot}/En.Exp.) suggests a better recovery of elastic energy (an increase of W_{el} in Equation 1) at high speed. The increase could be explained if the elastic structures (muscles and tendons) followed Hooke's Law (W_{el} proportional to square of force), and the forces exerted increased with running speed (Alexander and Vernon, 1975). We measured the maximal force (F_v) exerted at each step against the ground by hopping kangaroo and running man (Figure 2). The increase of F_v with speed does not seem sufficient to explain the increase of efficiency of positive work performed. In fact, in man, F_v remains almost constant over the range of speed, whereas

Figure 2. The maximal force, F_v, exerted by the feet against the ground as a function of speed for kangaroo hopping and man running. On the left-hand ordinate the corresponding acceleration of the body mass is given in G (1 G = 9.8 meters s^{-2}). Each point represents the average of all data obtained in two kangaroos (average body mass = 20.5 kg; Cavagna et al., 1977) and 11 men (average body mass = 73.5 kg; Cavagna et al., 1976). The vertical and horizontal lines give standard deviations; in each mean the number of items (sampled over intervals of 2 km h^{-1} from 8 to 30 km h^{-1}) was N = 2 ÷ 27.

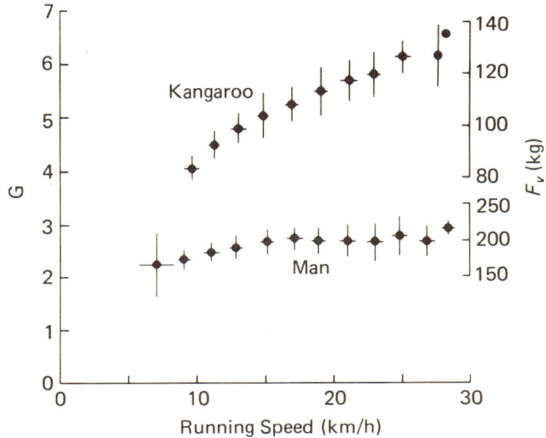

efficiency increases. In the kangaroo F_v increases by only 50%. According to Hooke's law, W_{el} would increase 2.25 times for this 50% increase in force ($W_{el} \propto (1.5)^2$). This could not explain the observed increase of the efficiency. Additionally, it is well known that the elastic structures within the muscles and the tendons do not follow Hooke's law. They become stiffer as the force applied to them increases (see, e.g., Jewell and Wilkie, 1958). Consequently, W_{el} increases much less than with the square of force.

The explanation of the increasing role of elastic storage and recovery of energy with increasing speed of locomotion may involve the contractile machinery instead of the more passive tendons. Some features of the contractile component are relevant in the process of elastic recovery of mechanical energy: (1) The contractile component generates the force necessary to put the elastic elements under tension; and (2) the contractile component is a site of elastic compliance (Huxley and Simmons, 1971).

In this paper we will discuss the relevance of these two properties of the contractile component to, and the effect of the speed of stretching on, elastic storage and recovery of energy.

Role of the contractile component as a tension generator

Storage of mechanical energy by muscle depends on the property of the active contractile machinery to resist lengthening with a force appreciably greater than the isometric force. This property is described by the well-known force–velocity relation. The force exerted by the contractile machinery increases steeply with increasing speed of stretching, reaching a maximum value that can be twice the force developed during an isometric contraction. Because the force exerted during stretching is greater than the isometric force, the mechanical energy stored in the elastic elements will also be greater.

At the reversal of a movement (i.e., between stretching and shortening), the speed of stretching falls to zero. On the basis of the force–velocity relationship alone, this would result in a fall of the force to the isometric value. Consequently, the additional mechanical energy stored during stretching (due to the force $> P_0$) would be lost. Fortunately, another property of the contractile component comes to the rescue. When the velocity of stretching falls to zero, the force developed by the contractile machinery does not fall to the isometric value immediately. In fact, Abbott and Aubert (1952) found that if a muscle is kept active at the stretched length, the force attained during stretching falls at a finite rate, which increases with the speed of stretching.

Abbott and Aubert's experiments were carried out on isolated muscles at 0 °C. During exercise the temperature of muscles is usually greater than 0 °C. Therefore, we studied frog sartorius at 0 and 20 °C to determine how the rate of fall of the force is affected by temperature. A typical tracing is given in Figure 3. The muscle was first tetanically stimulated at a length near its optimal length. It was then forcibly stretched using a Levin and Wyman ergometer (4 mm at 30 mm s^{-1}). It was kept active at the stretched length for a period of about 5 s. At the end of this period (arrow) stimulation was interrupted, and relaxation of the muscle occurred. One can see that during stretching the force attains a value, F', much greater than the isometric force, P_0. At the end of stretching the force falls from F' to P_0 at about the same rate at both 0 and 20 °C. However, during relaxation (after arrow in Figure 3) the force falls much more rapidly at the higher temperature. The relaxation rate has a $Q_{10} = 3$ to 4 (Buchthal et al., 1956).

Figure 3. Force–time tracings obtained at two different temperatures by stimulating tetanically in isometric conditions a frog sartorius (resting length = 3.7 cm, mass = 0.087 g) until the force reached the isometric value at that length (P_o) and then by stretching it (4 mm at 30 mm s^{-1}) so that the force attained a greater value (F') at the end of stretching. After stretching the muscle was kept active in isometric conditions for a period of about 5 s. Stimulation was then interrupted (arrow), and relaxation occurred. After relaxation the force settled at a value slightly greater than the initial because of the tension due to the stretched parallel elastic elements. At 20 °C fatigue of the muscle was relevant and responsible for most of the fall of the force taking place at the end of stretching. Note that temperature affects the relaxation rate (after the arrow) much more than the rate of decay of F'.

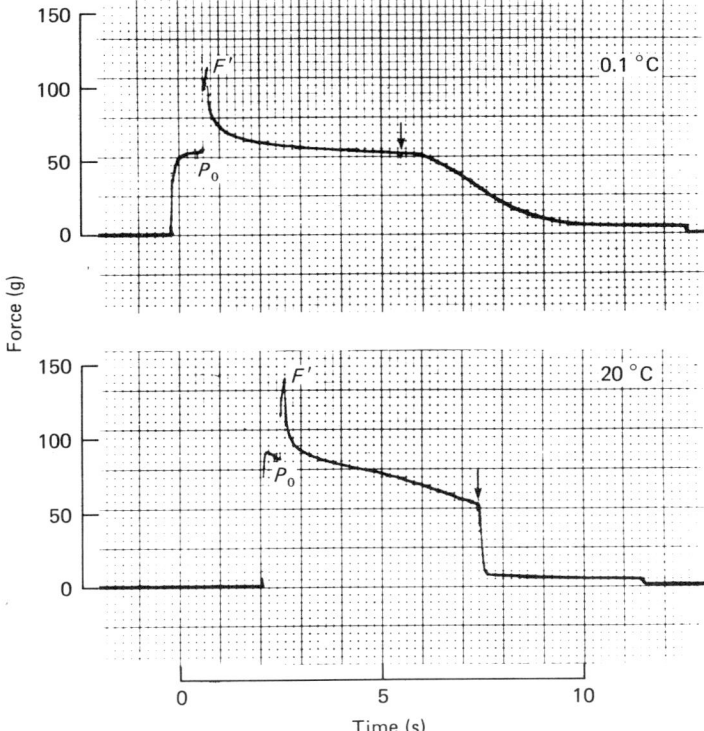

It seems possible that a detachment of the cross-bridges between actin and myosin may occur after stretching, as it does during relaxation. However, the causes responsible for the detachment appear to be different in the two cases. During relaxation the detachment is the result of a chemical reaction [rate constant g in Huxley's theory (Huxley, 1957)]. It is therefore highly affected by temperature. At the end of stretching the detachment may be due to the high stress to which some bridges are subjected during forcible stretching. This process seems to be less affected by temperature.

The finding that the force falls at the end of stretching at about the same rate at 0 °C and 20 °C has practical relevance. If the rate of fall were increased by temperature (in the same way as it is during muscle relaxation), there would be little possibility of recovering mechanical energy at the usual body temperature. Fortunately, it appears possible to recover mechanical energy at high or low temperatures as long as shortening immediately follows stretching. If an interval of time is left between stretching and shortening, the mechanical energy stored during stretching is dissipated into heat.

The effect of the speed of stretching upon the recovery of elastic energy

The force–length curves reproduced in Figure 4A were obtained during shortening of a tetanized gastrocnemius (frog on the left, rat on the right). The shortening took place so rapidly (2 to 4 msec) that the length change due to the cycling of the cross-bridges could be neglected [15 to 24 muscle lengths (l_0) per second for the frog, and 25 to 30 for the rat]. Thus, the recorded length change approximates the elastic recoil of muscle. This recoil is given on the abscissa as $\Delta l/l_0$, and the force is given on the ordinate as a stress: $P \, \delta l_0/M$ (δ = muscle density $\simeq 1$, M = muscle mass). The three curves refer to shortening from a state of isometric contraction, after stretching of the contracted muscle at a low speed, and at a greater speed. A minimum possible interval was left between stretching and release of the muscle. The length change imposed on the fully tetanized muscle during stretching was about 1 mm and occurred on the plateau of the isometric force–length diagram. In fact, the isometric force, P_0, was the same at the length where stretching began and at the length of release.

The three force-shortening curves are matched at a value of force equal to P_0. It can be seen that the average compliance exhibited by the elastic elements of muscle when the force falls from P_0 to zero is greater when shortening is preceded by stretching of the contracted muscle (Cavagna and Citterio, 1974). In addition, the present results show that the area below the curve is greater at high velocities of stretching. This is true for both frog and mammalian muscle, in spite of the different temperatures, 3 °C and 20 °C, respectively.

The mechanical energy released by the elastic elements of muscle can be calculated from the area below the stress–strain curves. This energy (ex-

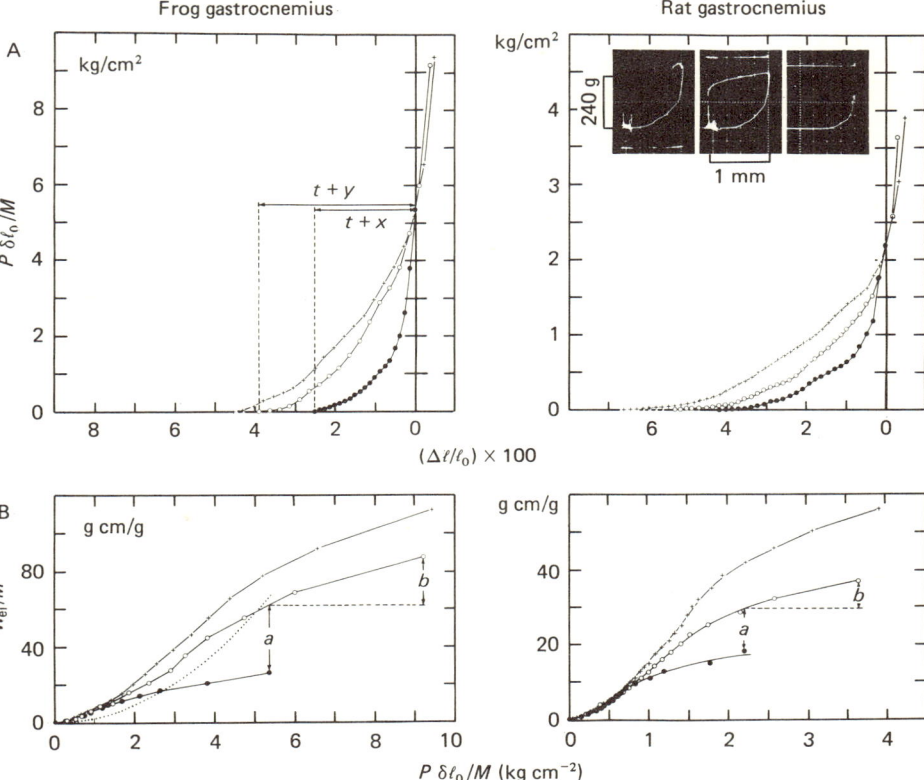

Figure 4. (A) Stress–strain curves of the elastic elements of muscle obtained during shortening at high speed. Left: 380 mm s^{-1} (+); 400 mm s^{-1} (o); and 250 mm s^{-1} (●). Right: 390 mm s^{-1} (+) and 330 mm s^{-1} (o and ●). Ordinate (stress) and abscissa (strain) as described in the text. Resting length (l_0) is the length where the muscle begins to show resting tension. Left: frog gastrocnemius (l_0 = 1.7 cm, M = 0.096 g, T = 3 °C); right: gastrocnemius of a 15-day-old rat (l_0 = 1.3 cm, M = 0.084 g, T = 20 °C). Release of the muscle took place during a state of isometric contraction (●) and after stretching at low speed [left: 0.4 mm s^{-1}; right: 0.9 mm s^{-1}, (o)] and at high speed [left: 7 mm s^{-1}; right: 6 mm s^{-1}, (+)].. When the force falls from P_0 to zero, the elastic recoil of the tendons (t) and the cross-bridges is ($t + x$) during release from a state of isometric contraction (●) and ($t + y$) after stretching at low speed (o). Insert shows the original tracings. (B) The mechanical energy released during shortening was calculated from the area under the stress–strain curves, plotted as a function of the stress. a indicates the additional mechanical energy released by the previously stretched muscle because of the modification of the elastic elements induced by previous stretching; b indicates the additional mechanical energy released because during stretching the force attains values greater than the isometric one. The dotted line indicates the mechanical energy that would be released if the full-points stress–strain curve (above) were a straight line as predicted by Hooke's law.

pressed per gram of muscle mass) is plotted as a function of the stress in Figure 4B. These graphs show that the mechanical energy released by the previously stretched muscle (open points and crosses) is greater than that released by the muscle during shortening from a state of isometric contraction (full points). The additional amount of energy released by the previously stretched muscle has been thought to be due to the greater forces attained during stretching (Hill, 1950). This can be as large as increment b. The present data show that an even larger amount of mechanical energy (increment a) can be released as a result of the modification of the elastic elements. In order to release the same amount of mechanical energy after stretching, *without* modification of the elastic elements, the full point curves in Figure 4 would have to be extrapolated to force values unattainable by the contractile component. The observed modification of the elastic characteristics of muscle allows the storage and release of a large amount of elastic energy without applying stresses that are too great to the muscles and tendons (a sort of "soft" spring).

Similar results have been obtained in frog sartorius, rat soleus, and rat extensor digitorum longus. Therefore, it seems that the increase in storage and recovery of elastic energy with increasing speed of stretching may apply to all vertebrate striated muscle, and may help explain the increase in elastic storage and recovery observed with increasing speed in running man and hopping kangaroo.

Figure 4 refers to experiments in which the muscle was stretched and released over the plateau of the isometric force–length diagram. If the same experiment is carried out at shorter muscle lengths, one observes that: (a) the force–length diagram of elastic elements obtained by releasing the muscle from a state of isometric contraction (full points in Figure 4) remains about the same (Jewell and Wilkie, 1958); but (b) the modification of the elastic elements induced by previous stretching is reduced, so that the curves obtained after stretching approach those obtained from a state of isometric contraction. This same phenomenon occurs if an interval of time is left between stretching and shortening (Cavagna and Citterio, 1974).

The contractile component as a site of elastic compliance

It is possible to discuss the experiments described above in terms of the properties of the cross-bridges between actin and myosin. Huxley and Simmons (1971) determined the force-shortening curve of a single fiber released during a state of isometric contraction (a) from the optimum fiber length (maximum overlap between actin and myosin) and (b) from a greater length (reduced overlap). The length change necessary to reduce the force to zero was the same in (a) and in (b) in spite of the initial force being much greater in (a) than in (b). This suggested that the elastic recoil of the fiber was totally due to the recoil of the bridges between actin and myosin; this was about 15 nm in each half sarcomere, corresponding to 1.5% of the total fiber length.

The experimental results described in Figure 4 may then be interpreted if one assumes that (1) at the end of stretching some cross-bridges are subjected to a force greater than the force they develop during an isometric contraction; (2) these cross-bridges are in an unstable condition having a probability of breaking that is greater the greater their strain; and (3) the elastic recoil of the tendons, t, taking place when the force falls from P_0 to zero, is the same when the muscle shortens immediately after stretching and when it shortens from a state of isometric contraction.

It follows that if the muscle is released immediately after stretching, the most strained bridges have the possibility of shortening, therefore delivering the mechanical energy stored during stretching. However, if the muscle is kept active at the stretched length (as in Figure 3), the most strained bridges would be the first to break. This would explain the observed fall of the force after stretching, and the reduction of compliance of the elastic elements during subsequent release (from P_0 to zero) (Cavagna and Citterio, 1974). If all the bridges were subjected to the same strain at the end of stretching, a detachment of a fraction of them would result in an increase of compliance during subsequent shortening rather than the observed decrease. In fact, in Huxley and Simmons's experiments the compliance of the elastic elements of the fiber was increased when release took place from a length at which there were fewer bridges.

For a given fall of the force, the elastic recoil of the most strained cross-bridges taking place when the muscle is released immediately after stretching (y) is greater than the recoil of the cross-bridges taking place when the muscle is released from a state of isometric contraction (x). The results given in Figure 4A show the elastic recoil of both the tendons (t) and the cross-bridges (x and y): $(t + y)/(t + x) = 1.5$ to 2 (i.e., the amount of shortening required to make the force fall from P_0 to zero is 1.5 to 2 times greater after stretching). The ratio y/x will obviously depend upon the value of t; for example, if $t = x$ (as suggested by Jewell and Wilkie, 1958, for frog sartorius), then $y = 2x$ to $3x$. In this case the total elastic recoil of the bridges would amount to 30 to 45 nm in each half sarcomere.

Rack and Westbury (1974) subjected an active muscle to stretch and shortening cycles and obtained evidence of an elastic extension of the contractile component up to 25 to 35 nm in each half sarcomere instead of 15 nm, as obtained by Huxley and Simmons (1971). The difference may be that in Huxley and Simmons's experiments the bridges "stretched themselves" in their attachment to the actin during the isometric contraction, whereas in the Rack and Westbury experiments (as in those by Cavagna and Citterio, 1974) the bridges were forcibly stretched after their attachment. In Rack and Westbury's experiments the force developed by the muscle oscillated between about 1.2 P_0 and 0.6 P_0 (calculated from Figure 3 of their paper), whereas in Cavagna and Citterio's (1974) experiments, as in the present ones (Figure 4), the force reached about 2 P_0 during stretching and then fell to zero during shortening: this greater change of force may imply a length change of the bridges even greater than 25 to 35 nm in each half sarcomere. Rack and Westbury also arrived at the conclusion that during stretching many cross-bridges "must be extended further than the 25 to 35 nanometers mentioned above."

The effect of the speed of stretching on the recovery of mechanical energy (Figure 4) is in agreement with the interpretation given above. In fact, both the extension of each bridge and the number of most strained bridges existing at the end of stretching are likely to increase with speed of stretching. The length change that each bridge will undergo during the period of time in which it remains attached to the actin will increase with the velocity of the movement of the filaments past each other. Additionally, the "oldest" bridges (which are about to break) will be able to undergo a maximum extension only if this occurs in a short time. A greater extension of the bridges will lead to a greater elastic recoil of the muscle. A greater number of bridges subjected to the maximum extension will lead to a greater mechanical energy released during this recoil (i.e., to a straighter force-shortening curve). The results given in Figure 4 show that this second effect of stretching is more relevant for the recovery of mechanical energy than the first one.

The presence of a greater number of more strained and therefore more unstable bridges (assumption 2 above), after lengthening at high speed, would also explain the finding that the force exerted by an active muscle after stretching falls with a rate that is greater the greater the speed of stretching (Abbott and Aubert, 1952).

Tendons or cross-bridges?

The mechanical energy released during rapid shortening of the whole muscle (W_{el}/M in Figure 4) is due to the recoil of both the tendons and the cross-bridges. However, the extra amount of energy released by the previously stretched muscle, when the force falls from the isometric value, P_0, to zero (a in Figure 4), is entirely due to the recoil of bridges if the recoil of the tendons (from P_0 to zero) is unaffected by previous stretching (assumption 3 above). This assumption is supported by the finding that the extra amount of energy a may be reduced almost to zero when the stretch–shorten cycle occurs at shorter muscle lengths. The tendons should behave equally at all muscle lengths independent of the overlap between actin and myosin. These considerations induce us to think that the amount of mechanical energy stored in the contractile component can be much greater than the elastic energy in the tendons.

This conclusion is in contrast to that reached by Alexander and Bennet-Clark (1977), who concluded that the elastic energy stored in the contractile component of running mammals (U) was a negligible fraction of the elastic strain energy in the tendons (U_t). These authors derived the equation:

$$U_t/U = (\lambda/2b)\,(\Delta l_t/l) \tag{2}$$

where λ is the sarcomere length, b is the elastic extension of each cross-bridge, Δl_t is the length change of the tendon, and l is the length of the muscle fiber. According to Alexander and Bennet-Clark, Equation 2 can be written as:

$$U_t/U = (70 \text{ to } 35)\,0.1\,(l_t/l)$$

by taking $\lambda = 2.1$, $b = 15$ to 30 nm, and $\Delta l_t = 0.1\,l_t$, and by assuming that "tendons stretch about 10% before breaking," and that "they are not much thicker than they have to be to withstand the forces their muscles exert on them." If we use a value of $b = 45$ nm (which seems reasonable from our previous discussion) and $\Delta l_t = 0.05\,l_t$ (because the force may attain at least $2P_o$ without breaking the tendons), Equation 2 becomes:

$$U_t/U = 1.2\,(l_t/l)$$

Preliminary measurements suggest that the length of the fibers in both gastrocnemii (of frogs and rat) is about one-third the length of the whole muscle, so that $U_t/U \simeq 2.4$. This value is much higher than we would expect from the data in Figure 4. How can we explain the discrepancy?

Equation 2 was derived assuming that both the tendons and the cross-bridges follow Hooke's law. However, the full point curves in Figure 4 show that the elastic elements of muscle (bridges plus tendons) do not follow Hooke's law (see also, e.g., Jewell and Wilkie, 1958). They store much less mechanical energy than expected on the basis of Hooke's law (dotted line in Figure 4).

Conclusion

The present experiments suggest that a great amount of mechanical energy can be stored in the contractile component of striated muscle, particularly when the active muscle is stretched at high speed. Because this result applies to the muscle of both the frog and the rat, it does not appear to have been a property of muscle that has changed during the evolution of vertebrates. This paper has also shown that it is necessary to take into account both the modification of the elastic elements due to previous stretching and the increase of the Young's modulus of the tendons with increasing force when calculating the amount of elastic energy stored in the tendons versus that stored in the cross-bridges. The analysis of records from kangaroo (hopping) and man (running) suggests that these principles apply equally in the locomotion of a "primitive" and a more advanced species.

References

Abbott, B. C., and Aubert, X. M. (1952). The force exerted by active striated muscle during and after change of length. *J. Physiol. (London) 117*:77–86.

Alexander, R. McN., and Bennet-Clark, H. C. (1977). Storage of elastic strain energy in muscle and other tissues. *Nature (London) 265:*114–7.

Alexander, R. McN., and Vernon, A. (1975). The mechanics of hopping by kangaroos (Macropodidae). *J. Zool. (London) 177:*265–303.

Buchthal, F., Svensmark, O., and Rosenfalck, P. (1956). Mechanical and chemical events in muscle contraction. *Physiol. Rev. 36:*503–38.

Cavagna, G. A., and Citterio, G. (1974). Effect of stretching on the elastic characteristics and the contractile component of frog striated muscle. *J. Physiol. (London) 239:*1–14.

Cavagna, G. A., and Kaneko, M. (1977). Mechanical work and efficiency in level walking and running. *J. Physiol. (London) 268:*467–81.

Cavagna, G. A., Thys, H., and Zamboni, A. (1976). The sources of external work in level walking and running. *J. Physiol. (London) 262:*639–57.

Cavagna, G. A., Heglund, N. C., and Taylor, C. R. (1977). Mechanical work in terrestrial locomotion: Two basic mechanisms for minimizing energy expenditure. *Am. J. Physiol. 233(5):*R243–61.

Dawson, T. J., and Taylor, C. R. (1973). Energetic cost of locomotion in kangaroos. *Nature (London) 246:*313–4.

Hill, A. V. (1950). The series elastic component of muscle. *Proc. R. Soc. (London) Ser. B 137:*273–80.

Huxley, A. F. (1957). Muscle structure and theories of contraction. *Prog. Biophys. Biophys. Chem. 7:*257–318.

Huxley, A. F., and Simmons, R. M. (1971). Mechanical properties of the cross-bridges of frog striated muscle. *J. Physiol. (London) 218:*59–60P.

Jewell, B. R., and Wilkie, D. R. (1958). An analysis of the mechanical components in frog's striated muscle. *J. Physiol. (London) 143:*515–40.

Margaria, R. (1938). Sulla fisiologia e specialmente sul consumo energetico della marcia e della corsa a varie velocità ed inclinazioni del terreno. *Reale Accademia Nazionale dei Lincei,* Memoria, 7:299–368.

Rack, P. M. H., and Westbury, D. R. (1974). The short range stiffness of active mammalian muscle and its effect on mechanical properties. *J. Physiol. (London) 240:*331–50.

24

Properties and distributions of the fiber types in the locomotory muscles of mammals

ROBERT B. ARMSTRONG

Mammalian locomotory muscles contain mixtures of fibers with different contractile and metabolic properties. The presence of these various *fiber types* within the same muscle permits it to generate forces over a broad range of shortening velocities with varying degrees of resistance to fatigue. Unlike the muscles of most lower vertebrates and invertebrates, which tend to have more homogeneous fiber compositions, mammalian muscles possess an amazing degree of versatility in meeting their contractile requirements. In fact, the heterogenous fiber composition in the skeletal muscles of mammals reflects the complexity of the innervating motor components of the nervous system.

Most of the research on mammalian skeletal muscle fiber types has been on Eutherian species from the more "advanced" orders. Comparatively little is known about the properties and distribution of the fiber types in the muscles of monotremes, marsupials, and insectivores. Representatives from these latter mammalian groups exhibit similar energetics (Taylor, this volume) and utilize the same basic mechanical principles (Heglund, this volume) as the so-called advanced mammals during locomotion; so it would be predicted they should possess muscle fibers with similar capabilities. The purpose of this paper is severalfold: (1) to summarize the contractile and metabolic characteristics of the fiber types as they have been defined in laboratory animals; (2) to survey representatives of the various mammalian subclasses and orders to determine if these fiber types are universally apportioned among mammals; (3) to seek general patterns of fiber type distribution within and among homologous muscles among mammals; and (4) to discuss briefly patterns of utilization of the fiber types during terrestrial locomotion.

Properties of mammalian skeletal muscle fiber types

Although the absolute shortening speeds and metabolic capacities of the fibers within a given muscle represent a continuous spectrum with no distinct points of demarcation (Burke et al., 1971), three well-defined types of fibers may be distinguished in the muscles of Eutherian mammals using enzyme histochemistry (Close, 1972; Ariano et al., 1973; Burke and Edgerton, 1975). Sections of muscles assayed for myofibrillar ATPase activity after alkaline preincubation (Padykula and Herman, 1955) show distinct dark or light staining intensities representing fast- and slow-twitching fibers, respectively (Peter et al., 1972). The oxidative capacities (mitochondrial densities) of the same fibers may be estimated by assaying serial sections of the muscle for NADH-diaphorase activity (Novikoff et al., 1961). With these two histochemical stains, fibers may be classified as fast-twitch–high-oxidative (FH), fast-twitch–low-oxidative (FL), or slow-twitch–high-oxidative (S) (Figure 1).

The physiological and biochemical properties of the three fiber types have been thoroughly described in laboratory animals (Close, 1972; Peter et al., 1972; Burke and Edgerton, 1975), and several of these characteristics are depicted in Figure 2. All of the muscle fibers within a given motor unit are of the same fiber type. Motor units composed of FL (FF or FG in Figure 2) fibers have fast contraction and relaxation speeds. Because of the large number of muscle fibers in each motor unit and the large cross-sectional area of the individual fibers, these units also generate high peak tensions. However, they fatigue rapidly during continuous use because of their low oxidative capacities (low mitochondrial densities and few capillaries). These motor units are primarily active during periods of intense muscular activity when high forces must be rapidly generated, and rely to a major extent upon anaerobic glycolysis for ATP production.

On the other end of the spectrum, motor units composed of S fibers have

Figure 1. Serial sections of laboratory rat *(Rattus norvegicus)* plantaris muscle stained for NADH-diaphorase (A), indicating the oxidative capacity of the muscle fibers, and for myofibrillar ATPase (B), indicating the contractile velocities of the fibers as fast (dark stain) or slow (light stain). Representative fast-twitch–high-oxidative (FH), fast-twitch–low-oxidative (FL), and slow-twitch–high-oxidative (S) fibers are identified on both photomicrographs. Note that the fibers are the same in the two sections. (× 113)

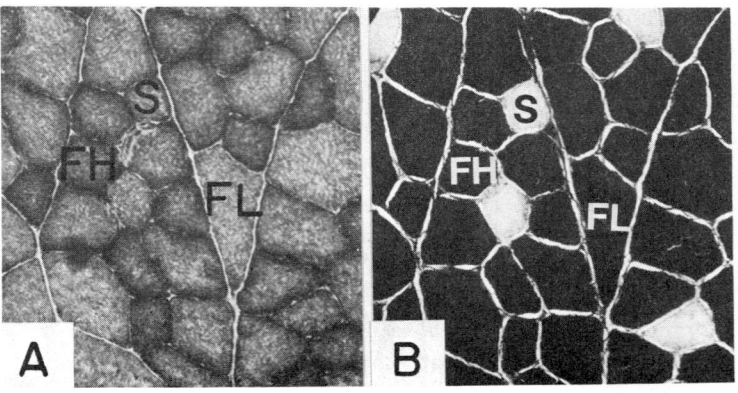

slow contraction and relaxation times, generate low tensions (primarily due to low innervation ratios and small cross-sectional areas), and are extremely resistant to fatigue during continuous use. These properties make these fibers particularly well-suited for postural maintenance, force development during low-intensity continuous cyclic exercise (e.g., locomotion), and manipulation requiring fine motor control. The characteristics of the FH (FR and FOG in Figure 2) fibers are intermediate to those of the FL and S, so

Figure 2. Diagram summarizing some important features of the organization of motor units in the medial gastrocnemius muscle of the cat. Shading in the muscle fiber outlines denotes relative staining intensities found for each histochemical reaction (identified in the FF unit fibers). Motor unit type nomenclature: FF – fast twitch, fatiguable; FR – fast twitch, fatigue-resistant; S – slow twitch. Histochemical profiles: FG – fast twitch, glycolytic; FOG – fast twitch, oxidative-glycolytic; SO – slow twitch, oxidative. These two systems are essentially interchangeable. (These three fiber types described by Burke and Edgerton correspond to the FL, FH, and S fibers, respectively – author's note.) (From Burke and Edgerton, 1975).

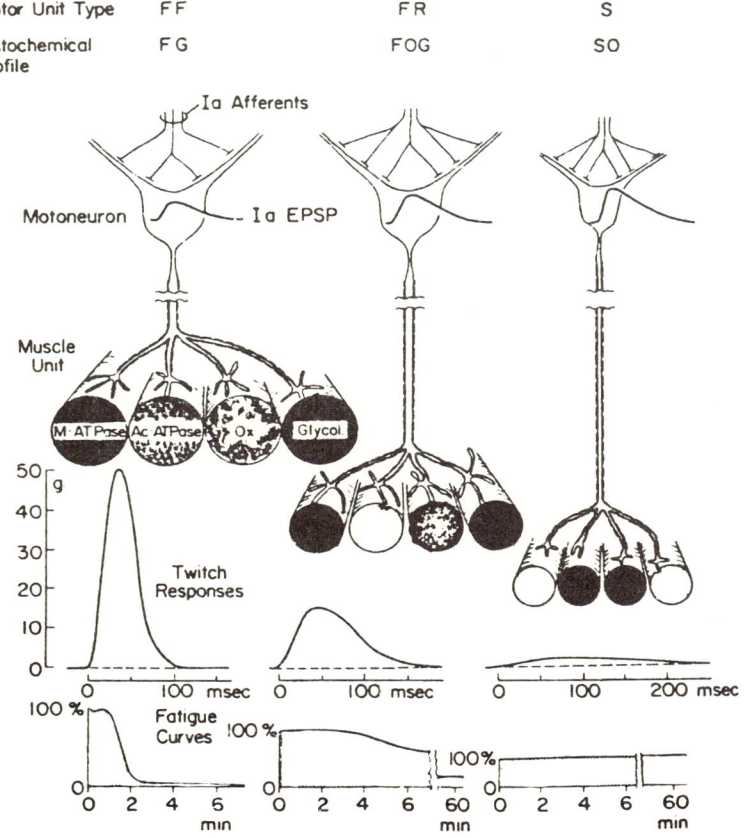

that within a muscle there is a continuous spectrum of motor units with varying contraction speeds and metabolic potentials.

As indicated, the different properties of the fiber types permit them to accomplish specific tasks in the muscle during the maintenance of posture, during locomotion, and in the performance of a variety of other specialized motor activities. Determination of fiber type composition thus provides a means for establishing the functional constraints under which a muscle must operate, and for predicting probable patterns of muscle fiber utilization during locomotion (Gollnick et al., 1973; Armstrong et al., 1974; Burke and Edgerton, 1975; Armstrong et al., 1977b; Sullivan and Armstrong, 1978).

Before comparing the fiber type composition of the muscles from the mammalian groups it is important to point out that the basic fiber types, as identified from histochemistry, do not necessarily have the same absolute contractile speeds or metabolic capacities. Rates of shortening of muscles, regardless of fiber composition, tend to be inversely related to body size. (Close, 1972). For example, the data in Table 1 show that within animals, muscles composed primarily or entirely of S fibers (SOL) have slow shortening speeds compared to muscles with a preponderance of fast fibers (EDL). However, the homologous muscles of larger animals have slower absolute shortening velocities than those of smaller animals, even though the fiber type composition as estimated from histochemistry is similar. The fast muscle of cat, for example, has about the same intrinsic shortening speed as the slow muscle in mouse. Thus, the myofibrillar ATPase stain is indicative of relative fiber contractile speeds within animals, but not across species.

The same is true of the metabolic capacities of the fibers. FH fibers may be identified both in bat flight muscle and in rat gastrocnemius muscle, for example, but these particular fibers in bats have about four times the oxida-

Table 1. *Fiber compositions and intrinsic shortening velocities of extensor digitorum longus (EDL) and soleus (SOL) muscles of mouse (Mus), rat (Rattus), and domestic cat (Felis)*

Animal	Body weight (kg)	Muscle	S fiber composition[a] (%)	Sarcomere Shortening speed[b] (μm s^{-1})
Mouse	0.02–0.03	EDL	9[c]	61
		SOL	67[c]	32
Rat	0.2–0.3	EDL	3	43
		SOL	84	18
Cat	2.0–3.0	EDL	14	31[d]
		SOL	100	13

[a]Ariano et al., 1973; [b]Close, 1972; [c]unpublished observations from our laboratory; [d]quadriceps muscle.
Adapted from Close, 1972.

tive capacity (as indicated by quantitative measurement of the activity of the Krebs cycle enzyme, succinate dehydrogenase) as those in rat muscle (Ianuzzo and Armstrong, 1976; Armstrong et al., 1977a). Thus, fiber typing with histochemistry permits a qualitative estimation of the composition of muscles and is useful for predicting muscle function in a given animal, but does not allow a description of shortening speed or metabolic potential in absolute quantitative terms.

We can now address the question: Are the three basic fiber types present in all mammalian groups, or do the conservative mammalian subclasses (Monotremata and Marsupialia) and orders (Insectivora) possess fibers with "primitive" characteristics? From the representatives of the various groups we have studied, it appears that no so-called primitive characteristics can be identified. The basic fiber types are readily apparent in the muscles of animals from these groups. For example, comparison of the fibers in the gastrocnemius muscle of two hopping mammals of about the same body size, one a rodent (springhare, *Pedetes* sp.), the other a marsupial (rat kangaroo, *Bettongia* sp.), shows the three fiber types are remarkably similar in the two animals (Figure 3). Figure 4 contains photomicrographs of gastrocnemius muscle sections from the spiny anteater (*Tachyglossus aculeatus*), the American opossum (*Didelphis virginiana*), and the tenrec (*Tenrec ecaudatus*). Again, the three fiber types may readily be identified. The only animals we have studied that do not possess distinct fast- and slow-twitch fibers are in the order Chiroptera (*Myotis*, Armstrong et al., 1977a; and *Eptesicus*, unpublished observations from our laboratory). These bats have no S fibers in their locomotory muscles.

Fiber type distribution in mammalian muscles and muscle groups

Several interesting points were previously noted in a study of fiber composition in the hind limb muscles of terrestrial Eutherian mammals (guinea pig, rat, cat, and lesser bushbaby, *Galago senegalensis*). Within the major extensor, antigravity muscle groups (i.e., triceps surae in the leg and quadriceps in the thigh), the deepest muscle (soleus and vastus intermedius, respectively) in the group was primarily composed of S fibers, whereas the other muscles contained only small proportions of S fibers. On the other hand, in the antagonistic muscle groups (flexors), all of the muscles possessed relatively low proportions of S fibers. Collatos et al. (1977) recently described similar patterns in the arm muscles of cats. This fiber distribution appears to reflect the upright position and locomotory mode of the animals because the high resistance to fatigue of the S fibers makes them particularly well-suited for maintenance of posture and efficient production of muscular force during locomotion. Indeed, these patterns stand in marked contrast to the fiber distribution in the muscles of the slow loris (*Nycticebus coucang*), which typically moves through tree branches in either the upright or a suspended position. Predictably, both extensor and flexor muscles in this animal possess high proportions of S fibers (Ariano et al., 1973).

We (Armstrong, Marum, Seeherman, and Taylor, unpublished observations) have now extended these observations to include all of the locomotory muscles in a variety of mammalian species. As demonstrated in Figure 5, terrestrial mammals, regardless of their taxonomic position, exhibit a predictable pattern of fiber distribution in their limb extensor and flexor muscles. In the extensor muscle groups, the deepest muscle (in Figure 5, medial head of triceps brachii muscle) is primarily composed of S fibers, whereas the more peripheral synergistic muscles have fewer S fibers. The

Figure 3. Serial sections of samples from gastrocnemius muscles of the springhare *(Pedetes cafer)*, a rodent (1), and the rat kangaroo *(B. lesueuri)*, a marsupial (2). The sections are histochemically stained for NADH-diaphorase (A), which is indicative of oxidative capacity, and for myofibrillar ATPase (B), which is indicative of contractile velocity. Fast-twitch–high-oxidative (FH), fast-twitch–low-oxidative (FL), and slow-twitch–high-oxidative (S) fibers are identified on both photomicrographs. Note that the same fibers are identified on both sections. (× 139)

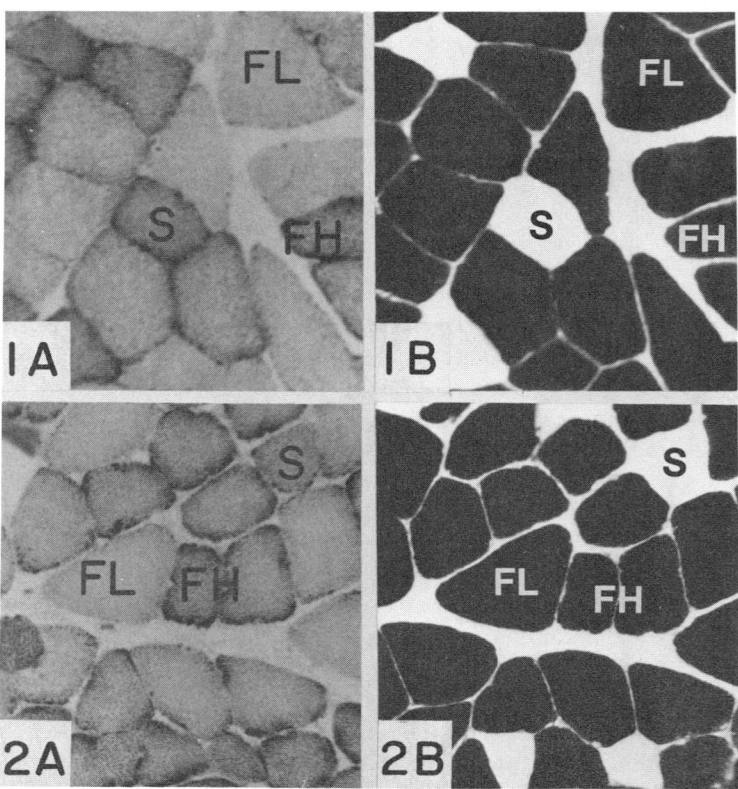

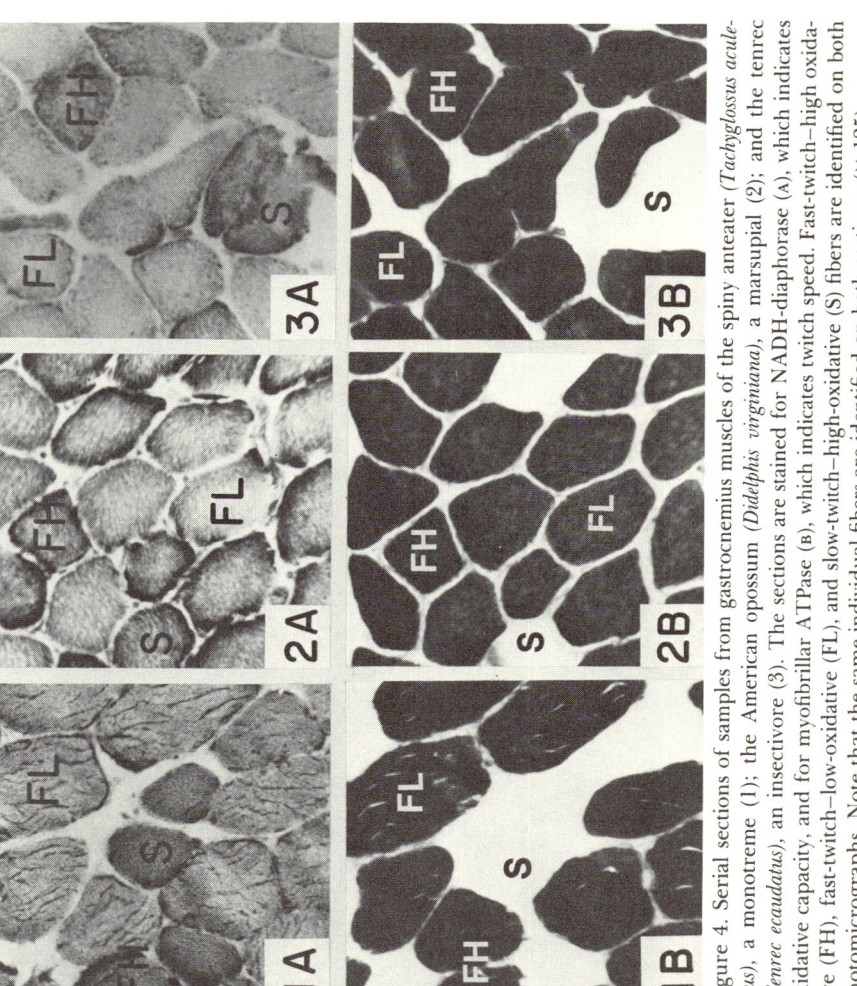

Figure 4. Serial sections of samples from gastrocnemius muscles of the spiny anteater (*Tachyglossus aculeatus*), a monotreme (1); the American opossum (*Didelphis virginiana*), a marsupial (2); and the tenrec (*Tenrec ecaudatus*), an insectivore (3). The sections are stained for NADH-diaphorase (A), which indicates oxidative capacity, and for myofibrillar ATPase (B), which indicates twitch speed. Fast-twitch–high oxidative (FH), fast-twitch–low-oxidative (FL), and slow-twitch–high-oxidative (S) fibers are identified on both photomicrographs. Note that the same individual fibers are identified on both sections. (× 125)

flexor muscle groups normally possess fewer S fibers than the extensors, and do not include a deep, predominantly slow muscle. This general pattern exists in all of the antigravity muscle groups and their antagonists, respectively, in both front and hind limbs in upright terrestrial mammals.

Figure 5 also illustrates the general distributions of fibers within locomotory muscles. The two values presented for the long head of triceps brachii muscle for each animal indicate the percentages of S fibers in the deep and

Figure 5. Slow-twitch–high-oxidative (S) fiber composition of the major extensor and flexor muscles of the forearm in opossum *(Didelphis virginiana)*, laboratory rat *(Rattus norvegicus)*, African lion *(Leo leo)*, and horse *(Equus caballus)*. Flexors: biceps brachii and brachialis muscles. Extensors: triceps brachii muscles; long, lateral, and medial heads. Intensity of the shading is proportional to the percentage of S fibers in the muscle cross sections. In every animal, the medial head of triceps brachii muscle contains the highest proportion of S fibers. The numbers in the long head of triceps brachii muscle indicate the stratification of S fibers in the muscle. (Not all muscles that appear in transverse sections through the arms are included.)

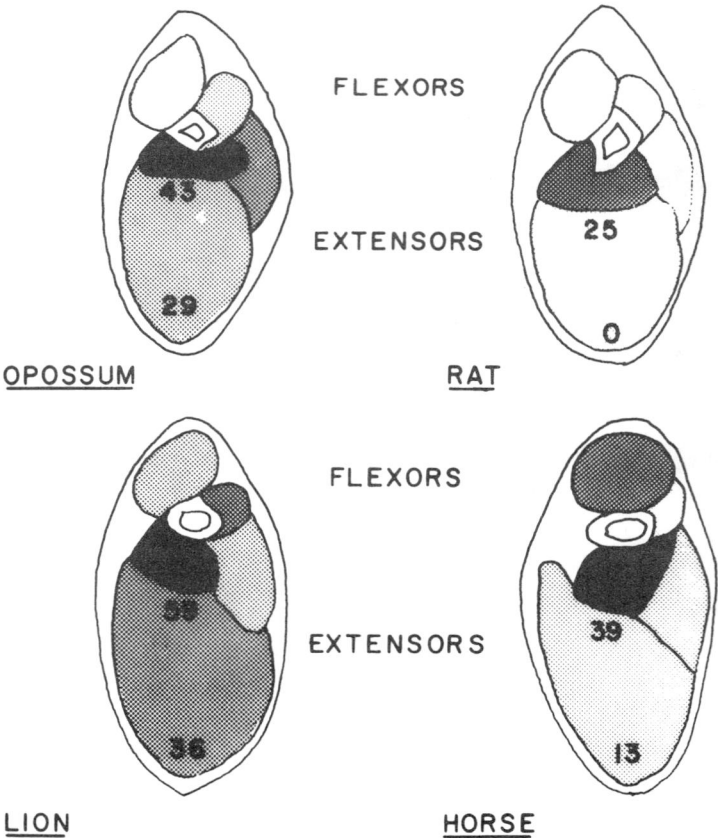

superficial portions of the muscle cross section. This stratification of S fibers is more prominent in muscles with a predominance of fast-twitching fibers. Slow muscles (e.g., medial head of triceps brachii muscle in Figure 5) are more uniform in fiber distribution. Thus, both muscles and parts of muscles that are closest to the bone have the greatest populations of S fibers. These patterns are found in mammals from all taxonomic groups.

Muscle fiber recruitment during locomotion

The stereotyped pattern of fiber placement that has been described suggests the existence of a basic functional design that animals can use for efficiently increasing the forces generated from a muscle or muscle group during locomotion. It has been demonstrated that motor units composed of S and FH fibers are recruited before the FL units during muscular activity requiring low to moderate force generation (Henneman and Olson, 1965; Burke and Edgerton, 1975). Thus, for most daily activities the FL units are not utilized, but form a reserve for times when the animal must generate bursts of intense muscular activity. The fibers that are most readily recruited, the S and FH fibers, have good blood supplies and high mitochondrial densities, and are metabolically capable of prolonged contractile activity (Close, 1972; Burke and Edgerton, 1975). These fibers also lie deepest in the muscles and muscle groups, where they can most efficiently apply mechanical tension on the bony levers during contraction.

Under the proper experimental conditions, fiber utilization during steady state locomotion can be determined by following glycogen depletion in the fibers (Gollnick et al., 1973; Armstrong et al., 1974; Burke and Edgerton, 1975). Figure 6 summarizes the results from an experiment in which laboratory rats ran on a motor-driven treadmill at different speeds, and the glycogen loss in the muscles was used to determine fiber recruitment. The muscles studied were soleus, plantaris, and gastrocnemius, which constitute a major group of extensors of the foot. Soleus muscle is primarily composed of S fibers (85%), and plantaris and gastrocnemius muscles possess about 10% S fibers distributed as indicated in Figure 6A. This follows the general pattern of fiber distribution described in the previous section. With increasing running speed there is a peripheral recruitment of muscle fibers with successively lower oxidative capacities and faster contractile speeds, both within and among muscles in the group (Figure 6B). At the slower running speeds the S and FH fibers contribute nearly 100% of the required force, whereas at higher running speeds there is a successive recruitment of FL motor units. Keeping in mind the distribution of the fibers in the muscles, it is apparent that those that are most readily activated for force generation, the S and FH fibers, are positioned where they can most effectively exert force on the bony levers. Presumably, similar recruitment patterns exist in the muscles of other terrestrial mammals during locomotion, although, interestingly, discontinuities in the recruitment of FL motor units at the gait transitions have been observed in lion muscles as the animals increase running speed (Armstrong et al., 1977b).

Summary

The properties and distributions of the fiber types in the locomotory muscles of mammals accurately reflect the particular functions of the

Figure 6. Slow-twitch–high-oxidative fiber composition (A) and active cross-sectional area* of muscle during running at different speeds (B) in soleus (S), plantaris (P), and gastrocnemius (G) muscles of laboratory rats *(Rattus norvegicus)*. Other abbreviations: T (tibia) and F (fibula). (*Based on muscle fiber glycogen depletion; from Sullivan and Armstrong, *J. Appl. Physiol.*, 1978.)

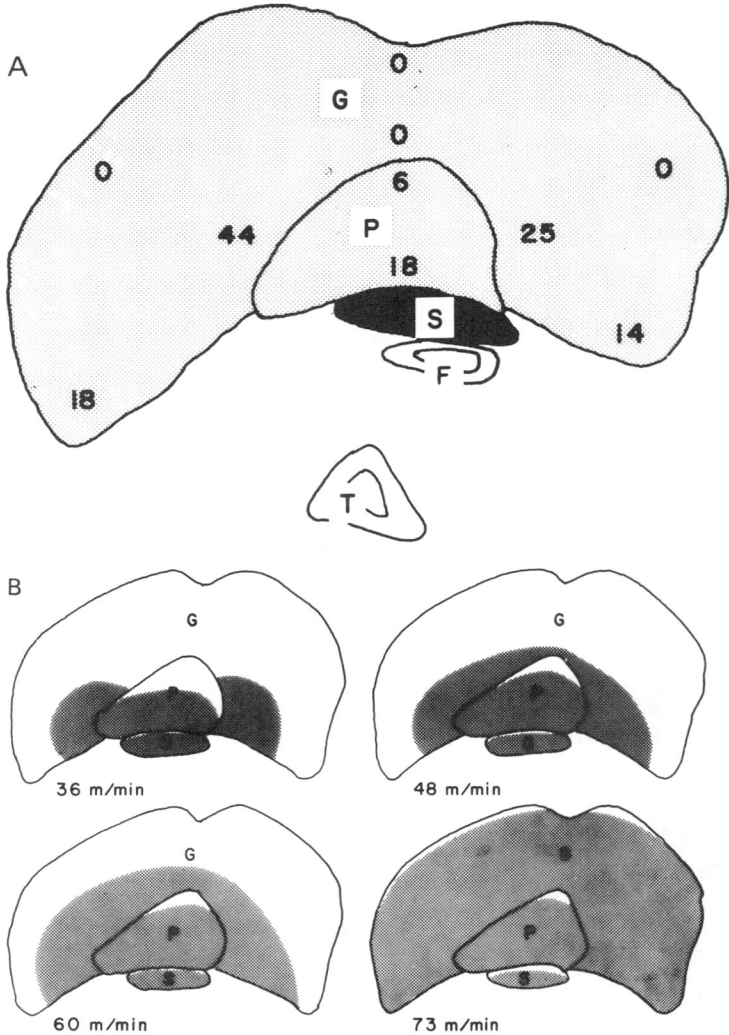

muscles, and appear to follow predictable patterns throughout the various mammalian groups. No conservative, or "primitive," characteristics are evident, and nearly all mammals have the same basic histochemical fiber types intermixed in their skeletal muscles in specific, stereotyped patterns. This leads to the conclusion that members of all mammalian groups possess muscular systems with similar flexibilities for meeting the varied contractile requirements involved in posture, locomotion over a wide range of speeds, manipulation of food, and other specialized muscular functions. Finally, the similarities in the properties and distributions of the muscle fiber types among mammals suggest that their neural motor control systems are essentially the same, and that deviations from the general distribution patterns reflect adaptations to specialized locomotory and behavioral modes rather than different taxonomic origins.

The author wishes to express his thanks to Ms. Patricia Marum for her help in the collection and analysis of the data included in this paper. This paper was supported by NIH grants 2 RO1 AM 18140 and 2 RO1 AM 18123.

References

Ariano, M. A., Armstrong, R. B., and Edgerton, V. R. (1973). Hindlimb muscle fiber populations of five mammals. *J. Histochem. Cytochem. 21:*51–5.

Armstrong, R. B., Saubert, C. W., IV, Sembrowich, W. L., Shepherd, R. E., and Gollnick, P. D. (1974). Glycogen depletion in rat skeletal muscle fibers at different intensities and durations of exercise. *Pflügers Arch. 352:*243–56.

Armstrong, R. B., Ianuzzo, C. D., and Kunz, T. H. (1977a). Histochemical and biochemical properties of flight muscle fibers in the little brown bat, *Myotis lucifugus. J. Comp. Physiol. 119:*141–54.

Armstrong, R. B., Marum, P., Saubert, C. W., IV, Seeherman, H. J., and Taylor, C. R. (1977b). Muscle fiber activity as a function of speed and gait. *J. Appl. Physiol. 43:*672–7.

Burke, R. E., and Edgerton, V. R. (1975). Motor unit properties and selective involvement in movement. *Exercise Sport Sci. Rev. 3:*31–81.

Burke, R. E., Levine, D. N., Zajac, F. E., Tsairis, P., and Engel, W. K. (1971). Mammalian motor units: Physiological–histochemical correlation in three types in cat gastrocnemius. *Science 174:*709–12.

Close, R. I. (1972). Dynamic properties of mammalian skeletal muscles. *Physiol. Rev. 52:*129–97.

Collatos, T. C., Edgerton, V. R., Smith, J. L., and Botterman, B. R. (1977). Contractile properties and fiber type compositions of flexors and extensors of elbow joint in cat: Implications for motor control. *J. Neurophysiol. 40:*1292–1300.

Gollnick, P. D., Armstrong, R. B., Saubert, C. W., IV, Sembrowich, W. L., Shepherd, R. E., and Saltin, B. (1973). Glycogen depletion patterns in human skeletal muscle fibers during prolonged work. *Pflügers Arch. 344:*1–2.

Henneman, E., and Olson, C. B. (1965). Relations between structure and function in design of skeletal muscles. *J. Neurophysiol. 28:*581–93.

Ianuzzo, C. D., and Armstrong, R. B. (1976). Phosphofructokinase and succinate dehydrogenase activities of normal and diabetic rat skeletal muscle. *Horm. Metab. Res. 8:*244–5.

Novikoff, A. B., Shin, W., and Drucker, J. (1961). Mitochondrial localization of oxidative enzymes: Staining results with two tetrazolium salts. *J. Biophys. Biochem.* 9:47–61.

Padykula, H. A., and Herman, E. (1955). The specificity of the histochemical method of adenosine triphosphatase. *J. Histochem. Cytochem.* 3:170–95.

Peter, J. B., Barnard, R. J., Edgerton, V. R., Gillespie, C. A., and Stempel, K. E. (1972). Metabolic profiles of three types of skeletal muscle in guinea pigs and rabbits. *Biochemistry* 11:2627–33.

Sullivan, T. E., and Armstrong, R. B. (1978). Rat locomotory muscle fiber activity during trotting and galloping. *J. Appl. Physiol.* 44:358–63.

25

Endocrines: problems of phylogeny and evolution

P. J. BENTLEY

For endocrine evolution is not an evolution of hormones but of the uses to which they are put; an evolution not, to put it crudely, of chemical formulae but of reactivities, reaction patterns and tissue competences.
P. B. Medawar (1953)

It would be truer to say that the evolution of the vertebrate hormones was nearly completed by Silurian times, and that the subsequent adaptive radiation of the vertebrates depended in part on changes in tissue reponses to them.
J. B. S. Haldane (1953)

The phyletic differences between groups of vertebrates indicate that the endocrine system has been subject to evolutionary change. A basically similar complement of hormones has been identified throughout the Vertebrata, but several chemical variations of a hormone may exist, and these often have a characteristic phyletic distribution so that putative pathways for their chemical evolution have been proposed. As pointed out by P. B. Medawar, however, the more fundamentally important aspect of endocrine evolution appears to have been the assumption of different functions by the hormones or, as put more precisely by J. B. S. Haldane, "changes in tissue responses to them."

The endocrine system provides a mechanism for the coordination of physiological and environmental events. Hormones are especially involved in the processes of reproduction, intermediary metabolism, and the regulation of the mineral content of the body. The physiological requirements of such processes differ depending on the particular environment, the animal's general way of life, and the physiological and morphological armory with which it is endowed.

The study of endocrinology in the marsupials and monotremes has a number of special attractions:

1. These animals have had the opportunity to evolve separately for a period of about 100 million years so that it is possible that they may have acquired some novel endocrine mechanisms.

2. All of the monotremes and most marsupials are confined to the Australasian geographical region where they occupy a variety of different habitats, some of which are quite unique.

3. The monotremes and marsupials also have a number of unique morphological and physiological characters which could influence the nature of the endocrine coordination.

The reproductive processes of monotremes and marsupials are especially characteristic of these mammalian orders; the former are egg-laying, whereas the latter invariably, after a short period of gestation, produce young in a relatively immature condition. The pituitary gland has a special role in regulating reproductive cycles in vertebrates, and Dr. H. Tyndale-Biscoe has provided a description of our, as yet, limited knowledge of its functions in marsupials. There are many parallels to pituitary function in eutherians, but there are also some differences, such as the nature of the regulation of the corpus luteum. This ovarian tissue has a special role in pregnancy, but this role appears to have been modified somewhat in marsupials. It is especially interesting to observe that prolactin, a pituitary hormone that appears to have assumed many different roles in the vertebrates, may contribute, uniquely, to the regulation of the function of the corpus luteum in marsupials.

An alternative systematic name for the Eutheria is the Placentalia, referring to the apparent absence of a placenta, or more specifically an allantoic placenta, in marsupials. The function of a placenta includes the nutritional sustenance of the fetus in utero and the physiological maintenance of pregnancy. The length of pregnancy and the ultimate size of the fetus may be expected to influence the role of the placenta so that it is, perhaps, not unexpected to observe in marsupials, with their relatively short period of gestation and small young, that placental function does not appear as prominent as in eutherians. Dr. M. Renfree, however, has shown that there is considerable morphological variability in the type of placentation and possibly its physiological role among the marsupials. Although different embryonic membranes may be involved, a placenta may still be important and, as in eutherians, may even be the site of formation of hormones concerned with pregnancy.

Steroids may play several special roles as hormones in vertebrates. Compared to the polypeptide hormones they show little chemical variability in their structures. There are, however, quantitative differences in their rates of synthesis and secretion. The biochemical mechanisms that determine such changes in the adrenal cortex have been described by Dr. M. Weiss. Synthesis of steroid hormones involves many distinct chemical steps involving different enzymes, but the basic process, whether it involves steroidal adrenocortical or sex hormones, appears to be basically similar in all vertebrates. There are, however, substantial quantitative differences in the final products that are secreted by some eutherians, marsupials, and monotremes. These patterns may be determined by variations in the properties of the enzyme systems that control the intermediate transformations of the steroid substrates. It is possible that the production of certain steroids may be advantageous to certain ways of life, and this could influence the evolution of the particular synthesizing enzymes involved.

The angiotensins are peptides that are formed from a plasma protein as a result of the action of an enzyme called renin. They can exert many effects, some of which may reflect hormonal roles. The renin–angiotensin system is present in most, but not all groups of vertebrates. In eutherian mammals it contributes to the regulation of aldosterone secretion from the zona glomerulosa of the adrenal cortex. This morphological specialization of this endocrine gland appears to be confined to mammals and so may represent a novel target organ for the angiotensins. As Dr. J. Blair-West has shown, the marsupials, like the eutherians, also appear to have utilized the renin–angiotensin system for regulating secretion of aldosterone. Although there is a considerable cross-reactivity between renin and its substrates from eutherians and marsupials, differences exist which presumably reflect variations in the chemical structures of these substances. Such evolutionary variation in these molecules, however, have apparently not resulted in any radical changes in the physiological role of angiotensin in controlling aldosterone secretion in marsupials.

The echidna *Tachyglossus aculeatus,* apart from being a monotreme, is a physiologically rather unique mammal. It possesses adrenocorticosteroid hormones, but, as described by Dr. C. Sernia, it can, apparently, under "normal" conditions survive without them. It is possible that the physiological emphasis of the functions of these steroid hormones may have changed in the echidna so that they are more concerned with the regulation of fat metabolism and responses to environmental stresses than mineral and carbohydrate metabolism. The echidna, rather remarkably, and in contrast to other mammals, lacks a specific cortisol-binding protein (CBG) in its plasma. This deficiency, however, does not appear to handicap the echidna, and indeed it could even be considered to be an adaptation that is consistent with its particular adrenocortical physiology. Information about other monotremes is sparse, but does suggest that it may be dangerous to make generalizations for this mammalian order based on observations in *Tachyglossus aculeatus,* which appears to be a rather specialized species.

The general functions of the adrenal cortex in marsupials appear to be rather similar to those observed in eutherians. As Dr. I. McDonald shows, however, there is considerable variability even within particular families of marsupials. The corticosteroids, it seems, may have somewhat different roles, even in quite closely related species. A unique example of the importance of the role of the cortisol-binding globulin in endocrine function has been observed in a small marsupial, *Antechinus,* in which declines in the level of this plasma protein appear to precipitate the death of the males following the breeding season.

It appears at this time that the endocrine function of more native species of marsupials than eutherians have been investigated. In the latter, observations have largely been confined to domesticated species. Dr. McDonald indicates that more extensive studies of native, wild, eutherians may well uncover far more variation than is usually thought to occur.

There appears to be general agreement that as far as their endocrine systems are concerned, marsupials and monotremes are certainly not "primitive" though they may be "different."

References

Haldane, J. B. S. (1953). Foreword. *Symp. Soc. Exp. Biol.* 7:ix–xxi.
Medawar, P. B. (1953). Some immunological and endocrinological problems raised by the evolution of viviparity in vertebrates. *Symp. Soc. Exp. Biol.* 7:320–38.

26

Pituitary–ovarian interactions in marsupials

C. H. TYNDALE-BISCOE and
SUSAN M. EVANS

The study of pituitary–ovarian interactions in marsupials is still in its infancy. Members of only two species, American opossum and tammar wallaby (*Didelphis virginiana* and *Macropus eugenii)*, have been hypophysectomized; and only in the latter has the effect on reproductive functions been examined and have pituitary hormones been measured (Hearn, 1974, 1975b). Apart from this species, all ideas about pituitary function in marsupial reproduction are inferences based on eutherian analogies; and the results from the tammar have taught us to be cautious about drawing such inferences.

Most marsupials are polyestrous, and the estrous cycle of all these comprises a luteal phase, associated with an active corpus luteum, and a follicular phase, which culminates in estrus and ovulation. In all except the kangaroos (Macropodidae), gestation is confined approximately to the luteal phase; the subsequent follicular phase and ovulation are suppressed during lactation but will recur if suckling is stopped. In the kangaroos gestation is longer, in most extending into the follicular phase so that estrus and ovulation occur at about the same time as parturition (Tyndale-Biscoe et al., 1974). The corpus luteum so formed does not complete development or induce a luteal phase so long as the young offspring suckles in the pouch. Nor does estrus or ovulation occur during this time. If the young is removed (RPY), the corpus luteum resumes its delayed activity, and in due course estrus and ovulation follow. If fertilization occurs at the postpartum estrus, development of the resulting embryo is arrested at the stage of a unilaminar blastocyst so long as the corpus luteum remains arrested. This phenomenon has been termed embryonic diapause, and the reproductive condition of the female has been termed quiescence.

The phenomena of corpus luteum quiescence, and the associated embryonic diapause, have stimulated most of the work on marsupial reproduction in the last 25 years, so that this paper will be very largely confined to pituitary—ovarian interactions of these phenomena, particularly in the tammar wallaby, *Macropus eugenii*, the species we selected for experimental work.

Annual cycle of the tammar wallaby

The largest wild population of tammars now occurs on Kangaroo Island, South Australia at latitude 36 °S, which has a maritime climate with most of the annual rainfall occurring during April to September, and the main flush of plant growth at the end of this time. The tammar has a highly regular annual cycle (Figure 1), which commences at midsummer (22 December). About 80% of females give birth and undergo postpartum estrus between late January and mid-February, one month after summer solstice, and they then suckle a pouch young until October, while concurrently carrying a diapausing embryo in the uterus. Pouch exit rather than birth thus coincides with the spring flush. From January to June unmated females will undergo repeated estrous cycles of 29 days, but any cycle begun after June becomes arrested at the stage reached 8 days after estrus and does not resume until midsummer. Young females that are weaned in October sometimes undergo their first estrus then but, like their mothers, do not complete the first cycle until midsummer. At this time the corpus luteum and associated embryo reactivate, and pregnancy is completed in 27 days.

It is thus clear that the tammar does not undergo a conventional anestrus during the second half of the year, and that seasonal quiescence is similar to lactational quiescence except that the proximate factor in its control is not the suckling stimulus. The natural resumption at the summer solstice implicates photoperiod, and we have shown that an alteration to a shorter day-length will induce premature reactivation of breeding activity (Sadleir and Tyndale-Biscoe, 1977).

Figure 1. Annual cycle of events in the reproduction of the tammar wallaby, *Macropus eugenii* on Kangaroo Island, South Australia.

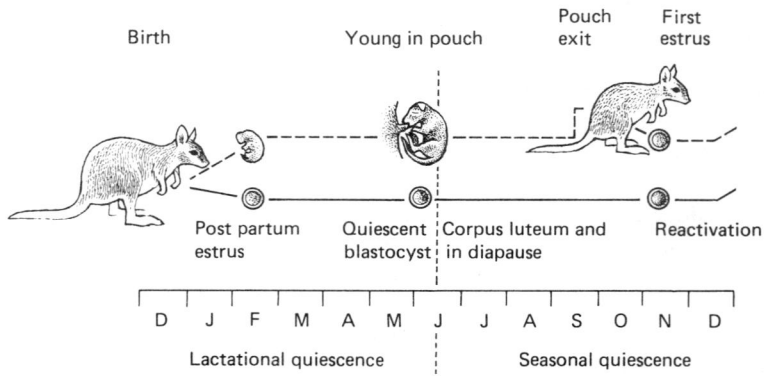

Role of the pituitary

The pituitary involvement in the control of reproduction in the tammar has been investigated by hypophysectomy, ovariectomy, and assay of pituitary gonadotropins, during the estrous cycle, lactation, and seasonal quiescence.

Hearn (1975a) has described the technique of hypophysectomy used on tammar wallabies and the general postoperative responses. Hypophysectomized animals survived in good health for four months without hormone therapy, but for most studies the animals were killed after shorter intervals.

Hypophysectomy performed at any stage of the estrous cycle or pregnancy prevents growth of follicles beyond 0.5 mm diameter, and, if preovulatory follicles of 3 to 4 mm were present at the time of hypophysectomy, neither estrus nor ovulation occurs subsequently, and the follicles become atretic. This is the normal effect of hypophysectomy in Eutheria, and the stage from antrum formation to ovulation is recognized as the gonadotropin-dependent phase of follicular growth. However, unlike that of eutherian mammals, the corpus luteum of the tammar after hypophysectomy develops normally and induces a luteal phase in the uteri; and if the female is pregnant, the embryo develops to full term (Hearn, 1974). This suggests that gonadotropins are necessary for follicular maturation and ovulation, but not for steroidogenesis by the reactivated corpus luteum.

Besides inducing the luteal phase, the corpus luteum affects follicle growth. Female tammars returned to estrus about 12 days after lutectomy performed during the first 12 days of the cycle, but after lutectomy on day 18 estrus occurred at the usual time 9 days later (Tyndale-Biscoe and Hawkins, 1977). Thus it would appear that the corpus luteum suppresses the whole of the gonadotropin-dependent phase during the first half of the cycle, but loses this capacity at about day 17.

The major steroid hormone secreted by the corpus luteum is progesterone. Its concentration in the circulation increases to a maximum during the second half of the cycle coincident with follicular growth (Lemon, 1972; Renfree and Heap, 1977; Hinds, unpublished results), which shows that progesterone cannot be exercising a feedback inhibition on the pituitary at this stage. Baird et al. (1975) have contrasted the human menstrual cycle, in which follicle growth is suppressed during the luteal phase, with the sheep estrous cycle; in the latter follicular growth occurs concurrent with the luteal phase, and ovulation follows its decline by 3 days. The tammar appears to share features of both these patterns; it resembles the human pattern in the first half of the cycle, when the corpus luteum inhibits follicle growth, and it resembles the ovine pattern subsequently when follicle growth can proceed in the presence of a highly active corpus luteum.

During lactational or seasonal quiescence, when the corpus luteum is arrested at the stage equivalent to day 8 of the cycle, it also prevents follicular growth, for estrus and ovulation occur about 12 days after lutectomy, the same as in the first half of the estrous cycle. This is further evidence that the tammar does not undergo a true anestrus. On the other hand, in other marsupials, such as the opossum, brush possum, and grey kangaroo, there is no quiescent corpus luteum, and the follicular phase is suppressed directly during lactation, as in many eutherian mammals. In these marsupials we may surmise that either gonadotropin secretion is depressed by the suckling stimulus, or enhanced prolactin secretion is inhibiting follicles directly, as has been suggested to occur in women.

In the tammar the quiescent corpus luteum immediately resumes development after hypophysectomy, inducing a luteal condition in the uteri, and the blastocyst reactivates. Subsequently estrus and ovulation do not occur, and the uteri and vaginae then do regress to a condition that resembles the true anestrus of other marsupials.

We have shown that the corpus luteum inhibition can be maintained after hypophysectomy with prolactin (Tyndale-Biscoe and Hawkins, 1977), and we conclude that prolactin inhibition, not luteinizing hormone (LH) stimulation, controls the corpus luteum. But the prolactin inhibition, once lifted, is no longer effective in suppressing the corpus luteum. When bromocriptine, an ergot derivative that depresses prolactin secretion, was administered to lactating females in a single dose, their corpora lutea reactivated, and 26 to 27 days later the females gave birth and underwent postpartum estrus and ovulation, while still suckling their first offspring in the pouch (Tyndale-Biscoe and Hinds, unpublished results, 1978). We can conclude from this experiment, first, that the corpus luteum is under a tonic inhibition, but once it escapes, by a transient fall in prolactin, it becomes independent of the pituitary; and, second, that in the tammar the gonadotropin-dependent phase of follicular growth is not inhibited by suckling as it is in the grey kangaroo and brush possum but by the quiescent corpus luteum itself.

At present we do not know how the corpus luteum inhibition of follicle growth is exercised, although it is presumably effected via the pituitary. Variations in plasma levels of progesterone and LH are not clearly correlated with the ovarian events except ovulation (see below), and estrogen and follicle stimulating hormone (FSH) are undetectable with present assays. Hearn et al. (1977) have suggested that the sensitivity of the hypothalamus to steroids may be enhanced by suckling, so that variations in circulating levels of steroids would not need to be invoked, but this idea has not yet been tested. Further understanding can only come about through a much better knowledge of the pituitary hormones in circulation.

Isolation of pituitary hormones

The anatomy of the marsupial pituitary conforms to the eutherian pattern (Green, 1951), and in the adenohypophysis the whole range of secretory cells has been identified tinctorially in *Didelphis* (Dawson, 1938), *Setonix*

(Henström, 1954), and *Macropus rufogriseus* (Ortman and Griesbach, 1958). In the last species Purves and Sirett (1959) also differentiated by bioassay the hormonal content of the separate rostral and caudal portions and correlated this with the distribution of cell types. Prolactin and thyrotropin activity were restricted to the rostral portion in which carminophils predominated, and growth hormone to the caudal in which orange G acidophils predominated; whereas ACTH and gonadotropin activity, predominantly LH, occurred in both portions.

The first isolation of a gonadotropin fraction was obtained from frozen tammar pituitaries by Hearn (1974). He did not differentiate between LH and FSH, but the material was subsequently found to be predominantly LH. Hearn developed a double antibody radioimmunoassay (RIA) to this fraction and with it measured plasma levels in female and male tammars during the breeding and nonbreeding season. In females the concentration remained within 2 to 5 ng equivalent NIH standard ml^{-1} plasma at all times of the year except for a marked but transient preovulatory peak at estrus (Hearn, 1974).

The question remained open whether marsupial gonadotropin is a single hormone, as it is thought to be in lower vertebrates, or two distinct hormones with separate functions as in Eutheria. Farmer and Papkoff (1974) separated two moieties from pituitaries of red kangaroos, which had biological properties similar to LH and FSH of Eutheria but were unable to make a definite separation. Using tammar wallaby pituitaries collected onto dry ice, Gallo et al., (1978) have now obtained two distinct fractions that on bioassay, chromotography, and amino acid composition resemble the FSH and LH of Eutheria and show like specificity.

In Eutheria androgen production by Leydig cells of the testis is specifically stimulated by LH and not at all by FSH, whereas in reptiles mammalian (eutherian) FSH is more potent in stimulating androgen production by the testis (Licht et al., 1977). It was therefore of considerable interest to find that wallaby LH stimulated androgen production by rat and by opossum Leydig cells, whereas the wallaby FSH stimulated turtle testes. Conversely, Cook et al. (1974) have shown that opossum testes can be stimulated to convert acetate to androstenedione and testosterone in vitro when incubated with human chorionic gonadotropin (hCG) and ovine LH, so that LH receptors in the testis of this marsupial can recognize eutherian LH.

In addition to the tammar, pituitaries from two species of grey kangaroos (*Macropus giganteus* and *M. fuliginosus*) have been extracted, and LH, FSH, prolactin, and growth hormone isolated and purified by Papkoff and colleagues. The purified gonadotropins of all three species were compared by our group with highly purified human ovine and rat FSH and LH in heterologous RIA's using antiovine LH and antiovine FSH as antibodies, and radio receptor assays using rat and kangaroos testes (Stewart et al., 1977; Evans and Sutherland, 1977). The results have confirmed the findings in the tammar that macropod marsupials have distinct LH and FSH molecules that possess antigenic determinants and receptor specificities similar to gonadotropins of eutherian mammals. Furthermore, the testes of the western grey kangaroo (M. fuliginosus) have been shown to possess distinct FSH and LH receptors that can differentiate between FSH and LH derived from eutherian and marsupial mammals.

Circulating levels of gonadotropins in female tammars

Measurements of LH and FSH in tammar plasma using these heterologous RIA's were similar to those obtained by Hearn with his homologous RIA.

During the estrous cycle, or pregnancy, LH levels remain low, from <0.2 ng NIH-LH-S19 ml^{-1} plasma (i.e., undetectable by this assay) to 1.0 ng ml^{-1}, except for large, transient peaks of 20 to > 50 ng ml^{-1}, associated with estrus and ovulation. FSH, however, remains undetectable (<50 ng NIH-FSH-S12 ml^{-1} plasma) at all times, except for a few females in which small elevations of 50 to 100 ng ml^{-1} were observed coincidentally with the LH peak.

Samples taken every 2 to 3 days throughout the year showed that in the lactating or seasonally quiescent female tammar, LH levels fluctuate between <0.2 ng ml^{-1} and 2.0 ng ml^{-1}. Hourly samples taken over 24 h, from two seasonally quiescent females showed that LH remained undetectable 85% of the time, with occasional small elevations up to 2.0 ng ml^{-1}, but FSH remained undetectable throughout.

These results from the tammar differ from what is known from eutherian species only in the apparent lack of FSH during all phases of the estrous cycle and the lack of alteration in LH levels between the breeding and nonbreeding seasons. Both these differences may be peculiar to the tammar, and, until several more species of marsupials with different patterns of reproduction have been investigated, it cannot be said that these are features that differentiate marsupials from Eutheria. Indeed, in male tammars both FSH and LH are readily detectable at all times, and LH occurs in higher concentration during the breeding season than during the nonbreeding season (P. C. Catling and R. L. Sutherland, personal communication) when the males associated with females.

Ovarian feedback on the pituitary

The levels of pituitary hormones circulating in the blood reflect dynamic interactions between the hypothalamus, the pituitary, and the target organ, in this case the ovary. A relatively simple way to examine the system is to remove the target organ, and hence the negative feedback loop, and then to introduce either single components of the organ or purified steroid hormones and observe the response by measuring the gonadotropins in circulation.

Despite our inability to detect FSH in intact female tammars, they are capable of secreting FSH in measurable amounts under experimental conditions. Plasma concentrations of FSH became detectable 3 to 8 days after bilateral ovariectomy and thereafter fluctuated between 100 and 600 ng ml^{-1} (Figure 2). Likewise LH concentrations rose after ovariectomy and thereafter remained elevated at 2 to 10 ng ml^{-1}.

In order to determine what part of the ovary is responsible for the negative feedback that maintains LH at low levels and FSH at undetectable levels in intact females, portions of the ovaries of females at ovariectomy were implanted under the skin of the pouch. In this site the subsequent fate of the graft can be monitored by palpation through the thin translucent skin. The tammar ovary contains a prominent mass of interstitial tissue that appears to be secretory (Renfree, this volume); so some animals received grafts of interstitial tissue and some received pieces of ovarian cortex. All the interstitial grafts and some of the cortex grafts deteriorated, and the gonadotropin profile of these females was the same as in the ovariectomized females. At autopsy the vaginae and uteri of these females were small and atrophic. In those females in which the cortex grafts became established and grew, the vaginae and uteri at autopsy were enlarged and vascular, indicating that they had been under estrogenic stimulation. In this group FSH and LH concentrations initially rose as in the ovariectomized females, and then between day 20 and day 40 declined to preoperative levels (Figure 2). This

Figure 2. Changes in the concentration of luteinizing hormone (LH) and follicle stimulating hormone (FSH) in the circulation of tammar wallabies after double ovariectomy (OVX) with and without grafts of ovarian cortical tissue.

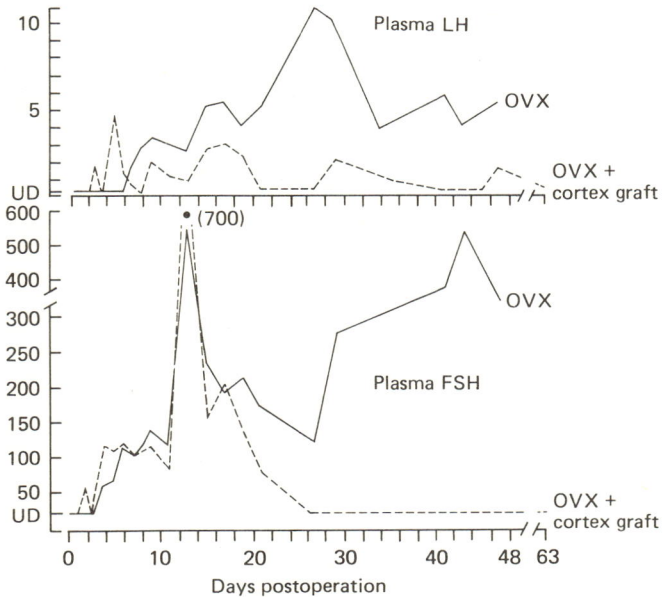

would suggest that the ovarian feedback is being exercised by steroid secretions from some component of the cortex, and that the interstitial tissue is not essential.

Progesterone and testosterone are present in the plasma of intact females, and estrogen has been detected, using a RIA sensitive to 2.5 pg ml^{-1} estradiol (Renfree and Heap, 1977). However, estradiol and progesterone, when injected subcutaneously in oil, depressed FSH and LH levels in ovariectomized females, whereas testosterone depressed LH only (Figure 3).

Conclusion

A tentative conclusion from these observations on the tammar is that there is a basic ovarian feedback on the pituitary, which maintains the low secretion rate of LH and undetectable FSH in intact females at all times of the year, and that it originates from the ovarian cortex. Superimposed on this is a secondary influence by the young corpus luteum, presumably on the pituitary, the effect of which is to suppress the gonadotropin-dependent phase of follicular growth in the ovary. As the corpus luteum matures, this effect wanes so that follicle growth can occur and proceed to ovulation.

Finally, the corpus luteum, which is not dependent on LH, is controlled by prolactin secretion from the pituitary, which in turn is influenced by either the suckling stimulus in the first half of the year or photoperiod in the second.

Our purpose is to discuss the physiology of "primitive" mammals, and

Figure 3. The effects of a single i.m. injection of oil, 700 μg progesterone kg^{-1} body weight and 5 μg estradiol kg^{-1} body weight in oil on FSH and LH concentrations in the plasma of an ovariectomized tammar wallaby. Values of LH that exceeded 9 ng ml^{-1} are indicated by numbers in the body of the figure.

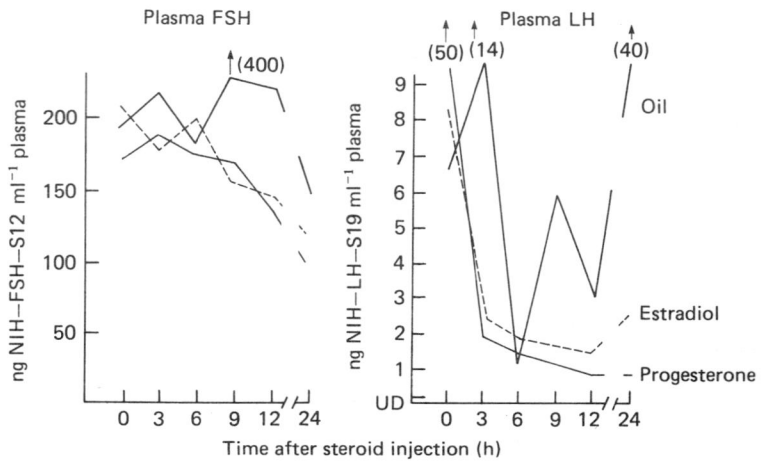

marsupials are included because of their long-separate evolution from eutherian mammals. However, in respect to pituitary—ovarian interactions there is remarkable similarity between what we now know about a few species of marsupials and current knowledge of eutherian mammals. The anatomy of the marsupial pituitary is very similar to the eutherian, and it secretes very similar gonadotropic hormones, which have similar chemical and immunological properties and bind to similar receptors in target organs.

What we know of the control of ovarian function has parallels in eutherian species, except for the unusual, if not unique, manner of pituitary inhibition of the corpus luteum. This latter phenomenon and its discovery provide an excellent justification for studying so-called primitive mammals. When John Hearn began to study the tammar pituitary in 1969, we expected that it would confirm the hypothesis of LH dependence of the corpus luteum for steroidogenesis, based on extensive knowledge in eutherian species. It proved to be quite different, and the results have helped us understand novel aspects of pituitary—ovarian interactions in eutherian mammals as well.

Thus we see marsupials as being contemporary mammals that have evolved different endocrine strategies rather than seeing them as relics from the dawn of mammals with an endocrine system much simpler than that of Eutheria.

References

Baird, D. T., Baker, T. G., McNatty, K. P., and Neal, P. (1975). Relationship between the secretion of the corpus luteum and the length of the follicular phase of the ovarian cycle. *J. Reprod. Fert.* 45:611–19.

Cook, B., Sutterlin, N. S., Graber, J. W., and Nalbandov, A. V. (1974). Gonadal steroid synthesis in the Virginia opossum, *Didelphis virginiana*. *J. Endocrinol.* 61:ix.

Dawson, A. B. (1938). The epithelial components of the pituitary gland of the opossum. *Anat. Rec.* 72:181–93.

Evans, S. M., and Sutherland, R. L. (1977). A heterologous radioimmunoassay for tammar wallaby luteinising hormone. *Theriogenology* 6:175.

Farmer, S. W., and Papkoff, H. (1974). Studies of the anterior pituitary of the kangaroo. *Proc. Soc. Biol. Med.* 145:1031–46.

Gallo, A. B., Licht, P., Farmer, S. W., Papkoff, H., and Hawkins, J. (1978). Fractionation and biological actions of pituitary gonadotrophins from a marsupial, the wallaby *(Macropus eugenii)*. *Biol. Reprod.* 19:680–7.

Green, J. D. (1951). The comparative anatomy of the hypohysis, with special reference to its blood supply and innervation. *Am. J. Anat.* 88:225–312.

Hanström, B. (1954). The hypophysis in a wallaby, two tree-shrews, a marmoset, and an orangutan. *Ark. Zool.* 6:97–154.

Hearn, J. P. (1974). The pituitary gland and implantation in the tammar wallaby *(Macropus eugenii)*. *J. Reprod. Fert.* 39:325–41.

Hearn, J. P. (1975a). Hypophysectomy of the tammar wallaby, *Macropus eugenii*: Surgical approach and general effects. *J. Endocrinol.* 64:403–16.

Hearn, J. P. (1975b). The role of the pituitary in the reproduction of the male tammar wallaby, *Macropus eugenii. J. Reprod. Fert. 42:*399–402.

Hearn, J. P., Short, R. V., and Baird, D. T. (1977). Evolution of the luteotrophic control of the mammalian corpus luteum. In *Reproduction and Evolution*, eds. J. H. Calaby and C. H. Tyndale-Biscoe, pp. 255–64. Canberra: Australian Academy of Science.

Lemon, M. (1972). Peripheral plasma progesterone during pregnancy and the oestrous cycle in the tammar wallaby, *Macropus eugenii. J. Endocrinol. 55:*63–71.

Licht, P., Papkoff, H., Farmer, S. W., Muller, C. H., Tsui, H. W. and Crews, D. (1977). Evolution in gonadotropin structure and function. *Rec. Prog. Horm. Res. 33:*169–248.

Ortman, R., and Griesbach, W. E. (1958). The cytology of the pars distalis of the wallaby pituitary. *Aust. J. Exp. Biol. 36:*609–18.

Purves, H. D. and Sirett, N. E. (1959). A study of the hormone contents of the rostral and caudal zones of the pars anterior of the wallaby pituitary. *Aust. J. Exp. Biol. 37:*271–8.

Renfree, M. B. and Heap, R. B. (1977). Steroid metabolism in the placenta, corpus luteum and endometrium of the marsupial, *Macropus eugenii. Theriogenology 8:*164.

Sadleir, R. M. F. S., and Tyndale-Biscoe, C. H. (1977). Photoperiod and the termination of embryonic diapause in the marsupial *Macropus eugenii. Biol. Reprod. 16:*605–8.

Stewart, F., Evans, S. M. and Sutherland, R. L. (1977). Radioimmunoassays and radioreceptor assays for wallaby follicle stimulating hormone. *Theriogenology 8:*176.

Tyndale-Biscoe, C. H., and Hawkins, J. (1977). The corpora lutea of marsupials: Aspects of function and control. In *Reproduction and Evolution*, eds. J. H. Calaby and C. H. Tyndale-Biscoe, pp. 245–52. Canberra: Australian Academy of Science.

Tyndale-Biscoe, C. H., Hearn, J. P., and Renfree, M. B. (1974). Control of reproduction in macropodid marsupials. *J. Endocrinol. 63:*589–614.

27

Placental function and embryonic development in marsupials

MARILYN B. RENFREE

The estrous cycle, pregnancy, and embryonic diapause

The pattern of the estrous cycle in the Marsupialia bears many similarities to the patterns in the Eutheria. The histological changes in the uterus and ovary during the follicular and luteal phases resemble those seen in many other mammals (see Pilton and Sharman, 1962; Sharman and Berger, 1969). Many of the steroid hormones and gonadotropins involved in reproduction in eutherian mammals have been found in marsupials, though particular features of their functions differ (Tyndale-Biscoe and Evans, this volume). Because the length of pregnancy is shorter than that of the estrous cycle, it has been suggested that the marsupial placenta has no endocrine influence on the duration of pregnancy (Sharman, 1970). The aim of the present paper is to review aspects of maternal–fetal physiology in the marsupials in relation to placental structure, and to show that the placenta has many complex functions, possibly including the synthesis and metabolism of hormones.

Marsupials may be monestrous, but the majority are polyestrous. The marsupial estrous cycle varies in mean length from 22 to 42 days, the average being 28 days (Tyndale-Biscoe, 1973). The life of the corpus luteum is not prolonged by the presence of a conceptus, although in the Peramelidae (bandicoots) the corpus luteum persists during lactation (Hughes, 1962; Gemmell, 1977). Because pregnancy is shorter than the estrous cycle in all except the swamp wallaby *(Wallabia bicolor)*, the young are born, depending on the species, at different stages of the luteal and follicular cycle. In one species, the red kangaroo *(Megaleia rufa)*, the timing of birth and estrus have been altered experimentally, suggesting that gestation and estrous cycle activity need not be closely linked (Clark, 1968). In the American opossums, *Didelphis virginiana* and *Marmosa mitis*, the numerous young are born toward the end of the luteal phase, and return to estrus is suppressed by the suckling stimulus. In contrast, in the tammar wallaby *(Macropus eugenii)* the single pouch young reaches the pouch at the end of the proestrous phase, and ovulation and fertilization occur postpartum (Figure 1). In this species, and in most of the Macropodidae, the fertilized egg enters a dormant phase at the blastocyst stage by the eighth day postcoitum (Tyndale-Biscoe, in press), and during this period of embryonic diapause both the blastocyst and the corpus luteum remain quiescent as long as the suckling pouch young is present. In *M. eugenii*, and in Bennett's wallaby, *M. rufogriseus fruticus*, photoperiodic influences control the diapause in the second half of the year after the winter solstice (Tyndale-Biscoe and Evans, this volume).

Morphology of marsupial placental membranes
Attachment and the fetal membranes

In all marsupials, as among eutherians, the yolk sac blood vessels vascularize part of the chorion, and in most the choriovitelline placenta so formed is the only functional organ of exchange between mother and fetus. Only in the bandicoots (Peramelidae), koala bear *(Phascolarctos)*, and possibly wombat *(Phascolomis)* does the allantois vascularize the chorion to form a placenta. (See Figure 2.)

In addition to the amnion, chorion, yolk sac, and allantois, there is in marsupials another membrane (an egg membrane) not found in any other mammals. This is the maternally derived shell membrane, a structure com-

Figure 1. Marsupial estrous cycles. In species such as the American opossum (*Didelphis virginiana*) (top), birth occurs toward the end of the uterine luteal (secretory) phase, and the presence of the numerous young in the pouch prevents the pro-estrous phase and the development of the follicle. Should the suckling young be lost, animals return to estrus about 7 days later. In the macropodids, such as the tammar wallaby (*Macropus eugenii*) (bottom), gestation occupies the whole length of the estrous cycle, and ovulation occurs a day after the birth of the single young. Suckling of the pouch young prevents development of the embryo beyond the blastocyst stage. (* − ovulation and fertilization.)

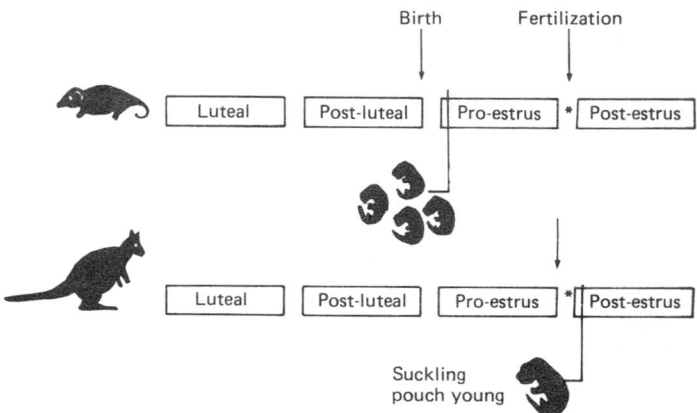

posed of ovokeratin, produced by the uterine endometrial glands (Hughes, 1974, 1977). The marsupial shell membrane attenuates but remains unruptured for two-thirds of gestation until the time of attachment. Once the shell membrane is lost, development is very rapid; attachment and invasion are presumably facilitated by its loss.

In macropodid marsupials the site of attachment is a simple interdigitation between maternal and fetal tissues, and the ectoderm cells of the yolk sac have numerous microvilli at their bases (Tyndale-Biscoe, 1973). In *Philander opossum* the ectoderm cells show incipient invasiveness at the margin of the vascular region of the yolk sac and penetrate folds on the endometrium. In both *Macropus eugenii* and *Philander* the ectoderm has numerous coated vesicles and inclusions, suggestive of absorptive activity (Enders and Enders, 1969; Tyndale-Biscoe, 1973). The endoderm of *Philander*, on the other hand, appears to be synthetically active. In *Bettongia* (Flynn, 1930), the cells of the bilaminar yolk sac are vacuolated, and "definite union" occurs in the region of the vascular omphalopleure. There is no direct evidence for Semon's (1894) and Hill's (1900b) idea that the vascular part of the yolk sac may be primarily important for respiration and that the nonvascular yolk sac may be the site of absorption from the uterine secretions. Biochemical evidence and ultrastructural data are not, however, inconsistent with this idea (Renfree, 1973b; Tyndale-Biscoe et al., 1974; Hughes, 1974).

Invasive properties

There is considerable variation in the degree of invasiveness of the marsupial yolk sac placenta (Table 1), and, as noted, in most species the shell

Table 1. *Types of implantation found in marsupial species*

Trophoblast not invasive	Trophoblast of yolk sac slightly invasive	Highly invasive chorionic villi
Phalangeridae	Phalangeridae	Peramelidae
Trichosurus vulpecula	*Schoinobates volans*	*Isoodon obesulus*
Pseudocheirus peregrinus		*Perameles gunnii*
	Dasyuridae	*Perameles nasuta*
Macropodidae	*Dasyurus viverrinus*	*Echymipera*
Macropus rufogriseus	*Sminthopsis crassicaudata*	*rufescens*
Setonix brachyurus		
Potorous tridactylus	Macropodidae	
Macropus robustus	*Macropus eugenii*	
Macropus giganteus	*Bettongia cuniculus*	
Didelphidae	Phascolarctidae	
Didelphis virginiana	*Phascolarctos cinereus*	
	Vombatus ursinus	
	Didelphidae	
	Philander opossum	

Source: Hughes (1974).

No fusion — shell
may persist

Fusion — shell
disrupted

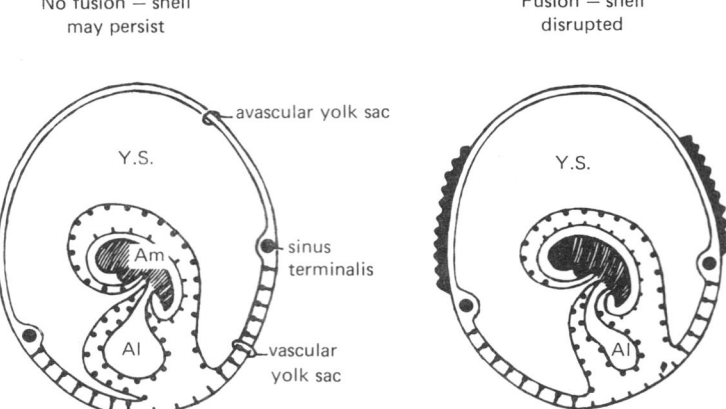

Didelphis

Dasyurus

Yolk sac placenta only

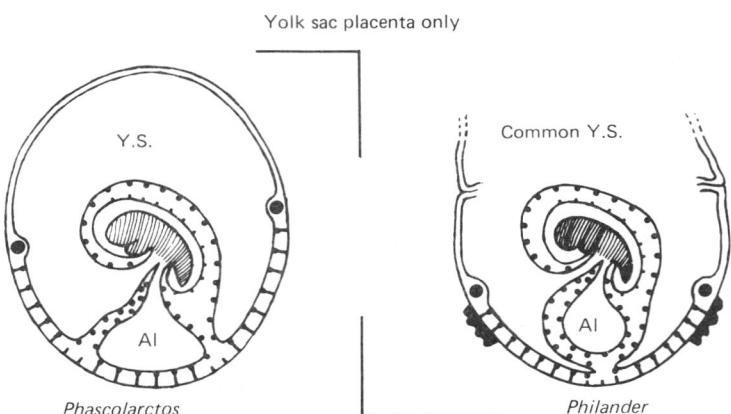

Phascolarctos

Philander

Yolk sac and allantoic placenta

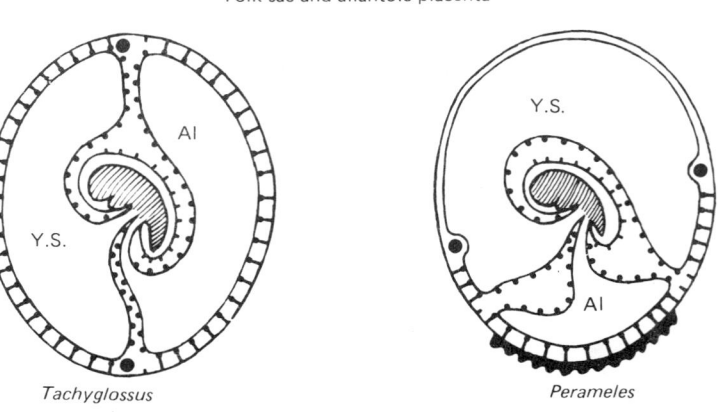

Tachyglossus

Perameles

membrane breaks down on attachment (Hughes, 1974; Tyndale-Biscoe et al., 1974). Attachment between fetal and maternal cells in the region of the sinus terminalis have been noted in *Schoinobates volans, Dasyurus viverrinus, Sminthopsis crassicaudata, Macropus eugenii, Bettongia cuniculus, Phascolarctos cinereus,* and *Vombatus ursinus,* as well as in *Philander* (Hughes, 1974; Padykula and Taylor, 1977); but it is unlikely that such close attachment occurs in *Didelphis* because the shell membrane remains intact (Tyndale-Biscoe, 1973). In *Potorous tridactylus* and *Trichosurus vulpecula* the yolk sac is not attached to the uterine epithelium at any point (Hughes, 1974), and no uterine invasion occurs in these species, or in *Pseudocheirus peregrinus, Protemnodon (Macropus) rufogrisea (-us),* and *Setonix brachyurus* (Sharman, 1961). However, in *Dasyurus,* in the marginal zone of the yolk sac there is an apparent syncytium formed between ectoderm cells, which invade the maternal capillaries, allowing extravasated blood to enter the yolk sac (Hill, 1900a; Tyndale-Biscoe, 1973). In early gestation in *Perameles* the vascular yolk sac is applied closely to the uterine epithelium, but the allantois subsequently contacts the chorion and fuses with it. The allantoic ectoderm completely fuses with the maternal luminal epithelium and "disappears" (Padykula and Taylor, 1977). This unusual feature of bandicoot gestation suggests that independent evolution of allantoic placentation occurred between marsupials and eutherians. The ectoderm of the chorion and the uterine epithelium are both syncytia at the time of fusion, and fetal capillaries come to lie close to the exposed maternal capillaries (Tyndale-Biscoe, 1973; Padykula and Taylor, 1977). Newborn bandicoots are relatively well developed at birth compared with other marsupials, although they have the shortest gestation period; it has been suggested that this family of marsupials may have the most efficient placenta of all (Tyndale-Biscoe, 1973; Padykula and Taylor, 1977; Taylor and Padykula, 1978).

Embryonic development and placental function
Phases of gestation
Although there have been many studies on the arrangements of the embryonic membranes of marsupials and on the reproductive tract, much less is known about the development of the embryo itself. Hill described aspects of early marsupial development in some Australian species (e.g., Hill, 1910),

Figure 2. Arrangements of the fetal membranes in five marsupials and in the monotreme egg after laying. The several patterns differ according to the proportion of the chorion that is vascularized, and hence respiratory; according to whether or not the allantois participates in this; and according to the region, if any, of the chorion that invades the uterine epithelium (from Tyndale-Biscoe, 1973, by permission, based on data from Enders and Enders, 1969, Flynn, 1923, Hill 1900a, and Semon, 1894). *Didelphis,* opossum; *Dasyurus,* native cat; *Phascolarctos,* koala; *Philander,* four-eyed opossum; *Tachyglossus,* echidna; *Perameles,* bandicoot. (Y.S., yolk sac; Am, amnion; Al, allantois.)

and several papers have described early intrauterine development of *Didelphis* (e.g., Hartman, 1916), but only one (McCrady, 1938) has described development from unfertilized egg to birth. Recently, Lyne and Hollis (1977) have begun detailed studies on the embryology of the bandicoot, and certain aspects of the embryology of the tammar wallaby are now known (Renfree, 1972a; Renfree and Tyndale-Biscoe, 1973, 1978; Tyndale-Biscoe, in press).

A feature of marsupial embryogenesis is the relatively slow growth of the blastocyst up to the time of attachment (Renfree, 1977). This period may be extended by embryonic diapause in the family Macropodidae. During the long free-vesicle or preattachment phase, the blastocyst lies free in the uterine lumen and must gain its nutrients and respiratory requirements by exchanges with the endometrium or luminal fluid. The duration of the attachment phase, when organogenesis occurs, is remarkably constant in marsupials over a wide range of body sizes (Figures 3 and 4).

The neonatal marsupial, although usually termed "altricial," has by comparison with the eutherian newborn a mixture of "altricial and precocial" features. At birth, the tiny, naked marsupial climbs to the pouch unassisted by the mother, using its well-developed forelimbs. Its lungs are fully functional, the nostrils are open, and the olfactory center of its brain is well developed. The mouth, tongue, and digestive system including liver and pancreas are sufficiently developed to cope with the change to a milk diet. By contrast, features, such as the eyes, the hind limbs, and the gonads remain undifferentiated; pouch and scrotum can be seen only after about 10 days postpartum in macropodids. The mesonephric kidney remains

Figure 3. Phases of gestation in marsupials. The preattachment or free-vesicle phase consists of a relatively long period of about two-thirds of gestation when no contact is made by the chorion to the maternal epithelium. The attachment phase, during which organogenesis is completed, is relatively short. Attachment coincides with the loss of the shell-membrane (from Renfree, 1977, by permission).

Gestation in marsupials

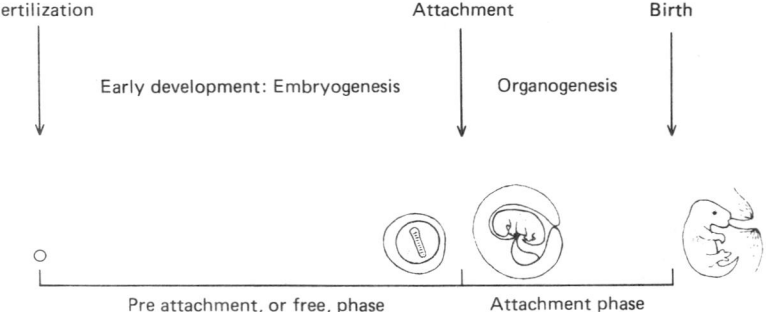

functional for the first few days after birth; the metanephros is differentiated but not functional immediately.

Endometrial responses during pregnancy, uterine and fetal fluids

In the tammar wallaby and in the quokka (*Setonix brachyurus*) there is a
marked response by the uterine endometrium during pregnancy, and a
number of uterine-specific proteins are elaborated (Renfree, 1972b, 1973a;
G. I. Wallace, personal communication). The proliferation of the uterine
glands is more prominent than in most eutherian mammals, and much more
so than in the rodent uterus (Enders and Given, 1977). The presence of
similar, pre-albumin, proteins in the uterine secretions and in the yolk sac
fluid indicates transfer, and it has been suggested that these molecules may
be important for activation of the delayed blastocyst (Renfree, 1973a; Tyndale-Biscoe, in press). Certainly their appearance correlates well with other
developmental events such as expansion of the blastocyst and onset of RNA
polmerase synthesis (Tyndale-Biscoe, in press). However, there is, as yet, no
direct evidence that these proteins derive from the uterine fluids or that
uterine-specific proteins are either responsible for initiation of growth after
diapause or contribute to the brief but rapid pregnancy.

The successful in vitro growth of marsupial embryos provides additional
data suggesting that these substances in the yolk sac fluid, and an intact yolk
sac membrane, are important for embryonic growth. New and his colleagues
have cultured embryos of the American opossum, *Didelphis virginiana* (New
and Mizell, 1972; New et al., 1977), in which the yolk sac is large, whereas
the allantois only enlarges (presumably filled with nitrogenous excretory

Figure 4. Relative duration of the preattachment phases in four different
families of marsupials. Numbers given are days of gestation. The period of
organogenesis is relatively constant in the four groups, which range in size
from less than 0.05 to over 30 kg.

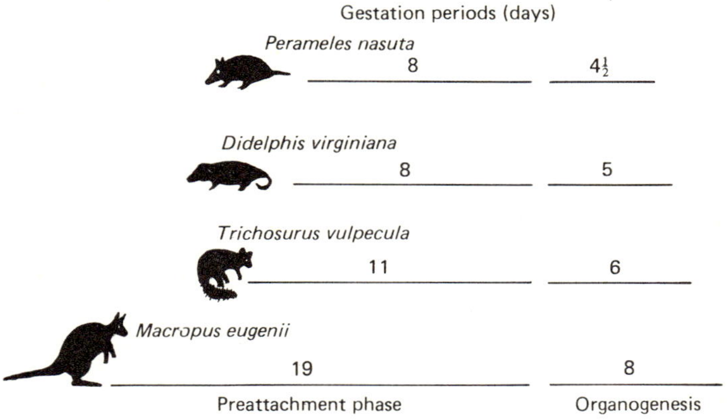

Gestation periods (days)

	Preattachment phase	Organogenesis
Perameles nasuta	8	$4\frac{1}{2}$
Didelphis virginiana	8	5
Trichosurus vulpecula	11	6
Macropus eugenii	19	8

products) during the last 3 days of pregnancy. These embryos grow successfully up to 30 h at all stages of development from the primitive streak to the
late fetus – the period of time that there would be an increase in concentration of substances in the yolk sac fluid. These experiments also show that the
integrity of the yolk sac membrane is important because the most successful
cultures occurred in those opossum embryos that had the yolk sac membrane intact, complete with its enclosed nutrient source of glucose, protein,
and amino acids.

Endometrial responses during pregnancy, role of the corpus luteum and stimulation by the placenta

There is apparently no unilaterally active (but systemically inactive) luteoly-
sin derived from the uterus because hysterectomy has no effect on the
corpus luteum of the brush possum or the opossum (Clark and Sharman,
1965; Hartman, 1925a). In lactating marsupials the corpus luteum is either
quiescent or absent. An exception is the bandicoots, in which the corpus
luteum of pregnancy is similar to that found in the first half of the estrous
cycle, but after birth and during subsequent lactation the life of the corpus
luteum is prolonged (Hughes, 1962). It has been suggested that the receptors on luteal cells are capable of responding to a luteotropic influence,
possibly prolactin (Tyndale-Biscoe and Hawkins, 1977). In contrast, in the
macropodids prolactin inhibits corpus luteum function, which implies that
these luteal cells are responding in an opposite manner to the peramelid
corpus luteum (Tyndale-Biscoe and Hawkins, 1977; Tyndale-Biscoe and
Evans, this volume).

The corpus luteum reaches a maximum size during the midluteal phase,
and declines toward the end of pregnancy. This pattern is mirrored by the
endometrium, which reaches the maximum weight during the time of placental attachment in the macropodids. In the nonmacropodids both gravid
and nongravid uteri respond to a similar extent, whereas in the macropodids the pregnant endometria are much heavier. In polyovular animals like
Didelphis, no assessment can be made of differences between uteri of the
same animal because both uteri are gravid, but a few values from nonpregnant animals suggest that the response is the same during the estrous cycle
and pregnancy (Renfree, 1975).

In the tammar wallaby *(Macropus eugenii)* the endometrium adjacent to the
corpus luteum is heavier than the contralateral, nonpregnant endometrium
whether or not pregnancy occurs (Renfree, 1972b; Renfree and Tyndale-
Biscoe, 1973). This effect is enhanced in the second half of pregnancy by a
factor derived from the embryo or its placenta (see Renfree, 1972b; Renfree
and Tyndale-Biscoe, 1973; Tyndale-Biscoe et al., 1974). The proliferation of
the endometrium is apparently dependent on the presence of an embryo or
the fetal membranes because the transfer of the blastocyst to a cyclic, nonpregnant animal or to the side contralateral to the corpus luteum initiates

similar endometrial proliferation. In the brush-tail possum, *Trichosurus vul-pecula,* a similar unilateral effect of the corpus luteum has been observed (von der Borch, 1963). The protein content of the secretion and weight of endometrium of the pregnant and nonpregnant uteri appear to be greater than in the macropodids, and there are fewer uterine-specific proteins (Tyndale-Biscoe and Cantrill,, unpublished results). In *Trichosurus* (1 to 2 kg body weight), however, the endometrium reaches a weight of 4 g and in *Didelphis* (2 kg body weight) a maximum of 12 g, both much greater quantities than the 0.5 g of the 5-kg tammar.

In the tammar during embryonic diapause, the proliferation of the endometrium can be initiated and maintained by injections of exogenous progesterone, but in this case both gravid and nongravid endometria respond. If progesterone injections are then stopped, only the uterus containing the embryo continues to show the secretory, enlarged endometrium, and the nonpregnant uterus declines to levels usually observed in the nonpregnant cycle (Renfree and Tyndale-Biscoe, 1973). In this experiment the ovaries were inactive, and the influence of the corpus luteum was replaced by the use of exogenous progesterone, so the continuation of endometrial stimulation after progesterone withdrawal must be related to the embryo in some way. This indirect evidence suggests that the placenta or embryo itself may be responsible for a hormonal stimulus (Renfree, 1972b).

This is supported by other evidence. In four species of marsupials, ovariectomy during the first few days of gestation results in death or loss of the embryos, but ovariectomy performed after days 6 to 9 (depending on the species) has no effect on the continuation of pregnancy, and the embryo continues to grow to full term (Hartman, 1925b; Tyndale-Biscoe, 1963, 1970; Sharman, 1964; Renfree, 1974). The uterus remains secretory, and embryonic development is not impaired, although parturition is inhibited. Thus, although the secretions of the corpus luteum are necessary for the initiation of development, pregnancy is able to continue in the absence of the corpus luteum as in many other mammals. In these studies of eutherian mammals it is usually assumed that the ability to maintain pregnancy without hormone replacement indicates placental hormone synthesis.

Steroids in pregnancy

Levels of circulating steroids during the estrous cycle are known for two species, the brush possum and the tammar wallaby. In the possum, levels change from 0.5 ng ml^{-1} before day 8 of the cycle, to reach a peak of 3 to 4 ng ml^{-1} at day 12, declining again by day 17 when the corpus luteum involutes (Thorburn et al., 1971; Tyndale-Biscoe, 1973). In the tammar, levels are much lower and change from 0.4 ng ml^{-1} at around day 17 to a maximum of about 1 ng ml^{-1} toward the end of the cycle just before parturition (Lemon, 1972). In the American opossum (*Didelphis*) progesterone concentration in the peripheral plasma during the luteal phase reaches 12 ng ml^{-1} (Cook et al., 1977), and in the bandicoot (*Isoodon macrourus*) levels of 20 ng ml^{-1} have been reported from animals during the early stages of lactation and 3 to 6 ng ml^{-1} in ovariectomized females and intact males (Gemmell, 1977).

Progesterone levels during pregnancy have been investigated in only one species, the tammar. Lemon (1972), using a competitive protein binding assay, has shown that levels are similar to that of the estrous cycle, and preliminary results using radioimmunoassay correlate well with these data (Renfree and Heap, 1977; Tyndale-Biscoe and Hinds, unpublished observations). Lemon showed that circulating levels of progesterone were significantly higher during the second half of pregnancy compared to the second half of the cycle, but in three animals ovariectomized at day 10 and measured at day 25 there was no rise in progesterone. The progesterone content of the corpus luteum shows a similar pattern to that in the peripheral plasma and suggests that most if not all of the circulating progesterone is derived from the corpus luteum (Renfree et al., in press). These results are not necessarily inconsistent with the idea of a placental endocrine effect, as only the gravid uterus responds; if the uterine differences are caused by hormones, their concentration must be very low because the contralateral uterus is unaffected.

Studies to characterize the steroids of the placenta are currently under way. In the quokka (*Setonix brachyurus*) the yolk sac placenta from three animals was shown to convert ^{14}C-progesterone in vitro, although the conversion was extremely low (<1%) (Bradshaw et al., 1975). In the tammar, the conversion of pregnenolone to progesterone is also very low, although the yolk sac placenta certainly is able to metabolize a number of steroid precursors; it actively converts androstenedione to compounds including 5α-androstanedione and androsterone (Renfree and Heap, 1977, and unpublished observations). The endometrium is able to produce estradiol-17β from estrone, and the corpus luteum actively converts pregnenolone to progesterone during quiescence and at all stages of pregnancy.

Hormones involved in parturition

Ovariectomy performed after about day 6 allows pregnancy to proceed uninterrupted, but parturition is inhibited. Lemon (1972) suggested that the high levels of progesterone she observed may be necessary for birth in macropodids. This does not appear to be the case from recent investigations in the tammar. It was found that if lutectomy is performed at day 17 or 21, parturition occurs in about half of the experimental animals, whereas if it is carried out at day 23 or 25, parturition is successful in both groups (Young and Renfree, 1979). It must be assumed that there has been sufficient stimulation and preparation of the birth canal by the secretions of the corpus luteum (progesterone and relaxin) prior to this time to allow parturition. Tyndale-Biscoe (1963, 1969, and personal communication) has shown that relaxin activity occurs in marsupial corpora lutea, and in the quokka, relaxin injections after ovariectomy may result in successful parturition.

The maternal pituitary also has an important function in parturition, but the precise factors involved have not yet been defined. Although hypophysectomy during pregnancy permits development of the corpus luteum and the embryo to full term, the fetuses are found dead in the uterus 2 days after the expected birth (Hearn, 1973, 1974). Another important influence in parturition may derive from the fetal adrenals which are large in *M. giganteus* (Owen, 1834) and have differentiated by day 22 in the tammar (Renfree, 1972a). Cortisol is detectable in the fetal serum (about 10 ng ml^{-1}), and neonatal pouch young serum (about 30 ng ml^{-1}) (Catling and Vinson, 1976), suggesting that the adrenal cortex is capable of synthesizing and secreting corticosteroids and androgens at and before birth.

Conclusions

Taken together, the findings reviewed here provide indirect evidence that the placenta of the macropodids, but probably not the phalangerids, has an important role in the continued stimulation of the uterus and maintenance of secretory activity, possibly by an endocrine mechanism. Although ovariectomy during gestation does not interrupt pregnancy of nonmacropodids if performed after the luteal phase has been initiated (*Trichosurus, Didelphis*), the earliest time that removal of the corpus luteum can be tolerated is only 7 days before expected birth (i.e., days 5 to 6 p.c. (post-conception) in *Didelphis* and day 10 p.c. in *Trichosurus*), compared with about 21 days in the macropodids. It could be argued that the relatively greater provision of uterine secretions and the much greater proliferation of endometrium in these two species means that sufficient nutrient material is available after ablation of corpus luteum and that continued stimulation from its secretions is unnecessary.

Tyndale-Biscoe et al. (1974), in discussing this evidence, proposed a hypothesis for the determination of gestation length in marsupials. Because gestation is shorter than the estrous cycle, they suggested that in nonmacropodid species (in which birth occurs toward the end of the luteal phase), the length of gestation must be genetically determined by the female (i.e., the time of birth is related to the length of the luteal phase of the mother), whereas in the Macropodidae, in which gestation occupies the whole length of the estrous cycle, the duration of pregnancy is the sum of the luteal phase plus any fetal or placental stimulation of the uterus subsequent to the decline of the corpus luteum that allows gestation to continue (Tyndale-Biscoe et al., 1974). Recently, Merchant has shown that the conceptus can also influence ovarian events because pregnant tammars return to estrus significantly earlier than cycling animals (Merchant, in press). A diagrammatic representation of these ideas, based on actual progesterone levels in the tammar and brush possum, is given in Figure 5.

Thus, in the aspects of placental function and embryonic growth, the

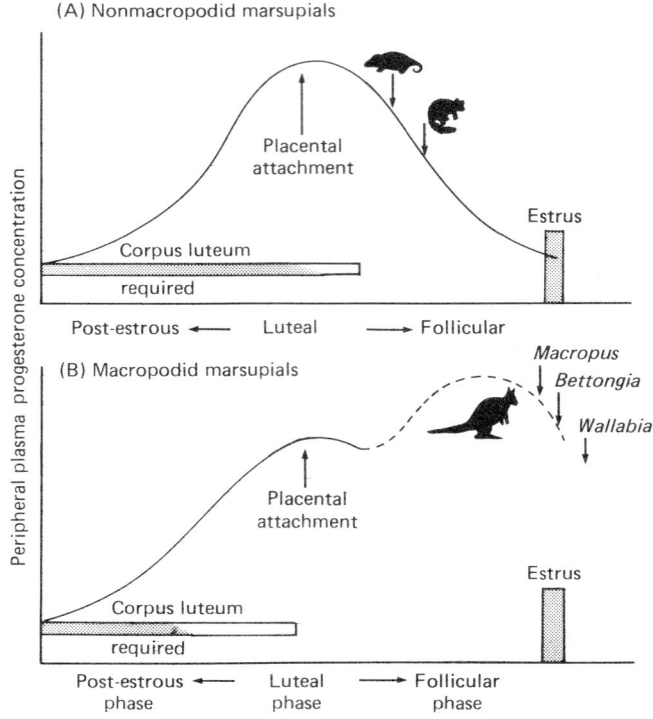

(A) Nonmacropodid marsupials

Peripheral plasma progesterone concentration

Placental
attachment

Estrus

Corpus luteum

required

Post-estrous ◄─── Luteal ───► Follicular

(B) Macropodid marsupials

Macropus
Bettongia

Wallabia

Placental
attachment

Estrus

Corpus luteum

required

Post-estrous ◄─── Luteal ───► Follicular
phase   phase   phase

Figure 5. A hypothetical scheme to illustrate the possible concentrations of ovarian steroid hormones during the estrous cycle and pregnancy in relation to the need for the corpus luteum, time of placental attachment, and birth. (A) Nonmacropodid marsupials, such as possums and opossums, in which hormone levels decline after midcycle; return to estrus is inhibited by the suckling pouch young. (B) Macropodid marsupials, such as the tammar, bettong, and swamp wallaby, in which hormone levels remain high after midcycle, despite decline in corpus luteum size. The dotted line shows the rise in progesterone which may be abolished by ovariectomy. Estrus occurs around the time of birth and is followed by a period of ovarian and embryonic quiescence during lactation. Stippling on the horizontal bar indicates the minimum duration of corpus luteum activity required to initiate and maintain pregnancy, and open bar shows the minimum time that the corpus luteum is required to allow successful parturition. Vertical stippled bar shows the time of estrus. Time of placental attachment is shown (↑), and time of birth (↓). Although the time of placental attachment is similar in both patterns, the ovary is required for relatively longer to maintain pregnancy in (A), and in (B) the extension of pregnancy to occupy the whole cycle results in birth occurring coincident with estrus. The temporal differences in the two patterns could relate to the duration of the placental effect.

marsupial pattern shares many of the features characteristic of eutherians. However, there are two alternative strategies – that of the marsupials giving birth to a small young which continues its development during a long lactation period in the pouch, and that of the eutherians producing a large young which is nurtured for extended periods by a complex placenta. Each produces an independent young at the end of lactation, and the continued existence of both modes of reproductive strategies attests to the adaptedness of both. As Kirsch (1977a,b), Parker (1977), and Pond (1977) all argue, the fact that marsupials do not "choose," in an evolutionary sense, to remain in the uterus to complete their development does not mean they are adaptively inferior. Indeed, because they display as wide a range of physiological, morphological, and ecological adaptations as the eutherians, it must be concluded that the marsupial represents an alternative pattern to the already wide range of vertebrate reproductive strategies.

I thank Dr. J. K. Findlay, Dr. A. P. F. Flint, and Dr. R. B. Heap for helpful suggestions. I also thank Dr. C. H. Tyndale-Biscoe, G. I. Wallace, and I. R. Young for allowing me to quote their unpublished results. Much of the work reported in this paper was supported by grants from the National Institutes of Health (HD-09387) and the Australian Research Grants Committee (D.I.-15759).

References

Bradshaw, S. D., McDonald, I. R., Hahnel, R., and Heller, H. (1975). Synthesis of progesterone by the placenta of a marsupial. *J. Endocrinol.* 65:451–2.

Catling, P. C., and Vinson, G. P. (1976). Adrenocortical hormones in the neonate and pouch young of the tammar wallaby, *Macropus eugenii*. *J. Endocrinol.* 69:447–8.

Clark, M. J. (1968). Termination of embryonic diapause in the red kangaroo, *Megaleia rufa*, by injection of progesterone or oestrogen. *J. Reprod. Fert.* 15:347–55.

Clark, M. J., and Sharman, G. B. (1965). Failure of hysterectomy to affect the ovarian cycle of the marsupial, *Trichosurus vulpecula*. *J. Reprod. Fert.* 10:459–61.

Cook, B., Sutterlin, N. S., Graber, J. W., and Nalbandov, A. V. (1977). Synthesis and action of gonadal steroids in the American opossum, *Didelphis marsupialis*. In *Reproduction and Evolution*, eds. J. H. Calaby and C. H. Tyndale-Biscoe, pp. 253–4. Canberra: Australian Academy of Science.

Enders, A. C., and Enders, R. K. (1969). The placenta of the four-eyed opossum (*Philander opossum*). *Anat. Rec.* 165:431–50.

Enders, A. C., and Given, R. (1977). The endometrium of delayed and early implantation. In *Biology of the Uterus*, ed. R. M. Wynn, pp. 203–43. New York and London: Plenum Press.

Flynn, T. T. (1923). The yolk-sac and allantoic placenta in *Perameles*. *Q. J. Microsc. Sci.* 67:123–82.

Flynn, T. T. (1930). The uterine cycle of pregnancy and pseudopregnancy as it is in the diprotodont marsupial *Bettongia cuniculus*. *Proc. Linn. Soc. N.S.W.* 55:506–31.

Gemmell, R. T. (1977). The structure and function of the corpus luteum of lactation of the bandicoot, *Isoodon macrourus*. *Theriogenology* 8:143.

Hartman, C. G. (1916). Studies in the development of the opossum, *Didelphis virginiana* L. I. History of the early cleavage. II. Formation of the blastocyst. *J. Morphol.* 27:1–83.

Hartman, C. G. (1925a). Hysterectomy and the oestrous cycle in the opossum. *Am. J. Anat.* 35:25–9.

Hartman, C. G. (1925b). The interruption of pregnancy by ovariectomy in the aplacental opossum: A study in the physiology of implantation. *Am. J. Physiol.* 71:436–54.

Hearn, J. P. (1973). Pituitary inhibition of pregnancy. *Nature (London)* 241:207–8.

Hearn, J. P. (1974). The pituitary gland and implantation in the tammar wallaby, *Macropus eugenii. J. Reprod. Fert.* 39:235–41.

Hill, J. P. (1900a). On the foetal membranes, placentation and parturition of the native cat (*Dasyurus viverrinus*). *Anat. Anz.* 18:364–73.

Hill, J. P. (1900b). Contributions to the morphology and development of the female urogenital organs in the Marsupialia. *Proc. Linn. Soc. N.S.W.* 25:513–32.

Hill, J. P. (1910). Contributions to the embryology of the Marsupialia with special reference to the native cat (*Dasyurus viverrinus*). *Q. J. Microsc. Sci.* 56:1–134.

Hughes, R. L. (1962). Role of the corpus luteum in marsupial reproduction. *Nature (London)* 194:890–1.

Hughes, R. L. (1974). Morphological studies on implantation in marsupials. *J. Reprod. Fert.* 39:173–86.

Hughes, R. L. (1977). Egg membranes and ovarian function during pregnancy in monotremes and marsupials. In *Reproduction and Evolution*, eds. J. H. Calaby and C. H. Tyndale-Biscoe, pp. 281–92. Canberra: Australian Academy of Science.

Kirsch, J. A. W. (1977a). The six-percent solution: Second thoughts on the adaptedness of the Marsupialia. *Am. Sci.* 65:276–88.

Kirsch, J. A. W. (1977b). Biological aspects of the marsupial–placental dichotomy: A reply to Lillegraven. *Evolution* 31:898–900.

Lemon, M. (1972). Peripheral plasma progesterone during pregnancy and the oestrous cycle in the tammar wallaby, *Macropus eugenii. J. Endocrinol.* 55:63–71.

Lyne, A. G., and Hollis, D. E. (1977). The early development of marsupials, with special reference to bandicoots. In *Reproduction and Evolution*, eds. J. H. Calaby and C. H. Tyndale-Biscoe, pp. 293–302. Canberra: Australian Academy of Science.

McCrady, E. (1938). The embryology of the opossum. *Am. Anat. Mem.* 16:1–233.

Merchant, J. C. (In press.) The effect of pregnancy on the interval between one estrus and the next in the tammar wallaby, *Macropus eugenii. J. Reprod. Fert.*

New, D. A. T., and Mizell, M. (1972). Opossum fetuses grown in culture. *Science (N.Y.)* 175:533–6.

New, D. A. T., Mizell, M., and Cockroft, D. L. (1977). Growth of opossum embryos in vitro during organogenesis. *J. Embryol. Exp. Morph.* 41:111–23.

Owen, R. (1834). On the generation of the marsupial animals with a description of the impregnated uterus of the kangaroo. *Phil. Trans. R. Soc.* 1834:333–64.

Padykula, H. A., and Taylor, J. M. (1977). Uniqueness of the bandicoot chorioallantoic placenta (Marsupialia: Peramelidae). Cytological and evolutionary interpretations. In *Reproduction and Evolution*, eds. J. H. Calaby and C. H. Tyndale-Biscoe, pp. 303–24. Canberra: Australian Academy of Science.

Parker, P. (1977). An evolutionary comparison of placental and marsupial patterns of reproduction. In *The Biology of Marsupials*, eds. B. Stonehouse and D. Gilmore, pp. 273–86. London: Macmillan.

Pilton, P. E., and Sharman, G. B. (1962). Reproduction in the marsupial *Trichosurus vulpecula. J. Endocrinol.* 25:119–36.

Pond, C. (1977). The significance of lactation in the evolution of mammals. *Evolution* 32:177–99.

Renfree, M. B. (1972a). Embryo–maternal relationships in the tammar wallaby *Macropus eugenii* (Desmarest). Ph.D. thesis, Australian National University, Canberra.

Renfree, M. B. (1972b). Influence of the embryo of the marsupial uterus. *Nature (London) 240:*475–7.

Renfree, M. B. (1973a). Proteins in the uterine secretions of the marsupial *Macropus eugenii. Dev. Biol. 32:*41–9.

Renfree, M. B. (1973b). The composition of fetal fluids of the marsupial *Macropus eugenii. Dev. Biol. 33:*62–79.

Renfree, M. B. (1974). Ovariectomy during gestation in the American opossum, *Didelphis marsupialis virginiana. J. Reprod. Fert. 39:*127–30.

Renfree, M. B. (1975). Uterine proteins during gestation in the marsupial *Didelphis marsupialis virginiana. J. Reprod. Fert. 42:*163–6.

Renfree, M. B. (1977). Feto-placental influences in marsupial gestation. In *Reproduction and Evolution,* eds. J. H. Calaby and C. H. Tyndale-Biscoe, pp. 325–32. Canberra: Australian Academy of Science.

Renfree, M. B., and Heap, R. B. (1977). Steroid metabolism in the placenta, corpus luteum and endometrium of the marsupial, *Macropus eugenii. Theriogenology 8(4):*164, Abstr.

Renfree, M. B., and Tyndale-Biscoe, C. H. (1973). Intrauterine development after diapause in the marsupial, *Macropus eugenii. Dev. Biol. 32:*28–40.

Renfree, M. B., and Tyndale-Biscoe, C. H. (1978). Manipulation of marsupial embryos and pouch young. In *Methods in Mammalian Reproduction.* ed. J. C. Daniel, pp. 307–31. New York: Academic Press.

Renfree, M. B., Green, S. W., and Young, I. R. (In press.) Growth of the corpus luteum and its progesterone content during pregnancy in the tammar wallaby, *Macropus eugenii. J. Reprod. Fert.*

Semon, R. (1894). Zoologische Forschungsreisen in Australien und dem malayischen Archipel. 2. Die Embryonalhullen der Monotremen und Marsupialier. *Denkschr. Med. Naturwiss. Ges. Jena 5:*19–58.

Sharman, G. B. (1961). The embryonic membranes and placentation in five genera of diprotodont marsupials. *Proc. Zool. Soc. London 137:*197–220.

Sharman, G. B. (1964). The effects of the suckling stimulus and oxytocin injection on the corpus luteum of delayed implantation in the red kangaroo. Proc. Second Int. Cong. Endocrinol. *Excerpta Med. Int. Cong. Ser. 83:*669–74.

Sharman, G. B. (1970). Reproductive physiology of marsupials. *Science (N.Y.) 167:*1221–8.

Sharman, G. B., and Berger, P. (1969). Embryonic diapause in marsupials. *Adv. Reprod. Physiol. 4:*211–40.

Taylor, J. M., and Padykula, H. A. (1978). Marsupial trophoblast and mammalian evolution. *Nature (London) 271:*588.

Thorburn, G. D., Cox, R. I., and Shorey, C. (1971). Ovarian steroid secretion rates in the marsupial *Trichosurus vulpecula. J. Reprod. Fert. 24:*139.

Tyndale-Biscoe, C. H. (1963). The effects of ovariectomy in the marsupial *Setonix brachyurus. J. Reprod. Fert. 6:*25–40.

Tyndale-Biscoe, C. H. (1969). The marsupial birth canal. In *Comparative Biology of Reproduction in Mammals. Symp. Zool. Soc. London 15:*233–50.

Tyndale-Biscoe, C. H. (1970). Resumption of development by quiescent blastocysts transferred to primed, ovariectomized recipients in the marsupial *Macropus eugenii. J. Reprod. Fert. 23:*25–32.

Tyndale-Biscoe, C. H. (1973). *Life of Marsupials.* London: Edward Arnold, 254 pp.

Tyndale-Biscoe, C. H. (In press.) Hormonal control of embryonic diapause and reactivation in the tammar wallaby. Ciba Fdn. Symp. *Maternal Recognition of Pregnancy,* ed. J. Whelan. Amsterdam: Elsevier.

Tyndale-Biscoe, C. H., and Hawkins, J. (1977). The corpora lutea of marsupials:

Aspects of function and control. In *Reproduction and Evolution*, eds. J. H. Calaby and C. H. Tyndale-Biscoe. Canberra: Australian Academy of Science.

Tyndale-Biscoe, C. H., Hearn, J. P., and Renfree, M. B. (1974). Control of reproduction in macropodid marsupials. *J. Endocrinol. 63*:589–614.

von der Borch, S. (1963). Unilateral hormone effect in the marsupial *Trichosurus vulpecula. J. Reprod. Fert. 5*:447–9.

Young, I. R. (1977). Relationship of hormonal and reproductive status to myometrial activity in the tammar wallaby. *Theriogenology 8*:207, Abstr.

Young, I. R., and Renfree, M. B. (1979). The effects of corpus luteum removal during gestation on parturition in the tammar wallaby, *Macropus eugenii. J. Reprod. Fert. 56*:249–54.

28

Adrenocorticosteroids in prototherian, metatherian, and eutherian mammals

MAGDA WEISS

The aim of this report is to evaluate the mechanisms of corticosteroid production by prototherian and metatherian mammals as compared with eutherian species. Emphasis will be on information that could give an insight into adaptive changes of mammalian evolution at a molecular level.

Our interest in Australian noneutherian mammals arose about 15 years ago because at that time little was known about their adrenocortical function, and because we anticipated that the information could serve to elucidate how adaptive responses in the three mammalian orders affected adrenal steroidogenesis. The adrenal cortex produces two important groups of steroid hormones: glucocorticoids (cortisol and corticosterone), which are concerned with intermediary metabolism, and mineralocorticoids (aldosterone), which regulate the retention of sodium in the body. The biosynthesis of these hormones from precursor material is via specific hydroxylating enzyme systems, the organization of which is genetically determined. It has been shown that the main sequence of these enzymic hydroxylations is the same in all vertebrates (Gottfried, 1964; Sandor, 1969). However, in humans inborn errors in enzyme function have been recognized, which can result in abnormal patterns of hormone production.

In keeping with the ultimate aim to trace evolutionary events, the forthcoming sections will deal with a number of selected parameters of adrenocortical function, which will be compared in the three mammalian orders.

Corticosteroids in blood

Adrenal venous blood

We first established the secretory patterns in seven marsupial species, selected to include representatives from each of the three living Australian superfamilies (Weiss and Richards, 1971). In the superfamily Phalangeroidea, which is the most diverse, members of all three families were studied, the brush-tailed possum (*Trichosurus vulpecula*) and the koala (*Phascolarctos cinereus)* (Phalangeridae), the wombat *(Vombatus hirsutus)* (Vombatidae), and the eastern grey kangaroo *(Macropus giganteus major)* (Macropodidae). In the superfamily Perameloidea the long-nosed bandicoot *(Perameles nasuta)* was studied, and in the superfamily Dasyuroidea the Tasmanian devil *(Sarcophilus harrisi)* and the Eastern native cat *(Dasyurus viverrinus)* (Dasyuridae) were studied. Descending the evolutionary scale, studies were carried out with the two living members of the class Monotremata, the echidna *(Tachyglossus aculeatus)* and the platypus *(Ornithorhynchus anatinus)*.

The secretion rates of the combined glucocorticoids (cortisol and corticosterone) in the three mammalian orders are presented in Table 1. From these data, it would be difficult to draw any conclusions as to possible phylogenetic differences among mammals. It is noteworthy, however, that the Phalangeroidae, which had generally lower corticosteroid secretion rates than Perameloidae or Dasyuroidae, had concomitantly lower adrenal weight to body weight ratios (Weiss and Richards, 1971). Furthermore, the echidna, which had the lowest corticosteroid secretion rate of all mammals, is able to survive bilateral adrenalectomy, if kept under stress-free conditions (McDonald and Augee, 1968).

It is well known that in mammals there exist large interspecies variations in the ratio of cortisol to corticosterone secretion. Eutherians such as man, sheep, cattle, dog, and hamster secrete mainly cortisol with smaller amounts of corticosterone. Rat, rabbit, and mouse produce corticosterone almost exclusively, whereas the guinea pig is predominantly a cortisol secretor (see Vinson and Whitehouse, 1970). The patterns of glucocorticoid secretion in marsupials are more uniform, in that all species investigated are predominantly cortisol secretors, and corticosterone is produced in lesser amounts (see Weiss and Richards, 1971). The highest cortisol to costicosterone ratio is found in the kangaroo (Weiss and McDonald, 1967) and the lowest in the quokka (Illet, 1969). In contrast, the nature of the major glucocorticoid in the two monotremes differs. In the echidna the predominant secretory product is corticosterone, whereas in the platypus it is cortisol. The ratio of cortisol to corticosterone in the former is about 0.2 (Weiss and McDonald, 1965), and in the latter about 5 (Weiss, 1973).

To evaluate the significance of these differences, it is essential to acquire a greater knowledge of the physiological role of these hormones in individual species. Since some species exhibit a reversal of the pattern under conditions of stress (Fevold and Drummond, 1976; Weiss et al., in press), it is possible that the major secretory product may reflect the most suitable adaptation of the individual to a specific metabolic requirement.

Peripheral blood

The peripheral levels of corticosteroids are dependent on a number of variables, such as adrenal secretory rate, pool size, and metabolic turnover rate. To avoid methodological differences that could also influence these

Table 1. *Secretion rates of glucocorticoids (cortisol and corticosterone) in eutherian, metatherian, and prototherian mammals (range)*

Eutheria		Metatheria		Prototheria	
Species[a]	Glucocorticoids (μg h^{-1} kg^{-1} body wt.)	Species	Glucocorticoids (μg h^{-1} kg^{-1} body wt.)	Species	Glucocorticoids (μg h^{-1} kg^{-1} body wt.)
Man	17–31	Possum	11–16	Echidna	0.5–1.3
Dog	32–95	Wombat	9–20	Platypus	14[b]
Cat	60	Kangaroo	16–48		
Sheep	20–55	Koala	20		
Rat	100–600	Bandicoot	37–55		
Rabbit	30–49	Tas. devil	80–106		
Guinea pig	89–165	Native cat	49–65		

[a]Dean (1962), Yates and Urquhart (1962).
[b]Data obtained from in vitro experiments (Weiss, 1973).

values, the data reported in Table 2 were obtained from the publication by Oddie et al. (1976), who employed the double derivative dilution method in their estimations. The large interspecies variations in glucocorticoid concentrations of the three orders could have important physiological implications with respect to differences in the animals' metabolic clearance rate and target tissue sensitivity. In this regard it is interesting to note that the glucocorticoid levels in the peripheral blood of echidna and two marsupials, wombat and koala, were similarly low (Table 2), despite the large differences in their respective adrenal secretion rates (Table 1). On the other hand, the peripheral blood aldosterone levels in the echidna fall in the range of eutherian and marsupial species (Table 2) (Weiss et al., in press), suggesting the possibility that the aldosterone receptor sites operate at fairly similar concentrations in all mammals.

Unusual steroids

An unusual feature of adrenocortical secretion, characteristic of marsupials, was the presence of 21-deoxycortisol. This steroid has not been isolated as a normal secretory product in eutherians, but it was found in the peripheral blood of patients with congenital adrenal hyperplasia (Wieland et al., 1965). We isolated 21-deoxycortisol in five marsupial species at concentrations that sometimes exceeded that of corticosterone (Weiss and Richards, 1971). In addition, the secretion rates of 11β-hydroxyprogesterone and 11β-hydroxyandrostenedione were also high relative to eutherians.

Table 2. *Peripheral blood corticosteroid levels in eutherian, metatherian, and prototherian mammals (means ± SD)*

Species	Combined cortisol and corticosterone (μg 100 ml^{-1})	Aldosterone (ng 100 ml^{-1})
Eutheria[a]		
Man	14.3 ± 6.4	6.5 ± 3.9
Dog	1.1 ± 0.5	2.1 ± 3.6
Cow	1.2 ± 0.7	2.9 ± 2.5
Sheep	0.6 ± 0.5	2.1 ± 1.7
Metatheria[a]		
Kangaroo	2.0 ± 0.2	3.2 ± 3.4
Wombat	0.1 ± 0.1	0.9 ± 1.4
Koala	0.2 ± 0.1	1.6 ± 2.2
Quokka	1.8 ± 0.7	6.6 ± 3.4
Native cat	1.9 ± 0.8	10.8 ± 1.6
Prototheria[b]		
Echidna	0.19 ± 0.04	10.2 ± 5.5

[a]Data from Oddie et al. (1976).
[b]Data from a collaborative study with C. J. Oddie.

The factors that would cause accumulation of 21-deoxycortisol include a deficiency in 21-hydroxylase, an inability to 21-hydroxylate an 11β-hydroxylated substrate, or some difference in the kinetic properties of the 11β-hydroxylating enzyme system. The first possibility can be excluded because cortisol is adequately produced in marsupials (Weiss and Richards, 1971). The second possibility has been tested, and it was shown that 21-deoxycortisol could be readily 21-hydroxylated (Weiss and McCance, 1974). The third possibility, that the properties of the 11β-hydroxylating enzyme system of marsupials could differ from that of eutherians, therefore seemed most likely. For testing this assumption, an in vitro approach was the most suitable.

In vitro studies

The biosynthetic pathways and the mechanisms involved in the formation of corticosteroids have been extensively studied by numerous workers, both in vivo and in vitro, and a reasonably clear picture has emerged concerning the sequence of steroid formation and the biochemical aspects of each step. It seems that in all vertebrates the pathways are basically similar, which may suggest that the enzyme systems for production of corticosteroids were stabilized very early in vertebrate evolution (Sandor, 1969).

More than 50 different steroids have been isolated from adrenal extracts of eutherian mammals. Many of these are biologically inactive by-products. Nevertheless, they indicate the complexity of enzyme systems that the adrenal cells possess. Some species also exhibit gross age-dependent changes in their adrenal enzyme activities. We observed such changes in guinea pigs and possums, in which the activity of the adrenal 5β-reductase increased markedly during the animal's maturation (Weiss and Ford, 1975, 1977). The physiological role of reductases in the adrenal remains to be established and merits further investigation.

Conversion products from [14]*C-progesterone*

Most of the conversion products formed by the platypus and echidna adrenal tissue were intermediates in the cortisol or corticosterone pathway (Table 3). However, as indicated, major quantitative differences appeared in the yields of conversion products by the two monotremes, particularly with respect to 11β-hydroxylated end products, cortisol, and corticosterone (Weiss, 1973). The very low 11β-hydroxylase activity of the echidna could well act as a limiting factor in the synthesis of corticosteroids, and could account for the low adrenal steroid secretory activity observed in this species (Table 1). The differences in enzyme activities of the platypus and the echidna may indicate the possibility that adaptation of adrenocortical function in the two genera evolved along separate lines appropriate with the demands of their different modes of life.

The adrenal conversion products formed from ^{14}C-progesterone were

Table 3. *Percent yield of conversion products from* ^{14}C-*progesterone by adrenal homogenates of three metatherian and two prototherian mammals (mean ± SEM)*

Conversion product	Possum[a]	Kangaroo	Antechinus[b]	Echidna	Platypus
Cortisol	52 ± 3.1	43 ± 5.2	38	0.3 ± 0.2	48 ± 11.2
Corticosterone	16 ± 2.7	2.8 ± 0.8	14	0.6 ± 0.3	6.1 ± 4.3
Aldosterone	0.7 ± 0.2	0.2 ± 0.1	0.1	<0.1	<0.1
11-Deoxycortisol	0.5 ± 0.2	2.0 ± 0.9	0.4	12 ± 2.4	0.8 ± 0.1
11-Deoxycorticosterone	—[c]	—	—	38 ± 7.5	—
21-Deoxycortisol	1.5 ± 0.3	2.9 ± 1.0	0.5		
11β-Hydroxyprogesterone	2.8 ± 0.6	0.5 ± 0.2	2.1	—	0.5 ± 0.2
17α-Hydroxyprogesterone	—	0.2 ± 0.1	1.8	5.2 ± 1.8	0.8 ± 0.3
Androstenedione	0.5 ± 0.2	0.6 ± 0.3	—	4.3 ± 1.2	—

[a]Immature animals. Adults yield 5β-reduced steroids (Weiss and Ford, 1977).
[b]Average of 2 animals.
[c]Levels undetectable, <0.1%.

investigated in three marsupials: possum, kangaroo, and marsupial mouse *(Antechinus swainsonii)*. Apart from 21-deoxycortisol no unusual products were isolated (Table 3). The ratios of cortisol to corticosterone in the three species varied and were in accordance with the values obtained from adrenal venous blood analyses (Weiss and McDonald, 1966, 1967).

The presence of 21-deoxycortisol in marsupials and the low yield of 11β-hydroxylated products in the echidna confirmed the in vivo findings and reinforced the view that the nature of the 11β-hydroxylating enzyme system in noneutherians may differ from that of eutherian mammals. We therefore considered it relevant to initiate investigations on the properties of this enzyme at subcellular levels.

Kinetic properties of 11β-hydroxylase
The 11β-hydroxylase system is located in the adrenal mitochondrial cell fraction, which was used for establishing the affinity (K_M) and activity (V) of the enzyme (Weiss and Vardolov, 1977). Table 4 lists the summarized data of the kinetic constants, obtained under identical experimental conditions, from representatives of each of the three mammalian orders. A comparison of values for marsupials and ox indicated a striking difference in the activities of the 11β-hydroxylases for 11-deoxycortisol, the value with ox being about 65 times higher than with possum and kangaroo. In contrast, the activities and affinities of the enzyme for 17α-hydroxyprogesterone were of a similar order of magnitude in all three species. Hence, with ox the K_M/V (which can be considered roughly as the inverse measure of enzyme efficiency) for 11-deoxycortisol substrate was about 600 times higher than that for 17α-hydroxyprogesterone, whereas with the marsupials the K_M/V was of a fairly similar magnitude for both substrates.

It has been previously established that both eutherian and marsupial adre-

Table 4. K_M and V values for 11β-hydroxylation of different steroid substrates by adrenal mitochondria from eutherian, metatherian, and prototherian mammals

Species	Substrate	K_M (μM)	V (pmol min^{-1} mg^{-1})	K_M/V
Ox	11-Deoxycortisol	38	2270	0.02
	17α-Hydroxyprogesterone	118	11	10
Possum	11-Deoxycortisol	86	36	2.4
	17α-Hydroxyprogesterone	147	59	2.5
Kangaroo	11-Deoxycortisol	714	32	22
	17α-Hydroxyprogesterone	294	101	2.9
Rat	11-Deoxycorticosterone	80	970	0.08
Echidna	11-Deoxycorticosterone	25	3	8.3

nals are capable of efficiently converting 21-deoxycortisol to cortisol (Weiss and McCance, 1974). Consequently the data in Table 4 suggest that the preferential intermediate in the biosynthesis of cortisol is 11-deoxycortisol in ox (as established by Hechter and Pincus, 1954) and 21-deoxycortisol in kangaroo, whereas in possum both intermediates are equally effective. Furthermore, the findings lead to the assumption that the adaptive changes in eutherians were directed to enhance the activity of 11β-hydroxylase for 11-deoxycortisol precursor, thereby rendering the pathway via 21-deoxycortisol redundant.

A comparison of the kinetic constants for 11β-hydroxylation of 11-deoxycorticosterone of the echidna with that of rat (Table 4) indicated that the affinities were similar, but that the activity of the 11β-hydroxylase with echidna was 1/300 that of rat (Weiss et al., in press). It was confirmed that besides ox and rat, the rate of 11β-hydroxylation of 11-deoxycortisol[*] or 11-deoxycorticosterone in other eutherians, such as human and guinea pig, is also about 200 to 1000 times higher than in marsupials (Weiss and Vardolov, 1977). Thus the activity of 11β-hydroxylase in the echidna, because it is about 1/10 that of marsupials, must be the lowest in the class Mammalia.

Another aspect relevant to the 11β-hydroxylase system is the enzyme substrate specificity. Experimental data from eutherians indicated that some species possess multiple 11β-hydroxylases, each being specific for a particular substrate. In the ox, for instance, three different 11β-hydroxylases have been isolated, which are specific for 11-deoxycortisol, 11-deoxycorticosterone, and androstenedione respectively (Hudson et al., 1976). Indirect evidence with human adrenal tissue points to a similar conclusion (Zachmann et al., 1971; Gregory and Gardner, 1976).

With the possum and kangaroo adrenal, the simultaneous addition of 11-deoxycortisol and 17α-hydroxyprogesterone, which can act as alternate inhibitors, showed a mutual competitive inhibition of 11β-hydroxylation in the range of 29 to 52%, indicating the presence of a single enzyme system (Table 5). No evidence of significant competitive inhibition was found with ox adrenals, indicating the presence of two substrate-specific 11β-hydroxylases. In conjunction with other studies described elsewhere (Vardolov and Weiss, 1978), it is concluded that in these marsupials, in contrast to the ox, both substrates are hydroxylated by the same non-substrate-specific 11β-hydroxylase system. The findings imply that substrate-specific 11β-hydroxylases could be a characteristic feature of eutherian adaptation. However, more extensive studies on a larger number of species remain to be carried out to give a fuller evaluation of this concept.

Effect of ACTH on adrenocorticosteroids in marsupials and monotremes

Corticosteroid levels rise in response to stress, and this response may aid in adaptation to the environment.

Our earlier data on adrenal corticosteroid secretion rates indicated that in marsupials and the echidna the steroidogenic response to ACTH is sluggish, and the maximum rise is lower than in eutherians (Weiss and McDonald, 1965; Weiss and Richards, 1971). This low response to ACTH may also be implicated in the low levels of corticosteroids observed in peripheral blood from conscious possums or echidnas (Khin Aye Than and McDonald, 1973; Sernia and McDonald, 1977; Weiss et al., in press). Because there is suggestive evidence that the 11β-hydroxylase system may respond directly to ACTH control (Kowal et al., 1970; Laury and McCarthy, 1970), it seemed relevant to investigate the effects of long-term ACTH treatment on 11β-hydroxylase activity in the possum and the echidna.

Compared with untreated possums, administration of ACTH for 4 days caused the K_M for 11β-hydroxylation of 11-deoxycortisol and of 17α-hydroxyprogesterone to decrease by factors of about 2.6 and V to increase by factors of about 3.6 (Vardolov and Weiss, 1978). Consequently, the rates of cortisol and 21-deoxycortisol formation after ACTH treatment increased 4 and 5 times, respectively. However, in spite of this rise, the rate of cortisol formation from 11-deoxycortisol remained in the possum 70 times less than that of untreated ox.

Table 5. *Competitive inhibition of 11β-hydroxylation of 11-deoxycortisol (S) and 17α-hydroxyprogesterone (17OHP) acting as alternate inhibitors by adrenal mitochondria of ox, possum, and kangaroo*

Species	Substrate (μM) S	17OHP	Product formed	Rate of product formation (pmol min^{-1} mg^{-1})	Inhibition of product formation (%)
Ox					
	19.2	−[a]	Cortisol	781	
	19.2	133	Cortisol	869	0
	−	20.2	21-Deoxycortisol	1.5	
	105	20.2	21-Deoxycortisol	1.4	8.5
Possum					
	67.1	−	Cortisol	15.9	
	67.1	182	Cortisol	9.4	41
	−	51	21-Deoxycortisol	15.2	
	200	51	21-Deoxycortisol	7.5	51
Kangaroo					
	67.1	−	Cortisol	2.9	
	67.1	200	Cortisol	1.4	52
	−	51.3	21-Deoxycortisol	14.5	
	200	51.3	21-Deoxycortisol	10.2	30

[a]Substrate not present.
Modified from Weiss and Vardolov (1977).

In the echidna ACTH administration for 4 days had a considerable trophic effect on adrenal growth, increasing the adrenal weight to body weight ratio by about 80%. The effect on the 11β-hydroxylation of 11-deoxycorticosterone showed a threefold rise in the enzyme's activity (V), whereas the affinity remained unchanged (Weiss et al., in press). The effect of short- and long-term ACTH treatment on the peripheral blood corticosteroid levels was also studied. The normal peripheral blood corticosteroid levels (ng 100 ml^{-1}) for cortisol, corticosterone, and aldosterone were: for untreated echidnas 58 ± 12, 130 ± 26, and 10 ± 5 (mean ± SEM), respectively; after short-term (30 min) ACTH treatment, 78 ± 8, 217 ± 7, and <1, respectively; and after long-term (4 days) ACTH treatment, 543 ± 231, 323 ± 57, and 3 ± 2, respectively (Weiss et al., in press). Thus prolonged ACTH treatment increased the levels of cortisol by about 800% and corticosterone by about 150%, thereby causing a reversal in the blood corticosteroid pattern. In this respect, the ACTH response of the echidna resembled that of rabbits, in which, after a similar treatment with ACTH, cortisol became predominant over corticosterone (Kass et al., 1954; Ganjam et al., 1972). Such shifts in the rate of synthesis of particular hormones may be important in withstanding conditions of continuous stress.

Discussion and conclusions

The adrenal gland of vertebrates appears to have undergone distinct evolutionary changes, which have become progressively more complex higher up the phyletic scale. However, irrespective of the complexity of adrenocortical morphology, the basic biosynthetic pathways for corticosteroid production have remained the same in all vertebrates. They may therefore be looked upon as "core" pathways which have a common ancestry. These "core" pathways may have provided the basis from which alternative pathways arose during the adaptive radiation of species. Although often quantitatively less important, they may have become a characteristic feature of a particular order or genus. Hence a comparison of the corticosteroid biosynthetic mechanisms of ox with those of kangaroos, both of which live in a grassland environment, could serve to show how molecular adaptations occurred in the two different streams, as well as provide information on the different ways in which homeostasis can be maintained.

This report suggests that in marsupials the alternative pathway, indicated by the presence of 21-deoxycortisol, arose as a result of a single non-substrate-specific 11β-hydroxylase enzyme system. In eutherians, on the other hand, the absence of 21-deoxycortisol could be due to adaptive changes which resulted in the formation of isoenzymes of 11β-hydroxylases, each exerting substrate specificity with varying degrees of activity. There is no doubt that the multiple substrate-specific 11β-hydroxylases bestow numerous physiological advantages upon the animal. The ones that come immediately to mind are (1) a finer control of biosynthetic pathways, (2) an individual response to physiological control systems (such as ACTH), and (3) a safeguard if one of the enzyme systems becomes defective (as can be seen in the adrenogenital syndrome). Furthermore, the higher enzyme efficiencies of eutherians, in conjunction with a faster ACTH response, would enable the animal to adjust more rapidly to stressful environmental changes.

In view of these considerations it would probably be justified to regard noneutherian mammals, with respect to steroidogenesis, as "enzymatically" disadvantaged over eutherians. However, in view of their seemingly divergent adaptational paths, the noneutherians should be regarded as an alternative stream rather than as primitive mammals.

References

Deane, H. W. (1962). The anatomy, chemistry and physiology of adrenocortical hormones. *Handbuch der experimentellen pharmakologie*, vol. 14, part 1, pp. 74–76. Berlin: Springer-Verlag.

Fevold, H. R., and Drummond, H. B. (1976). Steroid biosynthesis by adrenal tissue of snowshoe hares *(Lepus americanus)* collected in a year of peak population density. *Gen. Comp. Endocrinol. 28:*113–17.

Ganjam, V. K., Campbell, A. L., and Murphy, B. E. P. (1972). Changing patterns of circulating corticosteroids in rabbits following prolonged treatment with ACTH. *Endocrinology 91:* 607–11.

Gottfried, H. (1964). The occurrence and biological significance of steroids in lower vertebrates. A review. *Steroids 3:*219–42.

Gregory, T., and Gardner, L. I. (1976). Hypertensive virilizing adrenal hyperplasia with minimal impairment of synthetic route to cortisol. *J. Clin. Endocrinol. Metab. 43:*769–74.

Hechter, O., and Pincus, G. (1954). Genesis of adrenocortical secretion. *Physiol. Rev. 34:*459–96.

Hudson, R. W., Schachter, H., and Kinninger, K. W. (1976). Studies of 11β-hydroxylation by beef adrenal mitochondria. *J. Steroid Biochem. 7:*255–62.

Illet, K. F. (1969). Corticosteroids in adrenal venous and heart blood of the Quokka, *Setonix brachyurus* (Marsupialia: Macropodidae). *Gen. Comp. Endocrinol. 13:*218–21.

Kass, E. H., Hechter, O., Macci, I. A., and Mou, T. W. (1954). Changes in patterns of secretion of corticosteroids in rabbits after prolonged treatment with ACTH. *Proc. Soc. Exp. Biol. Med. 85:*583–7.

Khin Aye Than and McDonald, I. R. (1973). Adrenocortical function in the Australian brush-tailed possum *Trichosurus vulpecula* (Kerr). *J. Endocrinol. 58:*97–109.

Kowal, J., Simpson, E. R., and Estabrook, R. W. (1970). Adrenal cells in tissue culture. V. On the specificity of the stimulation of 11β-hydroxylase by adrenocorticotrophin. *J. Biol. Chem. 245:*2438–43.

Laury, L. W., and McCarthy, J. L. (1970). In vitro adrenal mitochondrial 11β-hydroxylation following in vivo adrenal stimulation or inhibition: Enhanced substrate utilization. *J. Endocrinol. 87:*1380–5.

McDonald, I. R., and Augee, M. L. (1968). Effects of bilateral adrenalectomy in the monotreme *Tachyglossus aculeatus. Comp. Biochem. Physiol. 27:*669–78.

Oddie, C. J., Blaine, E. H., Bradshaw, S. D., Coghlan, J. P., Denton, D. A., Nelson, J. F., and Scoggins, B. A. (1976). Blood corticosteroids in Australian marsupial and placental mammals and one monotreme. *J. Endocrinol. 69:*341–8.

Sandor, T. (1969). A Comparative survey of steroids and steroidogenic pathways throughout the vertebrates. *Gen. Comp. Endocrinol. suppl.* 2:284–98.

Sernia, C., and McDonald, I. R. (1977). Adrenocortical function in a prototherian mammal, *Tachyglossus aculeatus* (Shaw). *J. Endocrinol.* 72:41–52.

Vardolov, L., and Weiss, M. (1978). A study of steroid 11β-hydroxylation by adrenal mitochondria of marsupials. Part II. The effect of corticotrophin, metopirone and pH on 11β-hydroxylation of 11-deoxycortisol and 17α-hydroxyprogesterone by adrenal mitochondria of possum *(Trichosurus vulpecula). J. Steroid Biochem.* 9:47–52.

Vinson, G. P., and Whitehouse, B. J. (1970). Comparative aspects of adrenocortical function. *Adv. Steroid Biochem. Pharmacol.* 1:163–342.

Weiss, M. (1973). Biosynthesis of adrenocortical steroids by monotremes: Echidna *(Tachyglossus aculeatus)* and platypus *(Ornithorhynchus anatinus). J. Endocrinol.* 58:251–62.

Weiss, M., and Ford, V. (1975). The effects of age, gonadectomy and testosterone treatment on adrenal 5β-reductase activity in guinea pigs. *Int. J. Biochem.* 6:683–7.

Weiss, M., and Ford, V. L. (1977). Changes in steroid biosynthesis by adrenal homogenates of the possum *(Trichosurus vulpecula)* at various stages of sexual maturation, with special reference to 5β-reductase activity. *Comp. Biochem. Physiol.* 57B:15–18.

Weiss, M., and McCance, I. (1974). 21-hydroxylation of 21-deoxycortisone by adrenal glands of a marsupial and two eutherian species. *Comp. Biochem. Physiol.* 48B:1–13.

Weiss, M., and McDonald, I. R. (1965). Corticosteroid secretion in the monotreme *Tachyglossus aculeatus. J. Endocrinol.* 33:203–10.

Weiss, M., and McDonald, I. R. (1966). Corticosteroid secretion in the Australian phalanger *(Trichosurus vulpecula). Gen. Comp. Endocrinol.* 7:345–51.

Weiss, M., and McDonald, I. R. (1967). Corticosteroid secretion in kangaroos *(Macropus canguru major* and *M. (megaleia) rufus). J. Endocrinol.* 39:251–61.

Weiss, M., and Richards, P. G. (1971). Adrenal steroid secretion in the Tasmanian devil *(Sarcophilus harisii)* and the Eastern native cat *(Dasyurus viverrinus).* A comparison of adrenocortical activity of different Australian marsupials. *J. Endocrinol.* 49:263–75.

Weiss, M., and Vardolov, L. (1977). A study of steroid 11β-hydroxylation by adrenal mitochondria of marsupials. Part I. A comparison of 11β-hydroxylase activity and specificity for different steroid substrates by possum *(Trichosurus vulpecula),* kangaroo *(Macropus major)* and beef. *J. Steroid Biochem.* 8:1233–42.

Weiss, M., Oddie, C. J., and McCance, I. (In press). The effects of ACTH on adrenal steroidogenesis and blood corticosteroid levels in the echidna *(Tachyglossus aculeatus). Comp. Biochem. Physiol.*

Wieland, R. G., Maynard, D. E., Riley, T. R., and Hamwi, C. J. (1965). Detection of 21-deoxycortisol in blood from a patient with congenital adrenal hyperplasia. *Metabolism* 14:1276–81.

Yates, E., and Urquhart, J. (1962). Control of plasma concentrations of adrenocortical hormones. *Physiol. Rev.* 42:359–421.

Zachmann, M., Vollmin, J. A., New, M. I., Cirtius, H. C., and Prader, A. (1971). Congenital adrenal hyperplasia due to deficiency of 11β-hydroxylation of 17α-hydroxylated steroids. *J. Clin. Endocrinol. Metab.* 33:501–8.

29

The renin–angiotensin system in marsupials

J. R. BLAIR-WEST and ANGELA GIBSON

Renin is an endopeptidase present in the juxtaglomerular cells of the kidneys of all vertebrate species that have been tested, except elasmobranchs and cyclostomes. It is released into the blood from renal stores. Reninlike activity has also been found in other tissues, e.g., arterial walls, brain, uterus, placenta, and salivary glands in some eutherian species (Skeggs et al., 1974).

The only known physiological action of renin is the cleavage of angiotensin from renin substrate (Figure 1). This action can occur inside and outside the kidneys. There has been speculation as to whether the intrarenal role of the renin–angiotensin system (RAS) is more primitive than its extrarenal roles (Vane, 1974; Brown et al., 1972).

In view of the number of known variants of the angiotensin molecule and the examples of species specificity of renin–renin substrate reactions, the terms "renin," "renin substrate," and "angiotensin" have to be used generically to describe the enzyme, the substrate, and the physiologically active product of the reaction. Reviews of special aspects of the renin–angiotensin system are quoted in the text. This chapter is primarily concerned with the occurrence and function of this system in metatherian species.

Biochemistry of the renin–angiotensin system (RAS)

Progress in this subject has been reviewed extensively in publications edited by Page and McCubbin (1968) and Page and Bumpus (1974). Figure 1 summarizes the main elements of the RAS. The amino acid sequences of angiotensins in Figure 1 are those of man, horse, and pig; valine occurs in place of isoleucine in position 5 in other eutherian mammals, sheep, and cattle. This variation appears to have little effect on the biological activity of the angiotensins, but Khairallah et al. (1978) have reported that Ile^5-angiotensin II showed about 60% of the activity of Val^5-angiotensin II in stimulating aldosterone production from rat adrenal zona glomerulosa cells. The physiological functions of the RAS are virtually confined to the agonist properties of angiotensin II and III (see below).

The amino acid sequences of angiotensins from noneutherian species are unknown (except for the white leghorn fowl, a snake *Elape chimocophora*, and the Japanese goosefish). Biochemical and pharmacological properties of angiotensinlike substances from avian, reptilian, amphibian, and teleost renal and extrarenal sources indicate sequences that are different from known angiotensins (Taylor, 1977). Biological and radioimmunological tests (Best et al., 1974) showed that the C-terminal portion of angiotensin I in the eastern grey kangaroo (*Macropus giganteus* Shaw) is not the same as the sequence shown in Figure 1. The product of incubation of kangaroo renin substrate with various renin preparations was not reactive in radioimmunoassay that was sensitive to eutherian forms of angiotensin I. However, kangaroo angiotensin I could be converted to a reactive product in radioimmunoassay for eutherian angiotensin II. Assuming that renin splits a leu-leu bond in kangaroo renin substrate, a variation in kangaroo angiotensin probably occurs at the histidine in position 9. Other variations were also indicated. We plan to prepare a large quantity of kangaroo angiotensin I from a stock of renin

Figure 1. Biochemical relationships of the renin–angiotensin system.

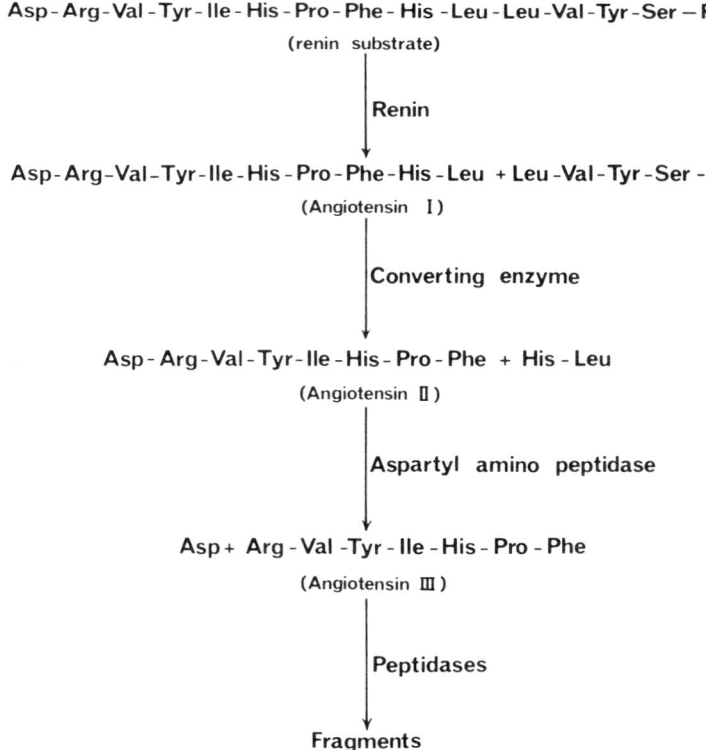

substrate and to determine the sequence of the decapeptide in collaborative studies with H. D. Niall and associates.

Other properties of the incubation product from several marsupial renin substrates are: (1) its formation in increasing amounts with increased incubation period and increased concentration of added renins, (2) similar blood pressure response and parallel dose-response curves to eutherian angiotensin in bioassay systems, and (3) that it is stable to boiling, dialyzable, and destroyed by α-chymotrypsin at pH 7.5 (Reid and McDonald, 1969; Simpson and Blair-West, 1971, 1972), indicating that the incubation procedure was detecting the reaction product of a renin–angiotensin system. Other biochemical evidence for the presence of renin and renin substrate in marsupial plasma was obtained by bilateral nephrectomy of an eastern grey kangaroo. After 5 days, renin was absent from plasma, and plasma renin substrate concentration rose from 300 to 3000 ng h^{-1} ml^{-1} (Simpson and Blair-West, 1972).

Plasma and kidney renins from all marsupials so far tested have reacted with sheep renin substrate (Reid and McDonald, 1969; Simpson and Blair-West, 1971). Simpson and Blair-West (1972) showed that marsupial plasma renins consistently reacted faster with sheep renin substrate than with kangaroo renin substrate at approximately the same concentration (Figure 2), indicating a higher affinity for the heterologous than the homologous substrate, findings similar to those for human renin (Skinner, 1967). Renin

Figure 2. Relationship between estimates of plasma renin concentration (PRC) obtained by incubation of plasma samples with kangaroo or sheep renin substrate. Results shown for eastern grey kangaroos (●), red kangaroo ( ), wombats (o), and wallabies – *Wallabia bicolor* ( ), *W. rufogrisea frutica* (□), *Thylogale billardierii* (■). (From Simpson and Blair-West, 1972.)

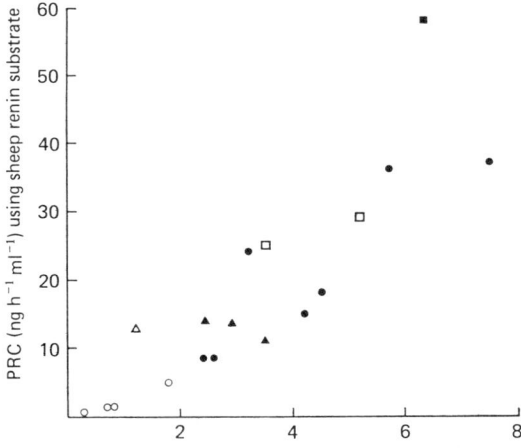

substrates from two species of kangaroo (*Macropus giganteus* Shaw and *Megaleia rufa*) and a species of wombat (*Vombatus hirsutus* Perry) reacted with sheep and pig renin preparations and with pepsin (Simpson and Blair-West, 1971). Pig renin was the most potent enzyme. The rate of reaction of kangaroo renin substrate with pig renin had an optimum at pH 5.4 compared with pH 6.5 for the reaction with sheep renin substrate. The maximal rate of reaction of pig renin with kangaroo renin substrate was one-tenth of the rate observed with sheep renin substrate, probably because of the amino acid sequence difference described above. The results, overall, indicate significant cross reaction between renins and renin substrates from marsupial and eutherian species, but structural differences in the components of the system seem to occur.

Functions of the renin–angiotensin system

The major extrarenal roles of the renin–angiotensin system (see Blair-West, 1976) are (1) regulation of blood pressure by peripheral vasoconstriction and possibly central actions in hypovolemic states, and (2) regulation of aldosterone secretion rate in sodium deficiency states in which angiotensin is essential for increased aldosterone biosynthesis. There is also strong evidence for an intrarenal role involving regulation of renal blood flow and glomerular filtration rate by local release of renin (Thurau, 1974). Other evidence suggests that angiotensin may affect reabsorption of sodium in the distal nephron segment (Thurau, 1974), and that angiotensin formed in the brain may stimulate thirst (Elghozi et al., 1977).

Knowledge of the biochemistry and functions of the renin–angiotensin system forms the basis of the methods used to detect and to measure the concentrations of renin in the plasma and kidneys of marsupial and other species.

Identification and assay of renin in marsupials

Renin in plasma is measured either as plasma renin activity (PRA – renin reacting with the homologous substrate in the plasma sample) or plasma renin concentration (PRC – renin reacting with added concentrated substrate after destruction of endogenous substrate). Tissue renin is measured by methods similar to PRC methods.

Johnston et al. (1967) detected renin activity in the kidneys of the North American opossum (*Didelphis virginiana*) by showing that infusion of kidney extracts increased blood pressure and stimulated aldosterone secretion. Sodium depletion increased the concentration of granules in juxtaglomerular (JG) cells, used as an index of renin content. Blair-West et al. (1968) detected renin activity at very low concentration in the kidneys of the eastern grey kangaroo (*Macropus giganteus*, Shaw) and the wombat (*Vombatus hirsutus*, Perry) by incubating homogenates with sheep renin substrate preparation. The kidneys of sodium-deficient wombats and kangaroos showed hypertrophy of the JG region, and the authors suggested that the apparently low renal renin concentrations could be due to species specificity of the heterologous renin–renin substrate reaction. Reid and McDonald (1968) showed that the JG complex of the kidney in the Australian possum (*Trichosurus vulpecula*) resembled the structure found in eutherians, but they did not demonstrate JG cell granulation characteristic of stored renin. Later (Reid and McDonald, 1969), Bowie-positive granules were identified in the walls of the afferent arterioles, and the degree of granulation was observed to increase during sodium depletion. Incubation of *Trichosurus* plasma alone (PRA) or with added sheep renin substrate (PRC) demonstrated significant levels of renin activity that were suppressed by ingestion of 0.9% NaCl solution and increased by administration of diuretics or by hemorrhage. Renin activity was found in kidney cortex. The authors were not able to measure plasma renin substrate concentration with heterologous renins or pepsin.

Simpson and Blair-West (1971, 1972) tested for the presence of a renin/angiotensin system in several marsupial species by analysis of plasma and kidneys. Renin substrate was shown to be present in the plasma of the three species tested, the eastern grey kangaroo (*Macropus giganteus*, Shaw), the red kangaroo (*Megaleia rufa*) and one species of wombat (*Vombatus hirsutus*, Perry), by incubation with sheep renin, pig renin, and pepsin. The range of concentrations was 270 to 460 ng ml^{-1} in animals with intact kidneys, which is similar to the levels in sheep (Simpson and Blair-West, 1971), rabbits (Romero et al., 1970), and rats (Carretero and Gross, 1967) and low relative to levels in man (Skinner, 1967). Bilateral nephrectomy of an eastern grey kangaroo abolished all renin activity in the plasma and increased renin-substrate concentration 10-fold after 5 days, which is similar to the effect of nephrectomy in other species, e.g., sheep (Blair-West et al., 1967).

Renin was shown to be present in the plasma of eastern grey kangaroos, a red kangaroo, wombats, two species of wallaby (*Wallabia rufogrisea fructica*, Ogilby and *Thylogale billardierii*, Desmarest), a tiger cat (*Dasyurus maculatus*), and a quokka (*Setonix brachyurus*) (Figure 3). Plasma renin concentrations (PRC) in wombats were uniformly low. Excluding that species, the PRC values were similar to those reported for *Trichosurus vulpecula* (Reid and McDonald, 1969), very high compared to sheep (Blair-West et al., 1971), and similar to the levels measured in man (Mitchell et al., 1970), using precisely the same procedure. These differences between metatherians and eutherians are more likely to be due to differences in affinity of the renins for sheep renin substrate than to difference in renin concentration.

Significant renin activity was demonstrated in the kidneys of all marsupials tested (Figure 4). Estimates of renal renin concentration were proportional to PRC. The lowest values were again observed in wombats. The renal values were consistently very low relative to the values in sheep and rabbits (Scoggins et al., 1970). The results of plasma and kidney assay suggest that the renal storage of renin in kangaroos and wombats is low relative to sheep or that renin is stored in an inactive form. The latter possibility is supported by findings that low pH activates renal renin in some eutherian species (Rubin, 1972; Boyd, 1972; Leckie, 1973). It has not yet been shown whether low pH activates renal renin in marsupial species.

Regulation and function of the RAS in marsupials

Almost all of the data on the regulation and functions of the RAS have been derived from studies in eutherian mammals. The major stresses affecting

Figure 3. Plasma renin concentration (●) using standard sheep renin substrate, and plasma renin activity (o) in the marsupial groups shown (B = *Wallabia rufogrisea frutica;* P = *Thylogale billardierii*). One sample was taken from each animal. Where both estimates were made on the same sample, results are joined by a solid line (from Simpson and Blair-West, 1971).

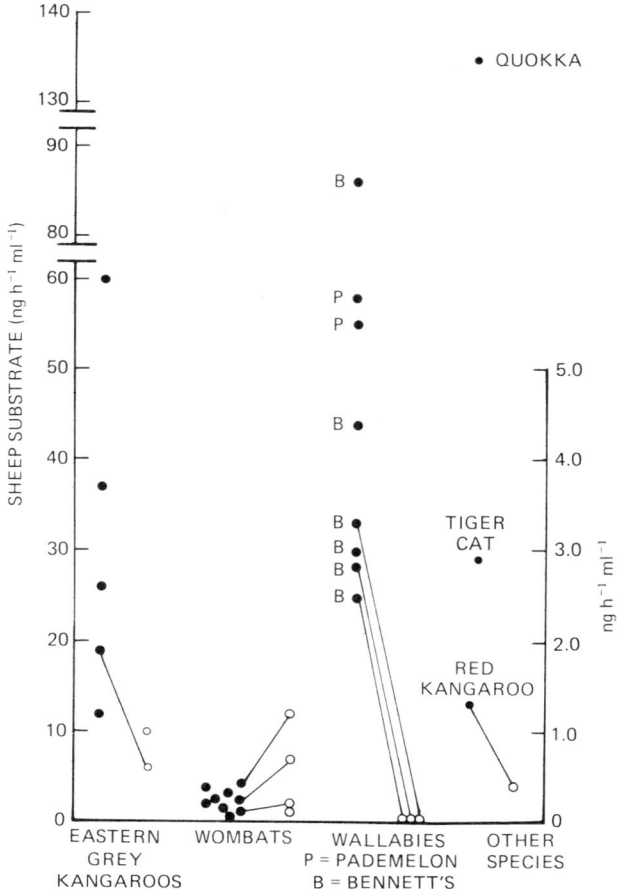

the release of renal renin into the blood involve changes of body sodium and fluid volume status. Briefly, the major proximate stimuli of increased renin release in mammals are: (1) reduced renal arterial blood pressure, (2) altered load or transport of sodium at the macula densa, (3) increased activity of the sympathetic nervous system, (4) reduced plasma potassium concentration, and (5) inhibition of the short-loop negative feedback of blood angiotensin II concentration. Blood levels of vasopressin are also implicated. Intrarenal prostaglandin production may mediate release of renin.

There have been no studies on the proximate stimuli of renin release in marsupials, but the effects of the stress of sodium deficiency and pharmacological findings are consistent with the operation of an eutherian-type renin–angiotensin system. Johnston et al. (1967), Blair-West et al. (1968), and Reid and McDonald (1969) all demonstrated changes in the juxtaglomerular apparatus consistent with stimulation of storage and release of renin in sodium-depleted marsupials. Reid and McDonald (1969) observed that high levels of plasma renin concentration in *T. vulpecula* were associated with appetite for salt, and these levels were reduced by voluntary ingestion of 0.9% NaCl solution. Sodium depletion by diuretics or blood loss also increased PRC. They concluded that the RAS is an important mechanism of sodium homeostasis in *Trichosurus* and operative within the usual range of stresses encountered by these marsupials.

Davis et al. (1967) reported that sodium deficiency increased the rate of aldosterone secretion in the American opossum *(Didelphis virginiana)*, and Johnston et al. (1967) found that injection of homologous renal renin extracts stimulated aldosterone secretion and raised blood pressure in the

Figure 4. Relation between plasma renin concentration and renal cortical renin concentration in four marsupial species (from Simpson and Blair-West, 1971).

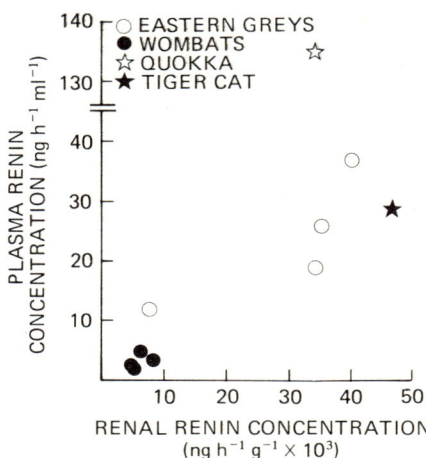

same species. In the same year, Coghlan and Scoggins (1967) reported that kangaroos and wombats from inland areas of Australia, naturally low in sodium content in soils and plants, had higher blood aldosterone concentration than animals of the same species from sodium-rich coastal areas. Later, Oddie et al. (1976) showed that blood concentrations of aldosterone, cortisol, and corticosterone in nine species of Australian marsupials were similar to the levels in introduced eutherians (sheep, cow, fox, dog, and man) with the exceptions of the wombat and koala.

Marsupial species were included in the detailed survey of the physiological adjustments of mammalian species to variations of sodium availability in natural Australian habitats (Blair-West et al., 1968; Scoggins et al., 1970). Rabbits, sheep, kangaroos, and wombats all showed an inverse relation between peripheral blood aldosterone concentration and sodium content of forage and renal excretion of sodium. The adrenal glands of kangaroos and wombats were heavier in animals from low-sodium habitats, and the zona glomerulosa area (which secretes aldosterone) was increased. The salivary glands of animals from sodium-poor areas showed hypertrophy of striated ducts. Renal renin contents were found to be uniformly low by the methods available at that time, but subsequent assays of plasma renin concentration in animals from different habitats indicate an inverse relation between plasma renin concentration and urinary Na concentration (Table 1).

In general, results of laboratory and field studies show that a renin–angiotensin system of the eutherian type, reactive to changes in Na status and correlated with the secretion of aldosterone in Na deficiency, is present in several metatherian species.

Table 1. *Plasma renin concentration (PRC) in eastern grey kangaroos* (Macropus giganteus) *and wombats* (Vombatus hirsutus Perry) *from various Australian habitats*

	PRC $(ng\ h^{-1}\ ml^{-1})$		Urinary Na concentration (mmol liter^{-1})
Kangaroos			
1. Snowy mountains	range	12–60	0.3–0.5
(N=5)	mean±SD	31±18	
2. Coastal range	range	6–24	6–71
(N=6)	mean±SD	15±6	
3. Sea coast	range	11–36	68–224
(N=6)	mean±SD	19±9	
Wombats			
1. Snowy mountains	range	0.6–5.0	0.3–0.5
(N=10)	mean ± SD	2.6±1.5	
2. Sea coast	range	0.3–0.4	1–13
(N=2)			

References

Best, J. B., Blair-West, J. R., Coghlan, J. P., Cran, E. J., Fernley, R. T., and Simpson, P. A. (1974). A novel sequence in kangaroo angiotensin I. *Clin. Exp. Pharmacol. Physiol. 1:*171–4.

Blair-West, J. R. (1976). Renin–angiotensin system and sodium metabolism. In *Kidney and Urinary Tract Physiology II*, vol. 2, ed. K. Thurau, pp. 95–143. Baltimore: University Park Press.

Blair-West, J. R., Coghlan, J. P., Denton, D. A., Scoggins, B. A., Wintour, M., and Wright, R. D. (1967). The renin/angiotensin–aldosterone system in sodium depletion. *Med. J. Aust. 2:*290–3.

Blair-West, J. R., Coghlan, J. P., Denton, D. A., Nelson, J. F., Orchard, E., Scoggins, B. A., Wright, R. D., Myers, K., and Junqueira, C. L. (1968). Physiological, morphological and behavioural adaptation to a sodium-deficient environment by wild native Australian and introduced species of animals. *Nature (London) 217:*922–8.

Blair-West, J. R., Coghlan, J. P., Denton, D. A., Funder, J. W., Scoggins, B. A., and Wright, R. D. (1971). Inhibition of renin secretion by systemic and intrarenal angiotensin infusion. *Am. J. Physiol. 220:*1309–15.

Boyd, G. W. (1972). The nature of renal renin. In *Hypertension '72*, eds. J. Genest and E. Koiw, pp. 161–9. Berlin: Springer-Verlag.

Brown, J. J., Chinn, R. H., Gavras, H. Leckie, B., Lever, A. F., McGregor, J., Morton, J., and Robertson, J. I. S. (1972). Renin and renal function. In *Hypertension '72*, eds. J. Genest and E. Koiw, pp. 81–97. Berlin: Springer-Verlag.

Carretero, O., and Gross, F. (1967). Evidence for humoral factors participating in the renin-substrate reaction. *Circ. Res. 20, 21 (Suppl. II):*115–27.

Coghlan, J. P., and Scoggins, B. A. (1967). The measurement of aldosterone, cortisol and corticosterone in the blood of the wombat (*Vombatus hirsutus* Perry) and the kangaroo *(Macropus giganteus). J. Endocrinol. 39:*445–8.

Davis, J. O., Johnston, C. I., Hartroft, P. M., Howards, S. S., and Wright, F. S. (1967). The phylogenetic and physiologic importance of the renin–angiotensin–aldosterone system. In *Proc. Third Int. Cong. Nephrol.*, Washington, 1966, vol. 1, pp. 215–25.

Elghozi, J. L., Fitzsimons, J. T., Meyer, P., and Nicolaidis, S. (1977). Central angiotensin in control of water intake and blood pressure. In *Hypertension and Brain Mechanisms, Progress in Brain Research*, vol. 47, eds. W. DeJong, A. P. Provoost, and A. P. Schapiro, pp. 137–49. Amsterdam: Elsevier.

Johnston, C. I., Davis, J. O., and Hartroft, P. M. (1967). Renin–angiotensin system, adrenal steroids and sodium depletion in a primitive mammal, the American opossum. *Endocrinology 81:*633–42.

Khairallah, P. A., Khosla, M. C., Bumpus, F. M., Tait, J. F., and Tait, S. A. S. (1978). Biological activity of angiotensins. *Proc. Fifth Meeting Int. Soc. Hypertension*, Paris, p. 136.

Leckie, B. (1973). The activation of a possible zymogen of renin in rabbit kidney. *Clin. Sci. 44:*301–4.

Mitchell, J. D., Baxter, T., Blair-West, J. R., and McCredie, D. A. (1970). Renin levels in nephroblastoma (Wilm's tumour). Report of a renin secreting tumour. *Arch. Dis. Child. 45:*376–84.

Oddie, C. J., Blaine, E. H., Bradshaw, S. D., Coghlan, J. P., Denton, D. A., Nelson, J. F., and Scoggins, B. A. (1976). Blood corticosteroids in Australian marsupial and placental mammals and one monotreme. *J. Endocrinol. 69:*341–8.

Page, I. H., and Bumpus, F. M. (1974). *Angiotensin.* Berlin: Springer-Verlag.

Page, I. H., and McCubbin, J. W. (1968). *Renal Hypertension.* Chicago: Year Book Medical Publishers.

Reid, I. A., and McDonald, I. R. (1968). Renal function in the marsupial *Trichosurus vulpecula*. *Comp. Biochem. Physiol. 25*:1071–9.

Reid, I. A., and McDonald, I. R. (1969). The renin–angiotensin system in a marsupial (*Trichosurus vulpecula*). *J. Endocrinol. 44*:231–40.

Romero, J. C., Lazar, J. D., and Hoobler, S. W. (1970). Effects of renal artery constriction and subsequent contralateral nephrectomy on the blood pressure, plasma renin activity and plasma renin substrate concentration in rabbits. *Lab. Invest. 22*:581–7.

Rubin, I. (1972). Purification of hog renin. Properties of purified hog renin. *Scand. J. Clin. Lab. Invest. 29*:51–8.

Scoggins, B. A., Blair-West, J. R., Coghlan, J. P., Denton, D. A., Myers, K., Nelson, J. F., Orchard, E., and Wright, R. D. (1970). The physiological and morphological response of mammals to changes in their sodium status. In *Hormones and the Environment*, eds. G. K. Benson and J. G. Phillips, pp. 577–602. London: Cambridge University Press.

Simpson, P. A., and Blair-West, J. R. (1971). Renin levels in the kangaroo, the wombat and other marsupial species. it. J. Endocrinol. *51*:79–90.

Simpson, P. A., and Blair-West, J. R. (1972). Estimation of marsupial renin using marsupial renin-substrate. *J. Endocrinol. 53*:125–30.

Skeggs, L. T., Dorer, F. E., Kahn, J. R., Lentz, K. E., and Levine, M. (1974). The biological production of angiotensin. In *Angiotensin*, eds. I. H. Page and F. M. Bumpus, pp. 1–16. Berlin: Springer-Verlag.

Skinner, S. L. (1967). Improved assay methods for renin "concentration" and "activity" in human plasma. *Circ. Res. 20*:391–402.

Taylor, A. A. (1977). Comparative physiology of the renin–angiotensin system. *Fed. Proc. 36*:1776–80.

Thurau, K. (1974). Intrarenal action of angiotensin. In *Angiotensin*, eds. I. H. Page and F. M. Bumpus, pp. 475–89. Berlin: Springer-Verlag.

Vane, J. R. (1974). The fate of angiotensin. In *Angiotensin*, eds. I. H. Page and F. M. Bumpus, pp. 17–40. Berlin: Springer-Verlag.

30

Physiology of the adrenal cortex in monotremes

C. SERNIA

Over the past 13 years information on the physiological function of the adrenal cortex in monotremes and especially the echidna, *Tachyglossus aculeatus*, has been steadily accumulating. In this paper the major findings in this area will be presented and compared to those on other mammals.

The echidna

In 1965, Weiss and McDonald reported that corticosterone and, in smaller quantities, cortisol are the major corticosteroids secreted by the adrenal cortex in the echidna. The presence of corticosterone as the main product, instead of cortisol which is more typical of mammals (Sandor et al., 1976), agreed with the nonmammalian morphology of the echidna adrenal gland (Wright et al., 1957). However, the concentration of corticosteroids in peripheral blood was only 1 to 2 ng ml^{-1}, and the highest secretion rate of corticosterone was 11.7 ng $kg^{-1}h^{-1}$. Furthermore, neither aldosterone nor free or conjugated urinary corticosteroids were detected in adrenal venous blood.

This unusually low adrenocortical activity raised the possibility that the echidna and other monotremes are less dependent on corticosteroids for the regulation of electrolyte and intermediary metabolism than the metatherian and eutherian mammals (Cahill, 1971; Rudman and Di Girolamo, 1971; Reid and Ganong, 1974; McDonald, 1977). In the following this possibility is discussed in the light of more recent experiments.

Metabolic consequences of adrenalectomy

Unlike eutherian and metatherian (see McDonald, 1977) mammals, in the absence of severe stress the echidna survives bilateral adrenalectomy indefinitely without developing the classical symptoms of adrenal insufficiency, i.e., profound disturbance of Na–K balance, hypoglycemia, acidosis, hemoconcentration, loss of appetite, and muscular weakness (McDonald and Augee, 1968). The only disturbance appears to be an inability to gain weight. Steroids from nonadrenal sources may be compensating for the loss of adrenal corticosteroids, but this seems unlikely because gonadectomy prior to adrenalectomy is without effect and extra-adrenal rests of steroidogenic tissue have not been detected.

A further pertinent observation made by McDonald and Augee (1968) relates to the lack of hypertrophy of the adrenal gland remaining after removal of the contralateral gland, even after 109 days. A possible explanation is that in low-stress situations, when the concentration of circulating corticosteroids is extremely low, the adrenals are not influenced by pituitary ACTH. However, the plasma concentration of corticosteroids in resting echidnas shows a diurnal rhythm, and the secretion of corticosteroids occurs as episodic pulses (Sernia and McDonald, 1977a). Furthermore, a normal plasma concentration of total corticosteroids (1.8 ng ml^{-1}) was measured in an echidna 2 months after unilateral adrenalectomy (Sernia, 1977). This recent evidence does not support the notion that a hypothalamic–adrenocortical negative feedback loop is absent, but it indicates that a single adrenal can increase its secretion to meet the low daily requirements of corticosteroids without an obvious increase in size.

In contrast to the absence of any serious metabolic disturbance when they are maintained in normal laboratory conditions, adrenalectomized echidnas cannot thermoregulate at low ambient temperatures (Augee and McDonald, 1973). When fasted and exposed to 5 °C they become hypothermic and die within 48 h, whereas intact echidnas in similar conditions maintain a normal body temperature until the depletion of their fat stores, after which they become torpid. Resistance to low ambient temperatures may be restored by treatment with intramuscular injections of cortisol acetate or by glucose injected intravenously during the period of cold exposure. These results, and the mild hyperglycemic action of glucocorticoids in intact echidnas, led Augee and McDonald (1973) to suggest that the maintenance of glucose concentration via gluconeogenesis is a major function of the glucocorticoids in the echidna – an idea that I will return to later.

The importance of the adrenal cortex is further implicated by the marked adrenal hypertrophy observed after repeated cold exposure (Table 1). Adrenal hypertrophy has also been found in echidnas dying (termed "Moribund" in Table 1) from parasitic infestation (two), impacted feces (one), or gradual emaciation, of unknown etiology, leading to death (two). Three echidnas receiving a weekly intravenous injection of a large dose of ACTH (5 IU kg^{-1}h^{-1} for 2 h) for 4 weeks also responded with a marked adrenal hypertrophy. These observations on adrenal weight changes indicate that in the echidna, as in other mammals, stressful stimuli elicit an adrenocortical response via stimulation of ACTH secretion (Yates et al., 1974).

Secretion of corticosteroids

Plasma concentrations and secretion rates of corticosterone, cortisol, and aldosterone have been measured in unrestrained, conscious echidnas. The methods were protein-binding and radioimmunoassay techniques for measuring plasma concentration and isotope dilution for estimating clearance rates (Table 2).

In agreement with the earlier observations of Weiss and McDonald (1965) and with the recent measurements in a single echidna by Oddie et al. (1976), the major glucocorticoids in peripheral plasma are corticosterone (1.7 ± 0.2 ng ml^{-1}) and, at lower concentrations, cortisol (0.8 ± 0.2 ng ml^{-1}). In spite of these low concentrations, the adrenals respond to stimulation by exogenous synthetic ACTH (Synacthen) with a large increase in secretion rate of both glucocorticoids, reaching a maximum with a dose of at least 1 IU ACTH kg^{-1}h^{-1}. This rate of ACTH injection is about 4, 15, and 160 times higher than the dosage of the same ACTH required for maximal adrenal stimulation in the brush-tailed possum (Khin Aye Than and McDonald, 1973), sheep (Bassett and Hinks, 1969), and man (Landon et al., 1964), respectively. Some metatherian species (McDonald, 1977; Bradshaw and McDonald, 1977) also require injection of ACTH at rates that are unlikely to reflect the secretion rate of endogenous corticotropin, but which are probably an indication of between-species differences in the affinity for eutherian ACTH.

Conditions that lead to adrenal hypertrophy also stimulate adrenocortical secretion. Surgery with ether anesthesia, disease, or handling echidnas unaccustomed to experimental procedures will elevate plasma glucocorticoid concentrations (Sernia and McDonald, 1977a). However, repeated exposure of normal echidnas to 5 °C has the unexpected effect of decreasing plasma

Table 1. *Adrenal weight in the echidna*

Condition	Adrenal wt/body wt[a] (mg kg^{-1})
Normal (laboratory held)	42.0 ± 8.1 (11)[b]
	41.0 ± 6.5 (7)[c]
	46.0 ± 8.5 (winter)[d]
Exposure to 4 °C	86.8 ± 10.0 (9)[d]
	64.0 ± 10.3 (5)
ACTH treatment	73.5 ± 5.4 (3)
Moribund	82.6 ± 8.2 (5)

[a]Data are means $\pm$ SD with number of animals in parentheses.
[b]Weiss and McDonald (1965), Weiss (1973).
[c]McDonald and Augee (1968).
[d]Augee and McDonald (1973).

Table 2. *Plasma concentrations, secretion rates (SR), and metabolic clearance rates (MCR) for the major corticosteroids in the echidna*

Treatment	Corticosterone Mean ± SD	No. of animals	Cortisol Mean ± SD	No. of animals	Aldosterone Mean ± SD	No. of animals
None						
Concentration[a]	1.7 ± 0.8	(12)	0.8 ± 0.2	(12)	5.4 ± 3.6	(8)
SR[b]	0.30 ± 0.18	(12)	0.44 ± 0.26	(11)	5.0 ± 4.9	(5)
MCR[c]	3.6 ± 3.2	(12)	10.5 ± 7.5	(11)	14.3 ± 3.0	(5)
Stress						
(disease, surgery)						
Concentration	4.2	(1)	4.4	(1)	17.6 ± 8.6	(5)
	10.0	(1)	3.4	(1)	(ether + surgery)	
ACTH injection						
(1–5 IU kg^{-1} h^{-1})[1]						
Concentration	10.9 ± 5.1	(12)	3.9 ± 2.1	(12)	53.8 ± 9.8	(5)
SR	2.18 ± 2.81	(13)	2.02 ± 1.77	(12)	35.0 ± 20.8	(5)
MCR	3.0 ± 2.6	(13)	8.7 ± 7.1	(12)	10.4 ± 3.8	(5)
Angiotensin injection						
(5 µg kg^{-1}h^{-1})						
Concentration					35.1 ± 10.9	(5)
SR					21.0 ± 9.4	(4)
MCR					11.5 ± 6.4	(4)

[a]ng ml^{-1} for corticosterone and cortisol; pg ml^{-1} for aldosterone.

[b]Measured as µg kg^{-1} h^{-1}.

[c]Measured as ml kg^{-1} min^{-1}.

Data from Sernia (1977); Sernia and McDonald (1977a); Sernia and McDonald (unpublished).

corticosteroid concentrations in spite of an actual increase in secretion rate (Figure 1). This seemingly contradictory situation is the result of an increased clearance rate in cold acclimatized echidnas, possibly a consequence of a higher hepatic and renal blood flow. It is also noteworthy that the secretory capacity, measured as the secretion rate at maximal ACTH stimulation, decreases after repeated cold exposure even though the adrenals hypertrophy (64.0 ± 10.3 mg kg^{-1} body weight). These results clearly indicate an involvement of the adrenal cortex in the adaptation to low ambient temperature. Further experiments along similar lines should eventually lead to an understanding of the metabolic role of the adrenal cortex in the hypothermic state of torpor in the echidna.

As with the glucocorticoids, the plasma concentration (5.4 ± 3.6 pg ml^{-1}) and the secretion rate (5.0 ± 4.9 ng kg^{-1}h^{-1}) of the mineralocorticoid, aldosterone, in unstressed echidnas are extremely low – perhaps the lowest among mammals (Oddie et al., 1976). Factors such as ACTH and angiotensin II, which stimulate aldosterone secretion in eutherian mammals (Reid and Ganong, 1974), have similar effects in the echidna (Table 2). However, there is no measurable renin activity in normal echidnas (Reid, 1971), and the electrolyte balance is undisturbed in adrenalectomized echidnas

Figure 1. Plasma concentration, secretion rate (SR), and metabolic clearance rate (MCR) of corticosterone during the infusion of saline or ACTH (1 IU kg^{-1}h^{-1}) in five echidnas. Measurements were made at room temperature before (clear bars) and after (hatched bars) a period of 3 to 4 weeks during which echidnas were subjected to a routine of 2 to 3 days at 5 °C with fasting alternating with a similar time at room temperature with feeding ad libitum. Data are expressed as mean ± SEM.

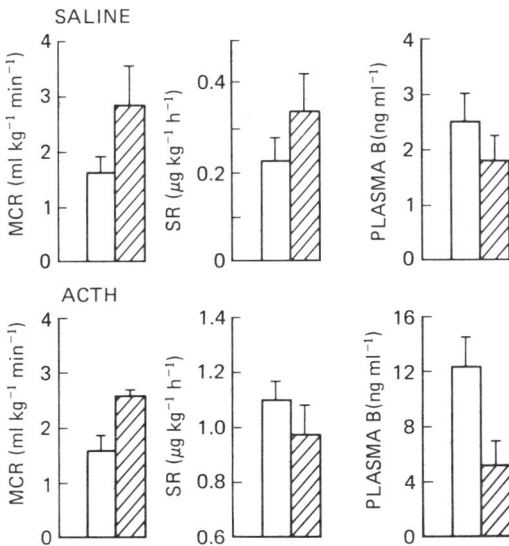

(McDonald and Augee, 1968). It therefore seems doubtful that the renin–angiotensin–aldosterone system is of major importance in the regulation of fluid and electrolyte balance in the echidna as it is in eutherian mammals.

Metabolic actions of the glucocorticoids

The intravenous injection of cortisol or corticosterone at a rate of 3 or 30 μg kg^{-1} h^{-1} (i.e., about the maximum secretion rate in normal echidnas or 10 times higher) has no effect on the plasma concentration of α-amino acids and urea and has only a slight hyperglycemic effect (Sernia and McDonald, 1977b). In contrast, plasma FFA (free fatty acids) concentration is increased in a dose-related manner. In long-term experiments where 0.2 mg kg^{-1} of cortisol or corticosterone acetate was injected (i.m.) daily for 5 days, essentially the same results were obtained. In addition, body weight, nitrogen intake, urinary nitrogen excretion, and fasting liver glycogen content did not change. Therefore, the major target of the glucocorticoids appears to be fat and not carbohydrate or nitrogen metabolism.

The failure of adrenalectomized echidnas to thermoregulate at low ambient temperatures, in spite of abundant fat stores, has been attributed by Augee and McDonald (1973) to a deficiency in glucose synthesis via gluconeogenesis. These results indicate that a failure to mobilize fat reserves is the primary deficiency, and the inability of adrenalectomized, cold-exposed echidnas to maintain blood glucose is probably due to a failure to meet an increased demand for glucose to compensate for the decreased availability of FFA. This explanation is consistent with the observation that normal, fasted echidnas exposed to low ambient temperatures do not maintain body temperature beyond the depletion of their fat stores (Augee and Ealey, 1968). A decreased potency in the lipolytic actions of ACTH, catecholamines, and growth hormone due to the absence of glucocorticoids may also be contributing to the disturbed fat metabolism in adrenalectomized echidnas. However, such hormone–hormone interactions have not yet been studied in the echidna except for cortisol–insulin antagonism (Sernia and McDonald, 1978).

Binding of glucocorticoids to plasma proteins

A corticosteroid-binding globulin (CBG), which binds the major glucocorticoids with a high affinity and renders them biologically inactive, has been found in all metatherian and eutherian mammals investigated (Seal and Doe, 1963; McDonald, this volume). However, both equilibrium dialysis and polyacrylamide gel electrophoresis of echidna plasma fail to show an analogous high-affinity, corticosteroid-binding protein. Thus the total concentration of blood glucocorticoids is biologically active, being either free or weakly bound to plasma proteins (Figure 2). In low-stress conditions, the secretion of and metabolic requirement for glucocorticoids is so low that the presence of a CBG, rather than its absence, would have been puzzling because it would have been difficult to envisage a physiological function for it. On the other hand, having glucocorticoids in a metabolically active form when an adequate adrenocortical response is essential (i.e., in stress) could provide a way of increasing the effectiveness of an adrenocortical system that has a low secretory capacity. Therefore, whereas the absence of a CBG is probably unique among terrestrial vertebrates, it seems appropriate for the limited but essential metabolic role of the adrenal cortex in the echidna.

Other monotremes

Very little is known of adrenocortical function in the long-beaked echidna *(Zaglossus bartoni)* and the platypus *(Ornithorhynchus anatinus)*.

Weiss (1973) found cortisone and cortisol as the major glucocorticoids in the platypus. Their combined plasma concentration was 140 ng ml^{-1}, and they were secreted (in vitro) at a rate of 640 ng mg^{-1} adrenal. The mean adrenal weight of three platypuses was 257 mg kg^{-1} body weight. In the long-beaked echidna, the adrenal weight in a nonlactating female has been measured by Dr. M. Griffiths at the very high value of 405 mg kg^{-1} body weight (personal communication).

This limited information suggests that adrenocortical function in these two monotremes is very different from that in the echidna and may show a level of secretory activity well within the range observed in eutherian mammals.

Figure 2. Distribution of cortisol in the plasma of the echidna, brush-tailed possum, and man (albumin-bound, ▥; globulin-bound, ■; unbound, ▢). The absence of CBG in the echidna is evident from the low percentage of globulin-bound cortisol and the failure to saturate the binding sites. (Reprinted from Sernia, 1978, with permission of *Australian Zoologist.)*

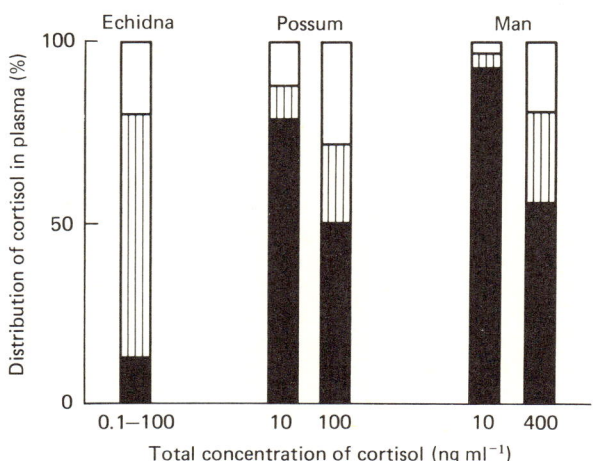

Conclusion

In the echidna, the secretion of corticosterone and cortisol as the major glucocorticoids and the response of the adrenal cortex to exogenous ACTH and to various stressful conditions show that in qualitative terms adrenocortical function in this monotreme is essentially mammalian.

The low secretion rates of glucocorticoids and aldosterone, the ability to survive adrenalectomy in low-stress conditions, the absence of a corticosteroid-binding protein, and the low diabetogenic effect of glucocorticoids indicate that the echidna is not particularly dependent on the corticosteroids for metabolic stability. However, insensitivity to the metabolic actions of glucocorticoids is not peculiar to the echidna and is found in both eutherian (e.g., dog, man) and metatherian (e.g., quokka, red kangaroo) mammals. Large differences in sensitivity may be present even within a single genus, as in the macropodid marsupials (McDonald, this volume). Indeed, the scanty information on the long-beaked echidna and the platypus indicates that, in spite of being represented by only three extant species, the monotremes also show the wide diversity in adrenocortical function seen in the Metatheria and the Eutheria. The low adrenocortical activity found in the echidna may be simply an expression of this diversity and need not reflect some ancestral mammalian condition.

References

Augee, M. L., and Ealey, E. H. M. (1968). Torpor in the echidna, *Tachyglossus aculeatus. J. Mammal. 49:*446–54.

Augee, M. L., and McDonald, I. R. (1973). Role of the adrenal cortex in the adaptation of the monotreme *Tachyglossus aculeatus* to low environmental temperature. *J. Endocrinol. 58:*513–23.

Bassett, J. M., and Hinks, N. T. (1969). Micro-determination of corticosteroids in ovine peripheral plasma: Effects of venipuncture, corticotrophin, insulin and glucose. *J. Endocrinol. 44:*387–403.

Bradshaw, S. D., and McDonald, I. R. (1977). Plasma corticosteroids and the effect of corticotrophin in a macropodid marsupial (*Setonix brachyurus*, Quay & Gaimard). *J. Endocrinol. 75:*409–18.

Cahill, G. F., Jr. (1971). Action of adrenal corticol steroids on carbohydrate metabolism. In *The Human Adrenal Cortex*, ed. N. P. Christy, pp. 205–39. New York: Harper & Row.

Khin Aye Than and McDonald, I. R. (1973). Adrenocortical function in the brushtailed possum *Trichosurus vulpecula* (Kerr). *J. Endocrinol. 58:*97–109.

Landon, J., James, V. H. T., Cryer, R. J., Wynn, V., and Frankland, A. W. (1964). Adrenocorticotrophic effect of synthetic polypeptide-β^{1-24}-corticotrophin in man. *J. Clin. Endocr. Metab. 24:*1206–13.

McDonald, I. R. (1977). Adrenocortical functions in marsupials. In *The Biology of Marsupials*, eds. B. Stonehouse and D. Gilmore, pp. 345–78. London: Macmillan.

McDonald, I. R., and Augee, M. L. (1968). Effects of bilateral adrenalectomy in the monotreme *Tachyglossus aculeatus. Comp. Biochem. Physiol. 27:*669–78.

Oddie, C. J., Blaine, E. H., Bradshaw, S. D., Coghlan, J. P., Denton, D. A., Nelson, J. F., and Scoggins, B.A. (1976). Blood corticosteroids in Australian marsupial and placental mammals and one monotreme. *J. Endocrinol.* 69:341–8.

Reid, I. A. (1971). Renin secretion in a monotreme *(Tachyglossus aculeatus. Comp. Biochem. Physiol.* 40A:249–55.

Reid, I. A., and Ganong, W. F. (1974). The hormonal control of sodium excretion. In *Endocrine Physiology*, ed. S. M. McCann, pp. 205–37. London: MTP and Butterworth.

Rudman, D., and Di Girolamo, M. (1971). Effect of adrenal cortical steroids on lipid metabolism. In *The Human Adrenal Cortex*, ed. N. P. Christy, pp. 241–55. New York: Harper & Row.

Sandor, T., Fazekas, A. G., and Robinson, B. G., (1976). The biosynthesis of corticosteroids throughout the vertebrates. In *General, Comparative and Clinical Endocrinology of the Adrenal Cortex*, vol. 1, eds. I. Chester-Jones and I. W. Henderson, pp. 25–125. London: Academic Press.

Seal, U. S., and Doe, R. P. (1963). Corticosteroid-binding globulin Species distribution and small scale purification. *Endocrinology* 73:371–6.

Sernia, C. (1977). Adrenocortical function in the echidna. Ph.D. thesis, Monash University.

Sernia, C. (1978). Steroid-binding proteins in the plasma of the echidna, *Tachyglossus aculeatus,* with comparative data for some marsupials and reptiles. In *Symposium on Monotreme Biology*, ed. M. L. Augee, *Aust. Zool.* 20:87–98.

Sernia, C., and McDonald, I. R. (1977a). Adrenocortical function in a prototherian mammal, *Tachyglossus aculeatus* (Shaw). *J. Endocrinol.* 72:41–52.

Sernia, C., and McDonald, I. R. (1977b). Metabolic effects of cortisol, corticosterone and ACTH in a prototherian mammal, *Tachyglossus aculeatus. J. Endocrinol.* 75: 261–9.

Sernia, C., and McDonald, I. R. (1978). The effect of cortisol on insulin sensitivity in the echidna, *Tachyglossus aculeatus* (Shaw). *Gen. Comp. Endocrinol.* 36:1–6.

Weiss, M. (1973). Biosynthesis of adrenocortical steroids by monotremes: Echidna *(Tachyglossus aculeatus)* and platypus *(Ornithorhynchus anatinus). J. Endocrinol.* 58:251–62.

Weiss, M., and McDonald, I. R. (1965). Corticosteroid secretion in the monotreme *Tachyglossus aculeatus. J. Endocrinol.* 33:203–10.

Wright, A., Phillips, J. G., and Chester-Jones, I. (1957). The histology of the adrenal glands of the Prototheria. *J. Endocrinol.* 15:100–7.

Yates, F. E., Maran, J. W., Cryer, G. L., and Gann, D. S. (1974). The pituitary adrenal cortical system: Stimulation and inhibition of secretion of corticotrophin. In *Endocrine Physiology*, vol. 5, ed. S. M. McCann, pp. 109–140. London: MTP and Butterworth.

31

Physiology of the adrenal cortex in marsupials

I. R. McDONALD

It is generally accepted that, in mammals, the steroidal secretion of the adrenal cortex is essential for the maintenance of life. The ultimately lethal effect of bilateral adrenalectomy is attributed mainly to the loss of mineralocorticoid secretion, causing renal sodium wasting and potassium retention. Death in the short term is a result of circulatory failure, due either to reduction of blood volume from the loss of salt and water or to potassium intoxication with cardiac arrest. The functions of the glucocorticoids are considered less critical and become apparent only in the long term if the renal salt wasting is ameliorated by an increase in salt intake or administration of a mineralocorticoid. If such adrenalectomized animals are exposed to an adverse environment, an important role of glucocorticoids is revealed. It includes mobilization, utilization, and replenishment of energy reserves, potentiation of autonomic nervous functions, and an influence on the reactivity of the immune system as well as the inflammatory response to injury.

Marsupial adrenal glands secrete the same glucocorticoids and mineralocorticoids as their eutherian counterparts, the major glucocorticoid being cortisol and the major mineralocorticoid being aldosterone (see Weiss, this volume). Secretion of these corticosteroids is stimulated by synthetic ($\beta 1-24$) and porcine adrenocorticotropin (ACTH), although the sensitivity of the marsupial adrenal is very much less than that of eutherian mammals (see McDonald, 1977). In this account of the physiological functions of these corticosteroids in marsupials, attention is drawn to elements of conformity and nonconformity with the accepted functions in eutherian mammals, as described in highly simplified form above.

Effects of bilateral adrenalectomy

The most often quoted element of nonconformity with the accepted role of the adrenal cortex in mammals is the report by Silvette and Britten (1936) and Britten and Silvette (1937) that, in the North American opossum, *Didel-*

phis virginiana, bilateral adrenalectomy was followed by a rise instead of the expected fall in serum chloride concentration. Later, Hartman et al. (1943) confirmed that the serum sodium concentration also rose after adrenalectomy and further reported that the opossum could survive the operation without any supportive treatment for up to 603 days. They concluded that "In the body economy of the opossum, the adrenal appears to be of less significance than in many other mammals." This, more than anything else, provoked the later, more extensive investigations into adrenocortical functions of the Australian marsupials that form the basis of this account.

The long-term survival of Hartman et al.'s adrenalectomized *Didelphis* was most likely due to the high (2%) salt content of their diet, which could have compensated for their renal salt wasting (see McDonald, 1977). Furthermore, the average survival of Britten and Silvette's (1937) adrenalectomized opossums was only 6 or 10 days, depending on whether or not they were fasted. All investigations of the effects of adrenalectomy in Australian marsupials (Anderson, 1937; Buttle et al., 1952; Reid and McDonald, 1968; McDonald, 1974) have demonstrated only a short survival period in the absence of salt or hormone supplementation. Death is preceded by evidence of renal salt wasting, hemoconcentration, and potassium retention, as in eutherian mammals. Therefore, it can be reasonably concluded that the adrenal glands are just as essential for survival of marsupials as of eutherian mammals. Furthermore, the critical factor would seem to be mineralocorticoid secretion.

Maintenance of adrenalectomized marsupials

There appear to be marked differences in the postadrenalectomy requirements for survival of different species of marsupial, although at present the comparisons are far from adequate.

Buttle et al. (1952) reported that the postadrenalectomy survival of the small wallaby, the quokka, *Setonix brachyurus* could not be prolonged by administration of salt, adrenocortical extract, or deoxycorticosterone, and concluded that this marsupial was unusually sensitive to the effects of adrenalectomy. At the other extreme, the much larger red kangaroo, *Macropus rufus,* can survive bilateral adrenalectomy in apparent good health indefinitely if provided with sufficient isotonic saline solution for self-selection to offset the renal salt wasting (McDonald, 1974). The brush-tailed possum, *Trichosurus vulpecula,* seems to be intermediate to the above two extremes. Although it exhibits an appetite for salt after bilateral adrenalectomy and can survive a week or more with only added salt in its diet, it is listless and does not thrive. This species behaves normally after adrenalectomy if it is given regular injections of both a glucocorticoid and a mineralocorticoid (Reid and McDonald, 1968).

Both the brush-tailed possum and the red kangaroo appear to thrive after adrenalectomy if they are given regular injections of the glucocorticoid cortisol alone. However, such animals always have an abnormally high plasma potassium concentration, and that of the red kangaroo maintained on salt alone is also high. Although administration of a mineralocorticoid will lower the plasma potassium concentration of such animals, omission of the glucocorticoid during mineralocorticoid administration results in a fall in plasma sodium concentration (Reid and McDonald, 1968; McDonald, unpublished). Therefore, it seems that glucocorticoids and mineralocorticoids may interact in the control of electrolyte balance in these marsupials.

Functions of the glucocorticoids

There are also intriguing differences in the apparent metabolic role of glucocorticoids in different species of marsupial.

Griffiths et al. (1968) reported that, in contrast to eutherian herbivores, the red kangaroo was completely insensitive to the nitrogen-mobilizing, diabetogenic actions of the glucocorticoid cortisone. Later, a similar insensitivity to these actions of cortisol – the glucocorticoid secreted naturally by this species – was also demonstrated (McDonald and Griffiths, unpublished). However, Khin Aye Than and McDonald (1974) found that the brush-tailed possum Trichosurus vulpecula was just as sensitive to these glucocorticoid actions as the laboratory rabbit and even more sensitive than the rat or dog. More recently, McDonald and Bradshaw (unpublished) have found that the quokka is, like the red kangaroo, insensitive to the nitrogen mobilizing, diabetogenic actions of cortisol. However, in yet another species of marsupial, the small shrewlike dasyurid Antechinus stuartii, injection of cortisol causes the expected negative nitrogen balance and hyperglycemia (Woollard, 1971; Barnett, 1973).

The lack of response of the two species of macropodid marsupial would seem to be at variance with the concept that a fundamental role of glucocorticoids is interference with the entry of glucose into cells, causing loss of nitrogenous constituents, particularly in lymphoid tissues (Munck, 1971), and induction of hepatic gluconeogenic enzymes (Cahill, 1971). It is also extremely curious that the adrenocortical secretion of cortisol by these marsupials is quite high. At present, we know of no significant metabolic action of cortisol in these two species of macropod. However, it has recently been shown that cortisol is diabetogenic in another wallaby, the tammar, Macropus eugenii (Cooley and Janssens, 1977) so that the lack of effect of cortisol on carbohydrate metabolism is not characteristic of all macropodid marsupials.

The immunosuppressive and anti-inflammatory actions of cortisol have been investigated in the small polyprotodont marsupial A. stuartii, a member of the family Dasyuridae, which includes such carnivores as the Tasmanian devil and the native cats. It has been shown that, in this species, treatment with cortisol causes a marked reduction in the splenic follicle index, the antibody response to injected antigen, and the concentration of the immunoglobulins G and M in the serum (Figure 1). These effects correlate with a significant mortality in the treated animals, compared with zero mortality in saline-treated controls. The animals die with evidence of a flare-up of latent viral, microbial, and parasitic infestations as well as hemorrhagic ulceration of the upper digestive tract (Bradley et al., 1975; Bradley, 1977). In a unique study of the role of stress in regulation of population numbers, Bradley (1977) has shown that the well-documented postmating total mortality of the male *A. stuartii* (Woolley, 1966) is primarily due to an increase in plasma free glucocorticoid concentration with associated immunosuppression. The males in the natural populations die with lesions similar to those found in laboratory-held animals treated with cortisol.

Metabolism and transport of corticosteroids

Two factors that may modify the relationship between corticosteroid production and plasma concentration or biological action of corticosteroid are metabolic clearance and high affinity binding to plasma proteins.

Metabolic clearance and metabolism of cortisol have been investigated in the brush-tailed possum, *T. vulpecula*. It has been shown that cortisol is hydrolyzed and conjugated in the liver in much the same way as in euthe-

Figure 1. Splenic follicle area index (A) and serum immunoglobulin (γG) concentration (B) in male and female *A. stuartii*. Each graph shows (left) values in males and females during the mating period and (right) the effect of cortisol acetate injected into males held captive for a month before the mating period began.

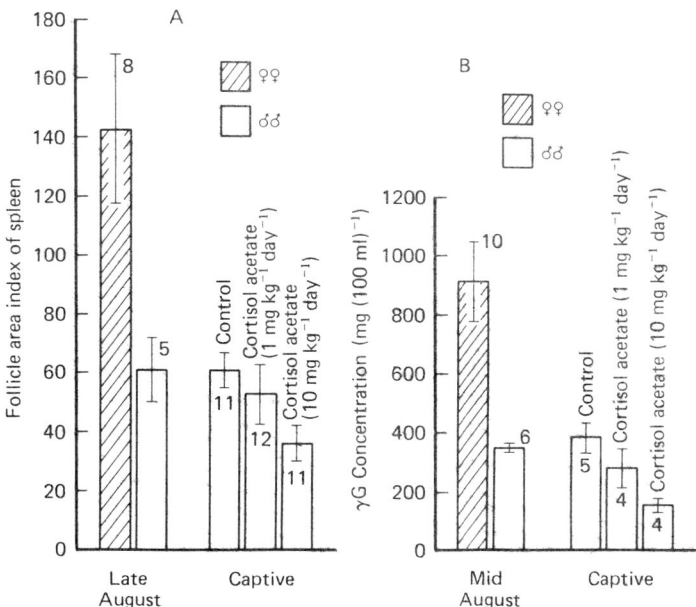

rian mammals, and that the major route of excretion of these metabolites is via the bile and digestive tract. This is in contrast to most eutherians, in which the renal route of excretion predominates (McDonald and Weiss, 1967). The metabolic clearance rate of cortisol in the conscious possum falls within the range quoted for eutherian mammals, if expressed as a function of body weight. The actual rate in individual animals may be significantly affected by environmental disturbance, which thus influence the relationship between secretion rate and plasma concentration (Khin Aye Than and McDonald, 1973).

The plasma of all marsupials so far investigated contains a high-affinity, limited-capacity corticosteroid binding protein with properties similar to the eutherian corticosteroid binding globulin (CBG). In some species, there appears to be more than one CBG (Bradley, 1977; Sernia, 1977).

An important role of CBG in determining the metabolic activity of the circulating glucocorticoids has been demonstrated in *A. stuartii*. In this species, the free, metabolically active glucocorticoid concentration in the plasma of males rises sharply by a factor of 10 or more during the single intensive mating period in late winter. This is due partly to a stress-induced stimulation of the adrenal cortex but mainly to a precipitous fall in the

Figure 2. Relationship between free, albumin-bound, and transcortin-bound glucocorticoid in blood plasma of male and female *A. stuartii* at the time of mating. The horizontal bars represent the range of total glucocorticoid concentration found in the plasma at that time. (Bradley, 1977)

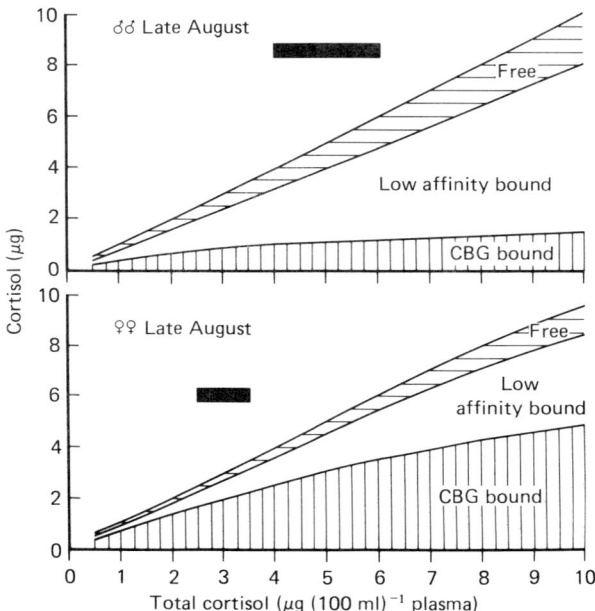

plasma CBG concentration to less than the total glucocorticoid concentration, so that virtually all the circulating glucocorticoid is free or only loosely bound to albumin. In the females, which exhibit much the same degree of adrenal hypertrophy and increase in total glucocorticoid concentration, plasma CBG remains high (Figure 2), and there is no evidence of the immunosuppression that results in the male mortality (Bradley et al., 1976; Bradley, 1977).

It has been shown that in the males, castration elevates plasma CBG concentration, whereas injections of testosterone or ACTH depress it (Figure 3). The fall in CBG of the males is, therefore, due to the synergistic actions of a rising plasma testosterone concentration before mating and increased ACTH secretion as a consequence of agonistic encounters and the stressful environment during the mating period (Bradley, 1977). The marked discrepancy between survival of males and females in this species can be attributed almost entirely to the difference in plasma CBG concentration, and illustrates clearly the important role it plays in limiting the potentially deleterious effects of excessive adrenocortical activity in a stressful environment.

These observations have an important bearing on the well-documented argument about the role of stress in the dramatic population crashes that occur in the microtine rodents (Christian et al., 1965), and they provide an

Figure 3. Effect of testosterone and ACTH on plasma maximum corticosteroid binding capacity (MCBC) of castrate male *A. stuartii*. Cortisol was administered with testosterone alone to suppress endogenous ACTH production. Note the added effect of ACTH. (Bradley, 1977)

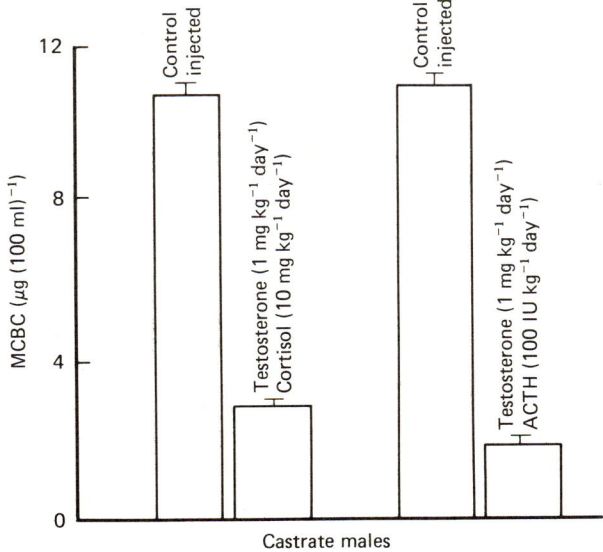

explanation for the lack of convincing evidence for a correlation between adrenocortical activity and mortality in natural populations. They further suggest that excessive adrenocortical activity, not adrenal exhaustion, could be the precipitating factor in such mortality.

Conclusion

The foregoing account indicates that, although there seems to be a basic similarity between adrenocortical functions in the marsupial and nonmarsupial therian mammals, there are considerable species differences and specializations in the marsupials. Because most data on adrenocortical functions in the nonmarsupial therians have been derived from study of inbred and long-domesticated species, it is not known if the diversity of corticosteroid actions found in the marsupials is unique to this group. It seems at least possible that a similar diversity may be found among the nondomesticated eutherian species as well. Investigation of such diversity, which must involve fundamental interactions between the steroids and their target cells, could have a significant bearing on our present concepts of the evolution of endocrine functions.

References

Anderson, D. (1937). Studies on the opossum *(Trichosurus vulpecula)*. II. The effects of splemectomy, adrenalectomy and injection of cortical hormones. *Aust. J. Exp. Biol. Med. Sci. 15:*22–32.

Barnett, J. L. (1973). A stress response in *Antechinus stuartii* (Macleay). *Aust. J. Zool. 22:*311–18.

Bradley, A. J. (1977). Stress and mortality in *Antechinus stuartii* (Macleay). Ph.D. thesis, Monash University, Victoria, Australia.

Bradley, A. J., McDonald, I. R., and Lee, A. K. (1975). Effect of exogenous cortisol on mortality of a dasyurid marsupial. *J. Endocrinol. 66:*281–2.

Bradley, A. J., McDonald, I. R., and Lee, A. K. (1976). Corticosteroid-binding globulin and mortality in a dasyurid marsupial. *J. Endocrinol. 70:*323–4.

Britten, S. W., and Silvette, H. (1937). Further observations on sodium chloride balance in the adrenalectomized opossum. *Am. J. Physiol. 118:*21–25.

Buttle, J. M., Kirk, R. L., and Waring, H. (1952). The effects of complete adrenalectomy on the wallaby *Setonix brachurus. Endocrinology 8:*281–90.

Cahill, G. F. (1971). Action of adrenal cortical steroids on carbohydrate metabolism. In *The Human Adrenal Cortex*, ed. N. P. Christy, pp. 205–39. New York: Harper & Row.

Christian, J. J., Lloyd, J. A., and Davis, D. E. (1965). The role of endocrines in the self-regulation of mammalian populations. *Rec. Prog. Horm. Res. 21:*501–78.

Cooley, H., and Janssens, P. (1977). Metabolic effects of infusion of cortisol and adrenocorticotrophin in the tammar wallaby *(Macropus eugenii* Desmarest). *Gen. Comp. Endocrinol. 33:*352–8.

Griffiths, M., Mackintosh, D. L., and Leckie, R. M. C. (1968). The effects of cortisone on nitrogen balance and glucose metabolism in diabetic and normal kangaroos, sheep and rabbits. *J. Endocrinol. 44:*1 – 12.

Hartmann, F. A., Smith, D. E., and Lewis, L. A. (1943). Adrenal functions in the opossum. *Endocrinology* 32:340–4.

Khin Aye Than and McDonald, I. R. (1973). Adrenocortical function in the Australian brush-tailed possum *Trichosurus vulpecula* (Kerr). *J. Endocrinol.* 58:97–109.

Khin Aye Than and McDonald, I. R. (1974). Metabolic effects of cortisol and corticotrophins in the Australian brush-tailed possum *Trichosurus vulpecula* (Kerr). *J. Endocrinol.* 63:137–47.

McDonald, I. R. (1974). Adrenal insufficiency in the red kangaroo *Megaleia rufa* (Desm.). *J. Endocrinol.* 62:689–90.

McDonald, I. R. (1977). Adrenocortical functions in marsupials. In *The Biology of Marsupials*, eds. B. Stonehouse and D. Gilmore, pp. 345–77. London: Macmillan.

McDonald, I. R. and Weiss, M. (1967). Turnover and excretion of cortisol in the Australian marsupial *Trichosurus vulpecula*. *Acta Endocrinol. (Kbh) Suppl.* 119:242.

Munck, A. (1971). Glucocorticoid inhibition of glucose uptake by peripheral tissues: Old and new evidence, molecular mechanisms and physiological significance. *Perspect. Biol. Med.* 14:265–89.

Reid, I. A., and McDonald, I. R. (1968). Bilateral adrenalectomy and steroid replacement in the marsupial *Trichosurus vulpecula*. *Comp. Biochem. Physiol.* 26:613–25.

Sernia, C. (1977). Adrenocortical function in the echidna. Ph.D. thesis, Monash University, Victoria, Australia.

Silvette, H., and Britten, S. W. (1936). Carbohydrate and electrolyte changes in the opossum and marmot following adrenalectomy. *Am. J. Physiol.* 115:618–26.

Woollard, P. (1971). Differential mortality of *Antechinus stuartii* (Macleay) nitrogen balance and somatic changes. *Aust. J. Zool.* 19:347–73.

Woolley, P. (1966). Reproduction in *Antechinus* spp. and other dasyurid marsupials. In *Comparative Biology of Reproduction in Mammals. Symp. Zool. Soci. London* 15:281–94.

INDEX